D0410075

T
be
da.

Glasgo
Libraries
names f

Stalin and the Scientists

STALIN
AND THE SCIENTISTS

A History of
Triumph and Tragedy
1905–1953

SIMON INGS

FABER & FABER

First published in 2016
by Faber & Faber Ltd
Bloomsbury House
74–77 Great Russell Street
London WC1B 3DA

Typeset by Donald Sommerville
Printed and bound in the UK by CPI Group (UK), Croydon CR0 4YY

All rights reserved
© Simon Ings, 2016

The right of Simon Ings to be identified as the author of this work
has been asserted in accordance with Section 77 of the Copyright,
Designs and Patents Act

A CIP record for this book
is available from the British Library

ISBN 978–0–571–29007–9

FSC
www.fsc.org
MIX
Paper from
responsible sources
FSC® C020471

Glasgow Life Glasgow Libraries	
SH	
C 006177062	
Askews & Holts	19-Oct-2016
509.470904 /sc	£20.00

2 4 6 8 10 9 7 5 3 1

For my children—

*Leo, whose every Christmas, more or less, has been spent
under a tree topped with a hand-fashioned cardboard Stalin,*

and Natalie, who supplied the glue for same.

Contents

CONTENTS

viii

Illustrations

Every effort has been made to trace or contact all copyright holders. The publishers would be pleased to rectify any omissions or errors brought to their notice at the earliest opportunity.

Preface

Come, brethren, let us look in the tomb at the ashes and dust, from which we were fashioned.
  Verse from the Orthodox burial service

This book – about more or less the whole of scientific life in the Soviet Union, from its birth until the mid-1950s – grew out of my fascination with someone other than Joseph Stalin.

Alexander Romanovich Luria's classic neuropsychological case study *The Mind of a Mnemonist*[1] was one of the first books of modern popular science: a slim book that shaped a genre. In it, Luria described the strange world of Solomon Shereshevsky, a man with a memory so prodigious it ruined his life.

Some years ago I met another Luria obsessive who wondered aloud if there was room for a new biography. I did some research and hit a wall. Luria's most astonishing achievement, in a career full of astonishing achievements, was his ability to lead a normal life. He betrayed no one, nor was he betrayed. He led a happy family life and enjoyed many close friendships with colleagues abroad. His work was as sound as it was brilliant. Luria's life and work are endlessly fascinating from a scientific point of view but, for a biographer, there is little to tell that has not already been told.

Yet here was a man – a Jew in a country goaded by the state into anti-semitism – who exposed himself again and again to political risk, who was repeatedly interrogated, sacked and admonished, whose work was forever being banned. Luria's career was an extraordinary demonstration of Winston

Churchill's adage that success consists of moving undaunted from failure to failure.

To unpick the ambiguities of Luria's quiet life, I realised I would need to explore Luria's world, and the more I explored, the more I came to appreciate the generation from which Luria sprang: young men and women who grew up in revolutionary times and whom Stalin and his government had not yet cowed into obedience.

So this has become a much bigger story: it describes what happened when, early in the twentieth century, a motley handful of impoverished and under-employed graduates, professors and entrepreneurs, collectors and, yes, charlatans, bound themselves to a failing government to create a world superpower.

Russia's political elites embraced science, patronised it, fetishised it and even tried to impersonate it. This process reached a head in 1939 when the supreme patron of Soviet science established a prize in his own name for scientific research, the Stalin Prize. At the same time, the 'supreme national scientific institution', the USSR Academy of Sciences, elected him an honorary member.

Suspected, envied and feared by the Great Scientist himself, scientific disciplines from physics to psychology, genetics to gerontology (a Soviet invention) sought to avert the many crises facing the country: famine, drought, soil exhaustion, war, rampant alcoholism, a huge orphan problem, epidemics and an average life expectancy of thirty years. Their work, writings and wrangles with the political authorities of their day shaped global progress for well over a century.

Tsar Alexander II, a successful war leader and diplomat, was an ambitious moderniser. Coming to power in 1856, he transformed Russia's military, its administration and its tax system, and spurred Russia's industrialisation. But Alexander's battle to push his country forward went disastrously awry in 1861, when he 'freed' the empire's 20 million serfs into poverty and homelessness. The tsar became a target of numerous murder plots and, after several narrow escapes, he was assassinated.

Nicholas II ascended the throne in 1894, by which time Russia's growing industrialisation had produced a revolutionary socialist movement. By 1905, following a string of embarrassing military defeats, support for the already unpopular government had dwindled. In St Petersburg, troops fired on a peaceful demonstration, sparking the 'liberal' Revolution of 1905.

The First World War brought another crisis, exposing Russia's dismal command of its natural resources. War losses and crop failures caused the economy to collapse and in St Petersburg – now called Petrograd – riots broke out.[2]

Nicholas II abdicated on 2 March 1917, and a shaky provisional government was declared. On 7 November,[3] led by Vladimir Ilyich Ulyanov (Lenin), the Bolsheviks seized power. But they were far from controlling the whole country, and a bitter civil war ensued. By 1922 Russia was devastated by battles, mass executions and, worst of all, by famine – a crisis that inspired many of the extraordinary scientific careers featured here.

Lenin's New Economic Policy, introduced in 1921, relaxed the revolutionary government's hold over the economy, reintroduced some limited private business and ushered in a period of extraordinary social and cultural change. Stalin's own son attended an experimental school run by psychoanalysts Sabina Spielrein and Vera Shmidt. Alexei Gastev, a poet and a leading architect of Russia's industrialisation programme, trained tens of thousands in the subtle art of production-line engineering. His colleague Isaak Spielrein, Sabina's brother, established a 'psychotechnical' community in Russia dedicated to the physical and psychological emancipation of the Soviet worker. Lev Vygotsky and his colleagues Alexander Luria and Alexei Leontiev embarked on a staggeringly ambitious project, rewriting psychology from first principles. The group's belief in practical, clinical experience led them from trauma wards to orphanages, and from the invention of the lie detector to expeditions in remotest Uzbekistan.

Following Lenin's death in January 1924, the Communist Party was torn apart by a bitter power struggle. Lenin's natural successors, including Trotsky, found themselves marginalised and ultimately destroyed by a man who had barely figured in the 1917 revolution. True, Joseph Stalin had been one of the Bolsheviks' chief operatives in the Caucasus. He had organised paramilitary units, incited strikes, spread propaganda and raised money through bank robberies, kidnappings, ransom demands and extortion. But his appointment in 1922 as General Secretary of the Communist Party was not considered significant at the time. What Stalin spotted straight away, however, was that the post gave him control over government appointments. Stalin built up a base of support, emerged victorious from the power struggle and went on to rule the Soviet Union for a quarter of a century, becoming one of the most powerful and murderous dictators in history.

Stalin scrapped Lenin's New Economic Policy and replaced it with five-year economic plans dictated from the top. This was good for some scientific disciplines, guaranteeing them virtually unlimited funds. For others it was a disaster.

Industrial development was pushed along at breakneck speed. Stalin's repressions created a vast system of labour camps managed by a government agency known by its acronym: Gulag. The convicts' labour, especially in Siberia, became a crucial part of the industrialisation effort. Around 18 million people passed through the gulag system, which became a sort of dark mirror of the state. Just as it boasted its own economy, the gulag boasted its own science base. Several scientists of world importance spent their careers in 'research prisons'.

In the end, only obedience mattered. Stalin believed that science should serve the state. 'Pure research' was not merely an indulgence. It was counterproductive. It was tantamount to wrecking. Even as he invested recklessly in Russian science, Stalin was arranging the sacking, imprisonment and murder of individual scientists. Ergonomists and industrial psychologists vanished without trace. Psychoanalysis was made illegal. Geneticists, botanists and agronomists languished in gulags across the Soviet Union.

Even more damaging was the state's approach to bureaucracy. Institutions were amalgamated with each other and centralised to the point where colleagues tore at each others' throats in an attempt to keep their jobs. Incredibly, no one thought of introducing a mandatory retirement age. Men conditioned to the acquisition and administration of power – we are talking about professors here, not ministers – clung on and on and on. Entire disciplines went to war with each other. Physiologists attacked psychologists. Laboratory pathologists denounced clinicians.

By the time Stalin died on 5 March 1953, the Soviet Union boasted the largest and best-funded scientific establishment in history. It was at once the glory and the laughing stock of the intellectual world.

Stalin and the Scientists is the story of politicians, philosophers and scientists who, over the course of half a century, found themselves intruding – or being dragged by main force – onto each other's turf. Tutting at this sort of thing comes naturally to us. Priests have no business in party politics. Scientists shouldn't laugh at religion. The developed world maintains very clear boundaries between these different kinds of discourse and it is not kind to those who stray off their own path and go skipping over the grass.

It was not always so. In Europe, by about the middle of the nineteenth century, it did seem possible that religion, philosophy, psychology, science and politics might achieve some sort of mutual understanding. Even the Bolsheviks weren't above reading religion in a psychological manner, so as to fold it into their own ideas of the good life.

What this meeting of minds required was that things be described completely in terms of their components. Imagine, for instance, that psychology is reducible to physiology, which is reducible to biology, which is reducible to chemistry, which is reducible to physics – this was pretty much the driving dream of mid-nineteenth-century scholarship.

Friedrich Engels, the German philosopher who cooked up the Marxist style of critical thinking called dialectical materialism, believed that, at some point in the future, all sciences would cohere to form one science, and this one science was bound to bring with it huge social benefit for mankind. In this respect, his thinking was absolutely, yawningly conventional.

Stalin and the Scientists describes what happened when this dream of science as a unified explanation of everything began to be eroded by scientists' own discoveries. It describes what this failure meant to a state that justified itself through science, and regarded its own science, Marxism, as the capstone of the whole nineteenth-century enterprise: a science of everything. It is, ultimately, the story of how impatient believers turned on the scientific community and demanded that the future happen right away.

No wonder they were impatient. No wonder they thought they could get away with it. The early twentieth century was a trans-formative and traumatic period. These were the years when the universe expanded out of all recognition. In 1917 the American astronomer Heber Curtis pointed out that the novae he observed in spiral nebulae were a hundred times farther away than novae in our own galaxy. In 1924 his close contemporary Edwin Hubble measured the distance to the nearer spiral galaxies. They were 2 million light years away. And the universe went on expanding. In 1922 and 1924, the Russian physicist and pioneer balloonist

Alexander Fridman showed that the universe need not be unchanging and that space itself could stretch: insights that led to the idea of the Big Bang.

The visible world was the least of it: in 1895, Guglielmo Marconi had sent longwave radio signals over a distance of a couple of kilometres, and since then hardly a year passed in which some researcher did not announce a new species of ray. The wildest of those claims were eclipsed in 1933, however, by Fritz Zwicky's discovery that a considerable fraction of the mass of the universe could not be seen at all. This missing mass became known as dark matter and today it is measured through its gravitational effects.

This was the moment the universe turned out to follow unexpected, even shocking laws. New areas of physics such as relativity theory and quantum mechanics were developed. Biology broke away from its descriptive origins and wrestled for years to reconcile the very different claims of natural selection and genetics. Everywhere you looked, you found unexpected inter-connections between the living and the nonliving, and between the very large and the very small. In 1917 William Harkins realised that nuclear processes are turning light elements into heavy ones, and that our whole world is made, quite literally, of stars. The following year the French biologist Paul Portier showed that the mitochondria powering our cells are direct descendants of bacteria.

This was the moment the world grew complex. In 1918 the English biologist Ronald Fisher used statistics to understand how large populations change over time. It took his peers decades to wrap their heads around his mathematics. Two years later in Germany Hermann Staudinger began Nobel prize-winning work on big molecules, and revealed for the first time the unimaginably intricate world of protein chemistry.

The world grew rich. Scientific investigations that had once been conducted in private and academic laboratories were now being funded by industry. We learned how to mass-produce. We learned how to throw our voices along wires. We learned how to fly.

The world grew healthy. People lived better, for longer. Medicine was changed out of all recognition by new forms of pain control, by germ theory and bacteriology, by lab-based chemical analyses, new diagnostic instruments and pharmaceuticals.

The world grew mindful. In 1894, the Spanish histologist Santiago Ramón y Cajal made the connection between neuronal growth and learning: insight that immeasurably enriched the study of physiology, even as the psychoanalysis developed by Freud, Jung and Spielrein was taking a quite different and utterly compelling 'top-down' approach to the mind.

To this ferment, Soviet scientists contributed the innovations, insights and discoveries that are the chief subject of this book. Though they had their intellectual beginnings as zoologists, psychologists, geologists and botanists, and were steeped in the classic descriptive traditions of nineteenth-century 'life science', my protagonists had complex lives that led them into entirely new areas of research. They operated on a heroic scale: from the biologist who took notes on the physiological effects of his own death sentence, to the botanist who delivered scientific lectures in a lightless underground cell while his wife, none the wiser, was sending food parcels to the wrong side of Russia; from the biologist who resorted to theft, fraud and kidnap to support his work, to the poet–ergonomist who built a machine – an actual machine, with pulleys and ropes – to churn out new forms of human being; and from the psychoanalyst who formulated the concept of the 'death instinct', to the zoologist who led an expedition to French Guinea to obtain a crossbreed of human and chimpanzee. (The film *King Kong* was released a year later.)

These were individuals who did pioneering research on extending human lifespan as well as on language, brain function and child development; founded the first management consultancy; explored the effect of living matter on rocks and minerals while building a model of the evolution of the biosphere; showed how Darwin's theory of natural selection could be reconciled with the findings of genetics; invented the modern conservation movement; and devoted decades of international exploration to amassing a seed collection that was one of the scientific wonders of the day. (During the Second World War, Adolf Hitler established a commando unit to seize this seed bank, hoping one day to control the world's food supply.)

The human cost of that peculiar, twisted, and ultimately tragic marriage between the state and its scientists was horrendous. Nevertheless, I hope I can make clear, and celebrate, what Soviet science managed to do for us.

PROLOGUE : Fuses (1856-1905)

For centuries the government has regarded knowledge as a necessary evil.[1]
  Vladimir Vernadsky

Every year, between 1550 and 1800, Russia conquered territory the size of today's Netherlands until, in the eighteenth century, it dawned on European writers that Russia had become larger than the surface of the visible moon.[2]

With no natural geographical boundaries to speak of, Russia's only means of defence was to control its neighbours, using them as buffers against a possible attack. To do that, it built up what by the nineteenth century was by far the largest standing army in Europe – an army that even in peacetime swallowed nearly two-thirds of the nation's annual budget. (Education and health took 7 per cent.)

It was a ramshackle sort of empire: a vast agglomeration of backward colonies, held together by military force and hootch.[3] There was no money for roads, let alone hospitals, let alone schools. The army itself suffered: there was no strategic railway network and in 1875, during Russia's military adventures in the east, the War Ministry in St Petersburg told a Russian commander that while he could have additional troops, he should not expect them for almost a year since they would 'have to walk from Europe to Asia'.

Without civic institutions, politics is devilishly hard to do. In Russia there were no institutions for reformers to reform: no councils, no unions, no guilds, no professional bodies, few

1

schools, few hospitals worth the name; in many places, no roads. Peter the Great, attempting to modernise his nation at the end of the seventeenth century, reached for a 'Prussian solution' to the problem of governing such a large, sprawling, uneducated mass. He saw to it that an elite was educated abroad, in Western Europe; on its return, this elite was expected to take up the reins of what was still an essentially feudal system. For the masses, modernisation consisted of containment, regimentation, curfew and exemplary punishment.[4]

By the late nineteenth century, Russian civic society had become more complex, and there was some political life to the place. But the dream of state power refashioning the land by fiat persisted. It appealed even to those whose ultimate vision was a stateless society. No one, from the tsars to their fiercest opponents, ever felt that they could trust Russia's illiterate and suspicious masses with the task of creating their own forms of government. Lenin's nannyish dream of a 'tutelary state' was fuelled by this mistrust, while Stalin's forced collectivisation of agriculture took tsarist megalomania to new depths. By the time dissidents were being dumped in mental hospitals under Leonid Brezhnev, state megalomania had entered realms of Alice-like surrealism. The communist ideal did not fail; it was never really tried, and the shadow of the Prussian solution hung, and still hangs, over all.

The pity of it was that Russia had never been poor. At the start of the twentieth century it was the world's largest grain exporter, and its fifth-largest industrial power. 'In terms of the size of its population Russia occupies first place amongst the civilised countries of the world,' boasted a government statistical annual in 1905.[5] On almost every other scale, though, Russia came last. In 1913, the empire had the lowest *per capita* income in Europe save for the Ottoman Empire. The average life expectancy – thirty years – put it about 150 years behind Britain and the United States.

What held Russia back was its form of government. For centuries the tsars had maintained tight bureaucratic control over the smallest details of national life. It was a good way to expand and settle a vast territory. It was a hopeless way to develop a big economy.

Western commentators bemoaned Russia's failure to adopt capitalism: without a free market, how would Russia ever emerge from its dark age? They were right, in a way: capitalism would have made all the difference to Russia, just as, over the preceding hundred years, it had made all the difference to Western Europe and turned a small island nation on its Atlantic periphery into the hub of an empire on which the sun famously never set.

The trouble lay in the iron physical limits imposed by Russia itself. Quite simply, whenever capitalism tried to penetrate Russia's heartland, it caught a cold and died.

Capitalism depends upon surpluses. Farmers grow surplus food to feed the cities; and factories in the cities make the machines the farmers need to grow more food. Out of this virtuous circle, a capitalist economy is born. But peasants in Russia had never produced those kinds of surpluses. Their communal style of agriculture was not geared to feed cities. It was geared so that people in the countryside wouldn't starve. We have good agricultural and climate data for Russia going back over a thousand years, from the year 873. In that span of time, Russia weathered a hundred hungry years and more than 120 famines.

Russia is as cold and as barren as it is big, and it gets colder, by tens of degrees, the further east you travel. When a shivering Ephikhodof quips in Chekhov's play *The Cherry Orchard*, 'Our climate is not adapted to contribute,' he is not kidding: a full third of Russia languishes in the permanent grip of ice and snow. Then there are its great rivers. Most of these flow northward, away from the potentially fertile lands of Central Asia, and into the Polar Sea. Three-quarters of Russia's population and its industry can directly access only about a sixth of its water.

Then there is the land shortage. Incredibly, in an empire with more surface area than the visible moon, Russia has insufficient fertile soil to feed its population easily. It relies on a narrow belt of fertile black earth which passes from the Danube through north-east Ukraine and just north of the Black Sea to Akmolinsk in the east. South of this belt rainfall is ten inches per year and much too arid for normal agriculture. To the north the rainfall is adequate but the soil is poor, the growing season short and the winter frosts very severe. That belt of fertile black earth that runs through the middle was already under the plough by the 1880s and could not hope to supply Russia's growing needs as its population boomed. (Between 1890 and 1913 cereal production actually shot up by more than a third – but it was promptly consumed.)

For enthusiastic students and reformers of agriculture it was clear that Russia desperately needed new strains of cereal and new varieties of fruits and vegetables that could flourish on the thinner, drier, colder lands outside the black-earth belt. Unfortunately, none of these Western-educated, academically trained enthusiasts were practical farmers, and none of them had the political clout to extend credit to millions of impoverished peasants. Their improved varieties of grain went unsold as Russian farmers persisted with their local, mongrel varieties. These were so heavily infested with weeds, it was generally assumed among farmers that seeds of wheat could grow into rye or wild oats.

Farming in Russia's north was not exactly primitive. It wasn't mere subsistence farming. It was *communal* – and it generated no significant surpluses whatsoever.

In the central and southern regions, by contrast, farming was arranged according to a feudal system in which serfs worked under the immediate rule of the nobility. This was an obvious and apparently fixable brake on the nation's progress, and even the notoriously repressive Nicholas I created committees to consider reform.

It was Tsar Alexander II who finally took the bull by the horns. In 1861 he released nearly half the population of the country from bondage, by freeing the empire's 22 million serfs. The transfer of land ownership was massive, as the nobility sold up and became absentee landlords. Very few used their generous state compensation payments to improve or modernise their estates. Meanwhile, emancipation gave the peasants 13 per cent less land than they had previously farmed. In the more fertile parts of the country, former serfs had to forgo up to half of what they had previously tilled.

Taxes were levied on land, and were inflated to take into account wages earned by the newly liberated young as they ran off to the towns and cities. But whatever these youths made was far more likely to be spent in bars and shops than be sent home to the communes. Moscow in particular, with its wild mix of contemporary buildings and ramshackle little homes, palaces and factories, crooked lanes and wide streets, squares and boulevards, drew young men from across the region. The communes emptied out: only women and the elderly remained, and agriculture declined. 'Peasants living in Moscow send scarcely half of the taxes,' ran one report from Klin District, north-west of Moscow, 'and the family struggles with need year-round, living half-starved. And those peasants who remain in the village, seeing around themselves decline in everything, cool toward agriculture and spend time in the taverns, which have grown to at least two in almost every village.'[6]

In 1861, in an attempt to stem the haemorrhaging of the countryside (and the food shortages that were becoming a constant of city life), the system of internal passports was strengthened. Fathers of delinquent sons now had a legal means to summon them back to the farm, 'and if you don't settle down, you scoundrel, I won't renew your passport. I'll have you brought home by the police, and when they've brought you home, I'll whip you with a birch rod myself, in the district administrative office, in the presence of honest people . . .'[7] The passport created a paper trail that followed you throughout life.

More or less admitting that the countryside had been made ungovernable, the government made much of the way the peasants would govern themselves through rural communes. For the young especially, this was not at all good news, given the way small rural communities traditionally exercised social control over their members. Public flogging was common.

All in all, the experience of emancipation convinced Russia's liberal elites, few as they were, that nothing short of the imposition of democracy could unpick the mess the country was in. In the cities, meanwhile, among crowds of underfed, underpaid, under-employed young men, an altogether more revolutionary mood was brewing.

On 13 March 1881 Tsar Alexander II rode in his carriage, as he did every Sunday, through St Petersburg to the neoclassical Michael Manege riding academy for military roll call. His route, via the Catherine Canal and over the Pevchesky Bridge, never varied. Among the pedestrians crowding the narrow pavements that day was Nikolai Ivanovich Rysakov, a young member of the Narodnaya Volya – the 'People's Will' movement, bent on igniting a social revolution by any means necessary. The means that day were contained in a small white package wrapped in a handkerchief.

After a moment's hesitation I threw the bomb. I sent it under the horses' hooves in the supposition that it would blow up under the carriage . . . The explosion knocked me into the fence.[8]

The explosion only damaged the tsar's bulletproof carriage, a gift from Napoleon III. The emperor emerged shaken but unhurt – and a second attacker, Ignacy Hryniewiecki, standing by the canal fence, threw his package at the emperor's feet.

The People's Will aimed to save Russia from tsarist autocracy. Their assassination of Alexander II ensured that the regime survived, and grew even more oppressive. The tsar's successor never forgot or forgave his father's ugly death: his legs torn away, his stomach ripped open, his face destroyed. Hope that the tsarist government might or could reform itself had been slim to start with. Under Alexander III it was utterly dashed.

Part One
CONTROL
(1905–1929)

And it seemed as though in a little while the solution would be found, and then a new and splendid life would begin; and it was clear ... that they still had a long, long road before them, and that the most complicated and difficult part of it was only just beginning.

ANTON CHEKHOV, 'Lady with Lapdog', 1899

Previous page: Soviet citizen science: in trials this air velocipede, invented by a worker in Moscow, averaged a speed of over 140 kilometres an hour.

1 : Scholars

In the late nineteenth century in Russia there existed something of fundamental importance – a solid, middle-class, professional intelligentsia which possessed firm principles based on spiritual values. That milieu produced committed revolutionaries, poets and engineers, convinced that the most important thing is to build something, to do something useful.[1]

Physicist Evgeny Feinberg on his mentor Igor Tamm

Head south-east out of Moscow in the morning, and by nightfall you will reach the city of Tambov and, in nearby woodland, a handsome, single-storey wooden structure that but for the modern signage might have sprung magically from the pages of a novel by Ivan Turgenev. It is a museum now. You can wander round Vladimir Ivanovich Vernadsky's study, his library and his living room. There is some simple information here about his life; his politics; how he anticipated, by over a century, James Lovelock's 'Gaia' theory of how living things and the planet's geology work as one system; rather less about his being the godfather of Russian atomic energy.

'To treat the needs of others as if they were one's own': Vladimir Vernadsky (seated, right) and some idealistic friends at St Petersburg University, 1884.

Vernadsky's father Ivan was a professor of economics and statistics in the Alexandrovsky Lycée. His first wife was Maria Shigaeva, one of Russia's early feminists and its first female economist. She died prematurely of tuberculosis in 1860, and in 1862 his father married again, to a distant relative of his first wife, Anna Petrovna Konstantinovich, who would be Vladimir's mother. She was a music teacher, a lively and warm personality, but she did not share the intellectual interests of Ivan's first wife.

In 1868, during a heated debate at the Free Economic Society, Ivan suffered a stroke. He resigned his post at the Lycée, and the family relocated from St Petersburg to Kharkov, where he ran the Kharkov branch of the State Bank. Vladimir's childhood here was a happy one, his memories beginning not with St Petersburg, but in the capital of the Ukraine, listening for hours to his opinionated, white-bearded uncle Evgraf Korolenko, who lived with the family.

In 1886, Vernadsky wrote to his future wife:

I recall dark, starlit winter nights. Before sleep, he loved to walk and, when I could, I always walked with him. I loved to look at the sky, the stars. The Milky Way fascinated me and on these evenings I listened as my uncle talked about them. Afterwards, for a long time I couldn't fall asleep. In my fantasies, we wandered together through the endless spaces of the universe . . . These simple stories had such an immense influence on me that even now it seems I am not freed of them . . . It sometimes seems to me that I must work not only for myself, but for him, that not only mine, but his life will have been wasted if I accomplish nothing.[2]

In 1876 the family returned to St Petersburg. Vladimir was now thirteen years old, and was scouring the bookshops for anything and everything to do with the home they had left. He taught himself Ukrainian and, since a lot of books about the Ukraine were in Polish, he taught himself that language as well.

As a university student in St Petersburg, Vladimir Ivanovich Vernadsky was 'very soft in appearance but very determined once he had set himself a goal'. Vladimir Posse, a medical student who went on to become a leading Marxist journalist, recalled that Vernadsky and Sergei Oldenburg, one of Vernadsky's closest friends, 'had already set themselves the goal not only of becoming professors but also members of the Academy of Sciences'.[3]

Vernadsky and his university friends were better off than most of their fellow students. Vernadsky, born in St Petersburg, Russia's imperial capital, and brought up in Kharkov, capital of the Ukraine, was among the wealthiest of them, having inherited from his father the 750-hectare estate of Vernadovka.

During one all-night conversation among the brotherhood, someone suggested they buy an estate together. It was going to be called 'The Haven'. The plan fell through, but it gave the group a name: Bratstvo Priutino or the 'Haven Brotherhood'. The influence of the novelist Leo Tolstoy on the group is palpable. In devoting their lives to the good of the Russian people, they swore (to quote Oldenburg's formula) 'to work and produce as much as possible, to consume as little as possible, to treat the needs of others as if they were one's own'.[4]

They attracted a lot of girls. Deprived of the right to a higher education, bright young women of their generation constantly sought whatever intellectual outlet they could.[5] They supported and helped run the Brotherhood's St Petersburg Committee of Literacy, preparing reading materials and reading lists, and setting up lending libraries.

Vernadsky's was not the only marriage to come out of this meeting of minds. The Brotherhood could be dreadful prigs, though: the wedding of Vladimir to Natalia Staritskaya, complete with frock coats, wedding gowns, engraved invitations and an orchestra, was boycotted by his abstemious friends.

Vernadsky, a mineralogist, had arrived to study at St Petersburg University at an opportune time. Mendeleev, Butlerov, and Dokuchaev were his mentors. Vasily Vasilievich Dokuchaev held the chair in mineralogy at St Petersburg University, and dispatched Vernadsky on fascinating, exotic, and sometimes dangerous scientific missions to various corners of the Russian Empire. Alexander Mikhailovich Butlerov was one of the pioneers of modern chemistry. More than ten years before Becquerel's discovery of radioactivity in France in 1896, Butlerov was arguing that the atom was divisible, and Vernadsky was witness to the lively debates he had with Mendeleev over this issue. Dmitry Ivanovich Mendeleev had formulated the periodic table of elements. When he lectured, the halls were packed. Listening to him, 'we entered a new and wondrous world . . . as if released from the grip of a powerful vise'.[6] From these men, Vernadsky acquired a view of an earth in constant flux, its elements flowing and spiralling through the earth's crust over geological time.

Like all his generation, Vernadsky went abroad to further his studies. He went to the University of Naples, and the world-renowned crystallographer Professor Arcangelo Scacchi, only to discover that the old man was succumbing to senility. He went on to Munich, and the laboratory of mineralogist Paul Groth. From his letters we know Vernadsky had a fine time there, a kid locked in an intellectual candy store.

In 1887 a son was born (George Vernadsky would later find modest fame as a historian in the USA). While Natalia returned to her family's dacha in Finland to look after the baby, Vladimir developed friendships and contacts that would shape his later career. In the summer of 1888, walking in the Alps, he had his epiphany: he saw that mineralogy, studied the right way, as a science of change and energy transfer, could connect cosmological history with the history of life itself. Vernadsky's fascination with earth's development at a cosmic scale would last him his whole career, though he worried that men like Groth would 'take me for a fantasiser'.[7]

Moving to Paris in 1889 ('really the most grandiose city I ever saw'), Vernadsky went to work at the Collège de France. To a

Russian, the Collège must have seemed an odd institution: there were no students as such, just professors (who were, however, obliged to deliver lectures), small labs and a staggeringly good library. Here, study was being given room to breathe. Researchers were well-resourced and their ideas were taken seriously. Here, leading a life of the mind was unlikely to land you in trouble with the authorities. It was a different kind of life.

In Vernadsky's homeland, universities were teaching institutions, not centres of research, and most certainly not intellectual melting pots. (The tsarist bureaucracy itself recruited from just a handful of expensive institutes, closed to everyone but the children of the aristocracy: the Corps of Pages, the Alexandrovsky Lycée, the Institute of Law.)

Repression of higher education was a fixed policy, dating from the thirty-year reign of Nicholas I. Inspectors watched the students, meting out punishments for scruffy uniforms or long hair. One student in Kiev University who appeared at a compulsory religious service without a proper uniform was thrown out of the church by an inspector and expelled from the university the next day.

When Nicholas I died in 1855, the government had tried undoing the harsh regime he had imposed. In Kiev, delighted Polish students marched through the streets in national dress. In Kazan, they wore animal skins. In Moscow and St Petersburg, students took to wearing peasant costumes, showing solidarity with the soon-to-be-liberated serfs. Appalled at what it had unleashed, the government promptly raised tuition fees, banned student assemblies, and reintroduced all the old rules on behaviour and uniform. This new repression lasted decades. Student-run organisations, 'reading rooms, dining halls, snack bars, theatres, concerts, balls, any meetings not having an academic character', were banned – and God forbid you should show any 'signs of approval or disapproval at lectures'. Punishments included admonitions, confinement in the *kartser* (the university jail) for terms ranging up to four weeks, suspension and expulsion.

Returning to Russia and a professorship at Moscow University, Vernadsky found little had changed in his absence. The city was dusty and provincial, and it literally stank. It was also oppressive. Middle-aged men in bowler hats passed by his home each morning as he set off to work. He used to offer them a cheerful greeting until one day, as he was leaving for a European trip, he spotted one of them tailing him through the railway station, and realised they were undercover policemen.

In the beginning of the 1890s [the police report runs] Vernadsky moved . . . to live in Moscow, where he continued his dubious acquaintance-ships, took an active part in evenings organised by students of Moscow university where he gave speeches about the necessity of coming together of professors and students for purposes of political education of youth and struggle with the present regime.[8]

Life at the university was dismal. It lacked even the most basic texts in his subject, and the mineralogical collection had not been catalogued since the 1850s – nor, for that matter, dusted. The place was riddled with corruption and the junior administrators were the worst of the lot, cutting up rooms meant for laboratories into unofficial student housing. Vernadsky had a good idea how this was all going to end, and spent time outside Moscow sorting out and refurbishing Vernadovka in case he lost his job.

Vernadsky reckoned that if push came to shove, and he and his family had to live there all year round, they could comfortably make do on what their acreage could provide. In the autumn of 1891, however, came the catastrophe that galvanised Vernadsky's political career. Famine struck Tambov and many places besides, razing the harvest across Russia's vital belt of black earth.

The crisis had been building for a year. In 1890, a dry autumn had delayed the sowing of winter cereals, and then winter had arrived much earlier than usual. It was dry, too: there was not enough snow to provide a blanket from the cold, so the winter crop froze to death.

Spring brought further trouble. Harvests in eastern Europe were equally dismal, and these countries had money to hand, so Russia's spring crop, instead of feeding its own mouths, was immediately snapped up for export.

Bread was scarce even in Moscow and St Petersburg. Lenin described the hunger bread of that time as 'a lump of hard black earth covered with a coating of mould'. In the countryside, people bulked out their dough and porridge with straw and weeds.

As 1891 wore on, conditions went from poor to calamitous. Five months passed without rain. The summer was far too hot and dry to plant out vegetables, but farmers had little choice but to chance it. Their plants withered and died. And after all that, a deluge: winter cereals planted that autumn were washed out of the soil by torrential rains.

Come the spring of 1892, farmers were watching in horror as the wind blew away their precious black earth in dust storms, 'concealing the sun's rays and turning day into night. Witnesses unanimously testified that the phenomenon had such a dreadful and frightening character that everyone expected "the end of the world",'[9] recalled soil scientist Per Zemyatchensky. Trains were halted by drifts of earth, and crops killed by blasts of dust. Swathes of the country were stripped of all vegetation; not even weeds remained. Farmers killed their livestock for food.

The catastrophe was epic. The commercial attaché in the British embassy in St Petersburg, E. F. G. Law, reckoned that the Russian government had 'to find the means of supplying a deficit of food to 35,500,000 people in sixteen provinces'. Even in the relatively well-off province of Tambov, the peasants lost over half their livestock. Vernadsky's estate manager wrote to tell his employer that they were selling their animals to the local gentry for a pittance, and about a quarter of them were already making 'famine bread', mixing their dwindling supplies of rye flour with hay, and even brick dust. They were knocking on the doors of Vernadovka for help.

Vernadsky did not immediately rush home. He realised that he could do more good by remaining in Moscow. A gifted bureaucrat, he assembled a relief effort among his friends. A retired neighbour, V. V. Keller, travelled tirelessly, informing him of the situation across the district. With another friend, L. A. Obolianinov, Keller visited Leo Tolstoy to study his methods of famine relief, and reproduced his organisation in Vernadovka. In Moscow, the historian Alexander Kornilov quit his government job to back the relief effort. The medievalist Ivan Grevs joined in; there were several future politicians, and even, under conditions of strict anonymity, the tsar's own uncle, Grand Duke Nikolai. This skilled, ad hoc administration made people's efforts count in a way liberal good intentions had never counted before. By July 1892, as the crisis eased, there were 121 famine relief kitchens in Tambov feeding 6,000 people; 1,000 horses had been saved and 220 more were gifted by lottery to horseless families.

And this raised a question: if a bunch of professors could do this sort of thing, why couldn't the government?[10]

1891 had given liberal opponents of the regime – the impotent *intellecty* parodied in the stories of Turgenev and Chekhov – a brief taste of civic power. They had enjoyed it, made the most of it, proved to their own satisfaction that they were worthy of it, and they wanted more. Their model response to the famine – scientific, rational and, in the best sense, bureaucratic – had given hope to Russia's demoralised educated class. Vernadsky and his friends had shown by example what it would be like for capable people to really participate in the running of their country. The vision spread. To realise that vision, however, required organisation.

The Union of Liberation was founded in July 1903 and campaigned publicly (and peacefully) for an end to autocracy.

Its tiny membership – just twenty liberals and radicals – held meetings in Vernadsky's apartment in Moscow. Vernadsky wrote to his wife: 'I consider that the interests of scientific progress are closely and inextricably tied to the growth of a wide democracy and humanitarian attitudes – and vice versa.'[11]

The difficulty was in attracting political support outside the tiny, well-heeled liberal coteries of Moscow and St Petersburg.

On Sunday 22 January 1905, more than 300,000 striking workers and their families walked towards the Winter Palace in St Petersburg, bearing icons and singing hymns. They came to petition the tsar for better labour conditions and an eight-hour working day. The Imperial Guard opened fire on them, leaving a thousand dead or wounded. Passing through the Alexander Gardens that day, suitcase in hand, was a young field geologist, B. A. Luri, Vernadsky's most promising student. Soldiers shot him twice in the back. In an angry article to a leading liberal newspaper, Vernadsky declared that 'one more victim has fallen in the long martyrology of the Russian intelligentsia'. But the time for writing stiff letters to the papers was long past.

Following the massacre, the students went on strike. The government, in an uncharacteristic gesture, polled the faculty councils on whether or not to resume classes. Perhaps they meant to give professors the impression that their opinions mattered. Whether they mattered or not, the professors spoke out. Not one university agreed to resume teaching. The councils declared that political reforms were necessary to secure peace in the universities. Vernadsky made a public appeal to his academic colleagues to break with tradition. They were independent scholars and teachers, not state hacks. They couldn't go on letting themselves be pushed around as if they were 'teaching on some godforsaken Philippine Island'.[12]

So the professoriate took a step that was blatantly illegal: they organised an Academic Union, declaring 'that academic freedom is incompatible with the existing system of government in

Russia', and by August had enrolled more than half the university teachers in the country.

All that remained was that some critical national event should give the Union the chance to show its mettle. Four months later, on 27 May, Admiral Togo of Japan launched an attack on the Russian fleet at Tsushima, and by 28 May over half the Russian fleet lay at the bottom of the Pacific. It was the Union's big chance – but they absolutely bottled it.

'All means may now be legitimately used to fight the danger represented by the continued existence of the present government.' So ran a dramatic appeal from the Union of Unions, a broad alliance of professional associations; the Academic Union was a member. 'Bring about the immediate elimination of the bandit gang that has usurped power and replace it with a constituent assembly . . . so that it will as quickly as possible end the war and the present regime.'[13]

But where was the Academic Union? The professors had stayed at home. Fearing the Union of Unions had become too radical, they had boycotted the congress.

Then the nation's students did something unexpected, something that triggered a chain reaction that exploded into the great general strike of mid-October, and the 1905 Revolution. They went back to college.

And they did not come alone. Following a series of open political rallies, curious workers began visiting the universities. No one was quite sure what to do with them. 'We had neither agenda nor speakers,' one student recalled. 'I began with a few words of welcome, suggested that we discuss the current political situation, and turned the meeting over to the floor. The ensuing discussion was utterly chaotic. Some of the volunteer speakers were wholly inarticulate.'[14]

News spread through the factories: the police were not interfering. The crowds grew. Whole factories turned up at the college gates without notice. Workers stood up to read their own poetry.

Come October it looked as though the buildings would literally collapse under everyone's weight.

In Moscow the chief of police warned that any meeting spilling onto the streets would be fired upon. Closing the university on 22 September, the rector, Sergei Trubetskoy, explained: 'University is not the place for political meetings. It cannot and should not be a public square, and by the same token a public square cannot be a university. Any attempt to turn the university into a popular meeting place will destroy it.'[15]

Given his responsibilities, Trubetskoy could do little else but close the university gates. But this pushed the workers into the path of the Moscow police, and street violence escalated sharply.

In an article dated 4 October, Lenin himself summed up, none too sympathetically, the impossible bind in which the professors were now trapped:

They closed the university in Moscow because they feared a bloodbath there. But by their action they caused a far greater bloodbath in the streets. They wanted to extinguish the revolution in the university, but they ignited the revolution in the streets. These professors have stumbled into a real vise.[16]

*

In June 1905 in Odessa, a general strike deteriorated into a mêlée between strikers, reactionaries and the local authorities. The crew of the battleship *Potemkin* mutinied in the harbour. Two thousand were killed, three thousand wounded.

On 8 October, Moscow's railroad workers struck. The strike became general: by 13 October Russia was paralysed. Two days later the tsar called his close advisors to the palace. The choice facing them was stark: concede major reforms or abdicate the running of the government to the military. Two days after that, Nicholas II pledged to endow a national assembly – the Duma – with legislative powers. Vernadsky and activists within the intelligentsia hurriedly constructed a political party to represent liberal opinion, the Constitutional Democrats – Kadets, for short. (Wags nicknamed them the Professors' Party.)

But this little, late experiment in parliamentary democracy was never going to work. The fractures that had opened up between the landowners and the people, the reactionaries and the radicals, the young and the old, had grown too wide, and the so-called 'liberal revolution' was ushered in with a bloodbath. 1905 led to far more casualties than the subsequent, decisive revolutions of 1917. A week of pogroms greeted the imperial manifesto, as those who depended on the status quo – small merchants, petty artisans and casual labourers – attacked those they believed had ruined them. Students, liberals and Jews were their particular targets.

In Moscow, in an attempt to restore order, the army moved in. In December 1905 they shelled the city, killing over a thousand people.

The peasants revolted. In an attempt to quell rural rioting, thousands were sentenced to death by military court-martial, and 14,000 were shot dead. Punitive expeditions scoured the Baltic, Poland and the route of the Trans-Siberian Railway sacking, burning and killing. 'Don't skimp on bullets,' came the order, 'and make no arrests.'

The gentry, in a panic, rushed to sell up. Land prices and rents collapsed.

There was hunger that winter; nothing like on the scale of 1891–2, but bad enough, and made far worse by the political chaos. In Tambov, Vernadsky, his son George and the Vernadovka estate's overseer organised famine relief, and took the opportunity to tell starving peasants about the forthcoming parliamentary elections. They were promptly arrested. George was free after a week but it took Vladimir a visit to the prime minister, Count Sergei Witte, to get his overseer's family out of jail.

In the capital, St Petersburg, 'an icy shudder of disenchantment with parties, politics and organisations pierced the hearts of all like a knife. Bitter quarrels arose. Finally no one gave a damn about anything . . . There was no faith in anything or anyone.'

This grotesque picture of the city comes from the poet Alexei Kapitonovich Gastev, returning from Stockholm in the spring of 1906:

Desperate, exhausted people stagger down magnificent Nevsky Prospekt. Miserable, useless, they commit suicide in the Neva and in the Fontanka and Mojka canals, they throw themselves from tall buildings onto the pavement, shoot themselves, take poison, even hang themselves on the crosses in the cemeteries. Suicides became so common that the newspapers finally took notice of only the most 'interesting' cases.[17]

The First Duma met first in April 1906. But it was a pale imitation of what the tsar had promised, and when the Kadets protested, it was promptly dissolved. In April 1906 Peter Stolypin became prime minister and set about expunging all trace of radical thought, dissolving the second Duma in 1907. As everyone involved started watching out for themselves, the Kadets quickly lost support. The Third Duma was a rubber-stamp affair, dominated by the landed gentry. Tsar Nicholas II was becoming ever more obsessed with his own royal privileges and, as he got rid of honest advisors, toadies and charlatans like Rasputin took their place.

The mood of national hopelessness is not hard to spot. The novels, poetry and plays of the period smack of apocalypse. Blok, Bely and Briusov all wrote as though the End Times were nigh, if they had not already arrived. Years before he produced his celebrated dystopian novel *We*, the satirist Yevgeny Zamyatin was writing stories that welcomed catastrophe as the quickest route out of Russia's political impasse.

On 20 November 1910 the writer Leo Tolstoy died. Though his political views were outdated, the young had regarded him with great affection. Students in universities all over the country held memorial services and passed resolutions promising to honour Tolstoy's memory by fighting against capital punishment. In

St Petersburg, thousands of students gathering for a demonstration on the Nevsky Prospekt were confronted by the city's police. When the pro-rector of the university, Ivan Andreev, tried to persuade the students to disperse; they lifted him on their shoulders and carried him along as they sang.

By the end of November, over 400 students had been arrested. The *Russian News*, Moscow's liberal newspaper, recorded how:

many police with rifles and fixed bayonets [were] ranged along the long corridors of the university; professors, along with small numbers of students, are escorted by armed police to lectures. Armed police stand in the lecture halls. Students storm in, trying to interrupt lectures with whistles, noxious gases, and the singing of revolutionary songs.[18]

One professor, accosted by a student, collapsed in hysterics and had to be carried away.

Police stormed Moscow University. With their institution now fatally compromised, its three highest officials decided to resign. With them went Vladimir Vernadsky and more than a third of the teaching staff. It looked as though the university was finished. By now, though, it hardly mattered. Both professors and students had better uses for their time.

Why fight the dismal state education system, when you could simply replace it? In Germany a network of well-funded research institutions, the Kaiser Wilhelm Gesellschaft, was being created. What if Russian entrepreneurs could do the same?

Moscow's professors came in the main from merchant families or had personal ties to local merchants and industrialists. They set up entrepreneurial organisations based on German and British models to lobby for private funding. They taught at private universities and on highly regarded women's courses, set up by liberal-minded industrial philanthropists. Many of the professors who resigned from Moscow University never went back. They enjoyed more freedom teaching outside the system, in institutions like Moscow City People's University, founded in 1908 on endowments from General Alfons Shaniavsky, a retired Polish officer and gold magnate who had already piled cash into women's medical education.

The state wouldn't employ the Shaniavsky's graduates the way it employed those who had earned a 'real' degree, but no one cared. Few imagined the state would last past the next winter. Courses in public administration, the cooperative movement, public health and education conscientiously prepared a new generation for a new era.

Having resigned from Moscow University, Vladimir Vernadsky was elected to the Imperial Academy of Sciences and moved to St Petersburg. As an able administrator, he nursed ambitious plans for learning, and in particular for the physical sciences. With his favourite student, Alexander Fersman, he lobbied the government to develop the country's mining potential. Annual expeditions with his students to remote regions of the empire identified and mapped sites rich in aluminium. But it was hopeless. In the summer of 1913, in Canada to attend an International Geological Congress, he wrote to his wife Natalia:

Over the past ten years the United States has made huge advances in science: today America gets along without the help of German universities, which not long ago was considered indispensable. When I involuntarily compare these years in Russia – the activity of Kasso and Company, that whole gang that constitutes our Imperial government – I feel depressed and uneasy.

Come the war, when a rich deposit of tungsten in Turkestan was put out of bounds because one of the grand dukes owned the land, another academician, the mathematician Alexei Krylov, went a lot further. If they lost the war, he snapped, the grand dukes weren't the only ones who stood to lose everything; the whole dynasty would go to the devil's mother.[19]

In the aftermath of the assassination of Austria's Archduke Franz Ferdinand, as Russia toppled towards war, a crucial weakness was revealed: Russia's utter dependence on trade with Germany. Its science, its industry, its very economy depended on good relations with the country it was fighting.

The hopelessness of the situation was movingly summed up on 18 July 1914 by Alexander Sergeevich Serebrovsky, later one of the Soviet Union's leading geneticists. He wrote in his diary:

I am a soldier now. The city was decorated with flags and so on. God, what a sea of unrestrained lies in all this . . . Many people I know were waiting for a manifesto and equal rights for nationalities. Jews were walking with the torah singing God, Save the Tsar. Some intellectuals shouted hurray when the tsar passed by.

All these expectations turned out to be crap. The tsar didn't have enough state wisdom. He could with one scratch of his pen affirm his throne for many years to come, create a faithful Russia for himself, and win the sympathy of the people, including the intelligentsia. They could really reach the unification of the people and the government. Oh, I know that I would have gone to the front with a totally different feeling and state of mind . . . They drove expectations to the highest tension, made even streetcar drivers take part in parades, and gave nothing . . .

I love Russia too much to wish it a victorious war now. A victorious war for Russia is like a return to a distant past. Losing the war is another step forward to the bright future.[20]

2 : Revolutionaries

Across the Earth, from corner to corner of the world revolutions howl. The war that was begun to delight the kings, the tsars, the presidents, has become a tornado that tears down imperial palaces, burns royal mantles, sends crowns flying and turns kings into dust. The world in which everything seemed to be arranged so beautifully has collapsed . . . And we want to be the newcomers; we shall lift the curtain from the cities, the streets, the workshops, the bazaars . . . We shall immediately set our factory of art humming.[1]

  Alexei Gastev

Karl Marx and Friedrich Engels liked to think of themselves as scientists. At least, they considered themselves philosophers of science.

But even that isn't quite right: German has no separate word for science; nor does Russian. The German term *Wissenschaft* (enquiry) and the Russian term *nauka* (knowledge) make no clear distinction between sciences and humanities, let alone between physical sciences and their 'soft' social-science cousins.

Friends and rivals: Vladimir Lenin plays chess with Alexander
Bogdanov on a visit to Maxim Gorky on Capri. Italy, April 1908.

That these languages lack a dedicated term for the physical sciences reflects their rich nineteenth-century cultural heritage. Back then it was still possible to believe that the universe was knowable, and that one kind of knowledge was, at least in principle, compatible with all other kinds of knowledge. People thought that the universe could not abide contradictions.

It was this dream of uniting all disciplines under one roof that captivated Friedrich Engels. He constructed an entire philosophy around the idea, called 'new materialism'. (It was a later philosopher, Georgy Plekhanov, who stuck the slightly baffling handle 'dialectical materialism' to it.)

The world needed a 'new' materialism because the 'old' materialism of Newton would no longer serve. No one seriously disputes that Newtonian mechanics is a scientific triumph: a productive, insightful model of how the physical world ticks. The trouble is, the more widely we apply it, the more alienated it makes us feel. Vital for Newton's success was the presence of organised religion to explain and describe some of the major gaps in his account of reality. (Isaac Newton was, unsurprisingly, a profoundly religious man.)

By the beginning of the nineteenth century, however, evidence against the 'revealed knowledge' of the Bible was piling up at an unignorable rate. Scientists and philosophers were having to look elsewhere for an explanation for non-Newtonian phenomena: things like love, and grief, and memory, and the colour green.[2]

Engels's philosophy is 'materialist' because it finds a home for these subjective experiences in the physical world; it doesn't imagine that minds exist in a separate, spiritual realm. It is 'dialectical' in the sense that all knowledge is obtained through reasoned argument and enquiry. Knowledge is provisional, because *we* are provisional. We're not the same people we were yesterday, and none of us will be around forever. So knowledge is being constantly re-reasoned, re-confirmed and re-believed.

Dialectical materialism is interested in how things change. Everything, according to dialectical materialism, has a past and a future, and this history matters if we are to understand how the world works.

Marx, with his lively interest in economics, history and psychology, bought into nineteenth-century 'scientism' in a major way, and he believed that, through Engels' more elegant 'new' materialism, it might be possible to extend the natural sciences into all spheres of life.

He believed, that is, in scientific government.[3] His idea was not to politicise science, but to extend science into politics, to the point where there could be no distinction between knowledge and policy. In 1894 Lenin wrote this about Marx:

The irresistible power of attraction that draws socialists of all countries to this theory lies precisely in the fact that it unites a rigorous and most lofty scientism (being the last word in social science) with revolutionism, and unites them not by chance, not only because the founder of the doctrine combined in his own person the qualities of a scientist and a revolutionary, but unites them in the theory itself intrinsically and inseparably.[4]

Lenin's vision of science found its embodiment in one remarkable man – a polymathic talent and close friend who founded the Bolshevik movement with him in 1904. It took Alexander Alexandrovich Bogdanov years to complete his most important philosophical work, *Tektology: Universal Organisation Science*, and by the time of its completion he found himself a renegade. Expelled from the Party, he became, in his own words, an '"official devil" who had to be "foresworn", who had to be "blown and spat upon" as in the ritual of Christening and who could be used to frighten unruly infants into obedience in matters of theory'.[5]

Bogdanov dreamt of a synthesis of all the disciplines, and sought to impose his scientist's philosophy over the whole of human experience.[6] In his Utopian novel *Engineer Menni*

(1913), set on Mars after the creation of a communist society, he imagined the triumphant birth of a 'Universal Science', compared to which 'the philosophy of former times was nothing but a vague presentiment'. Being universal, this science would describe everything in terms of everything else. This meant that everyone, however ill-educated, would be able to understand it.

It was capitalism that had fragmented scientific progress, shattering it as surely as God had shattered Babel, so that every discipline spoke a different, esoteric language. Separated from 'labour', scientists had come to believe that the more their discipline baffled outsiders, the closer it was to the truth of things. This pursuit of 'science for its own sake' was a tragic error. Come the dawning of a truly socialist society, practice and theory would once again be fused, and science could at last be put to the service of society. Real science was applied to the benefit of mankind, or it was nothing.[7]

Reduced to one line, what Bogdanov meant was this: *there is no such thing as pure science.* The next sixty years would test the truth of this assertion.

The careers of Russian revolutionaries, under close observation by the tsar's secret police, followed a predictable pattern. After a spell of exile in Central Asia or Siberia, one would generally flee to some foreign city: Geneva, or Paris, or London, some liberal place where one could compose inflammatory material and smuggle it back into Russia.

In 1883 a group of Russian exiles in Geneva established the Liberation of Labour group to educate Russian revolutionaries in the principles of Marxism. Lenin[8] visited them in 1895. On his return to Russia he was arrested, imprisoned for just over a year, then exiled to Siberia. He left Russia again in 1900. Through *The Spark*, a newspaper written abroad and distributed in Russian cities, Lenin made himself the centre of the Russian Social Democratic Workers' Party – a portmanteau grouping of revolutionary organisations. In 1902 the Marxist philosopher Leon Trotsky joined the paper's staff.

It took hardly a year for a major split to develop within the Social Democratic Workers' Party. Lenin, Bogdanov and the majority wanted Party membership limited to activists. A minority, including Trotsky, wanted to encourage broader support. The controversy, pitting purity against compromise, divided the Party into Lenin and Bogdanov's Bolsheviks (from *bolshoi*, meaning large), and Trotsky's Mensheviks (from *menshe*: smaller).

By 1905 the factions were so estranged that they held their congresses in separate cities – the Bolsheviks in London and the Mensheviks in Geneva – and they quite missed the 1905 revolution.

Exiled after the revolution petered out, Bogdanov hid out in Finland with Lenin and his family, and supported Lenin through his acrimonious split with Leon Trotsky. Lenin and Bogdanov were the closest of friends, and, with Lenin's wife Nadezhda Krupskaya, established a seemingly unbreakable bond.

But 1905 was also the year Bogdanov's thinking began to stray from what Lenin considered the straight path. The failure of the 1905 revolution had convinced Bogdanov that Russian workers were not ready for political power. Bogdanov's political education had begun while delivering lectures to factory workers; now he wanted the Bolsheviks to ally themselves to underground workers' organisations. Lenin's continued dabbling with the Duma seemed a waste of time. Increasingly, therefore, Bogdanov began to resemble the sort of puffed-up intellectual revolutionary that Lenin despised: the sort who joined the Party for the thrill of conspiracy, but had no patience or aptitude for the churn and grind of day-to-day politics.

On the other hand, Bogdanov was extraordinarily good at getting hold of money. His piratical raids on the funds of political opponents, and the districts harbouring them, had sustained the

Bolsheviks for years. And, as the two old friends began to bicker, Lenin found it harder and harder to fund his own projects.

Bogdanov was fascinated with the possibilities of mutualism, workers' power, and trade unions. Like Western Marxists (and unlike Lenin), Bogdanov was interested in people's alienation from the world and from each other. The individual 'I', Bogdanov argued, is a capitalist artefact, reflecting society's emphasis on private property and individual profit. In a socialist society, the self will exist only as part of the whole, and 'people will be immortal' in so far as they 'develop their "I" beyond the limits of the individual and more toward the workings of community.'[9]

What, he wondered, would it take to turn a society of the alienated 'I' into the socialist utopia of the 'we'? A keen student of Georges Sorel's book *Reflections on Violence* (which was translated into Russian in 1907), Bogdanov argued that workers needed a myth to inspire them to action.

Although subsequent events give Bogdanov's talk of 'we' overcoming 'I' a sinister edge, it is hard to see at first why his well-intentioned handwaving should have aroused his friend Lenin to such fury. But arouse him it did. Bogdanov's three-volume *Empiriomonism* – a fully worked-out programme for the creation of a new and truly revolutionary mutual culture – drove Lenin, who didn't have the cleanest mouth to start with, to a correspondence so extraordinarily abusive, Bogdanov returned Lenin's long letter (it took up three notebooks and was passive-aggressively titled 'A Declaration of Love') with a note saying that if Lenin wanted to remain his acquaintance, what he had written would have to be considered 'unwritten, unsent and unread'.

Even contemporaries thought 'Lenin was going slightly out of his mind' to be so exercised over Bogdanov's philosophising. But the more one understands the predicament of Bolshevism at the turn of the century, the more one realises what was at stake. The two men were fighting for the nothing less than the soul of the revolution.

For Marx and Engels, the history of science had been a story of constant improvement. Science pursues a complete explanation of the world. Like Zeno's tortoise, science will never entirely achieve this high goal. But the gap between what we know and what is actually out there will always be closing. Over time, science gets closer and closer to the Truth.

But while Marxists were enthusing over the scientific potential of Marxism, large sections of the scientific community were turning *away* from scientism, *away* from ever-more-reasonable explanations of things, and towards descriptions of the world that were counter-intuitive and mutually incompatible, and in some cases (notoriously so in quantum physics) literally made no sense.

If we want to draw a line in the sand and pick a date when the tide turned, we could do a lot worse than plump for 1898, the year 'Mme Curie threw the bomb . . . she called radium. There remained no hole to hide in,' wrote Henry Adams in his autobiography. 'Even metaphysics swept back over science with the green water of the deep-sea ocean and no one could longer hope to bar out the unknowable, for the unknowable was known.'[10]

The unprecedented energy of Mme Curie's invisible rays cracked scientism wide open. Science ceased to be merely a matter of what you observed; now you had to consider seriously *how* you were observing. You didn't have to be a quantum physicist to hit this problem. You could as easily be attempting to induce mutations in fruit flies. You could be a physiologist studying salivation in dogs.[11]

Among those who surfed this tidal change in scientific thinking, one of the more eloquent was the Austrian Ernst Mach. He is best remembered today as the man who first accurately measured the speed of sound, and the popular assumption that he was a physicist is a fair one – but not quite accurate. Mach's most interesting work pushed him into an unlabelled zone between physics, physiology and psychology.[12] His work on optical illusions, and the physiology of sensation, and how our senses cleverly but somewhat imperfectly register the world, led him to a devastating conclusion about science.

Mach argued that science makes no pronouncements about ultimate reality. All scientists can do is select the most elegant explanations for their results. These results are useful, but they aren't *true*. Indeed, the same body of knowledge can lead scientists to several, equally valid conclusions!

Marxism was supposed to be the capstone for a certain sort of nineteenth-century science. Now Mach had come along and demolished the whole edifice. Needless to say, Lenin hated Mach's 'empiriomonism'. As far as he was concerned, it was a recipe for disaster: a retreat from the real world into idealist handwaving. His response was strident: 'For the *sole* "property" of matter with whose recognition philosophical materialism is bound up is the property of *being an objective reality*, of existing outside our mind.'[13] This is not so much an argument as a howl.

Mach's views (bolstered, it seemed, by the discoveries of X-rays and the electron) were extremely worrying for Lenin. They seemed to flatly contradict the idea that matter is real.

But Bogdanov approached Mach's arguments much more phlegmatically. He understood – as Lenin did not – that Mach was writing *for scientists*, about the ever-narrowing but ever-present gap between our experience and the world as it is. Mapping that ever-narrowing gap did not make the gap wider: Lenin's fears were unfounded.

Bogdanov found that gap to be a fascinating place. In exploring it, he refined what we mean by the word 'experience'. Say that, for a moment, you put this book aside and reach through the open window of your room to pluck a rose. You are now 'experiencing' the rose in two ways. You can smell the flower, see it, prick your thumb on a thorn. And there is also *your* experience of the rose. Its beauty. Its transience. Its colour, so close to the colour of your child's lips, or your favourite ice cream, or what have you.

Bogdanov called these two orders of experience *objective* and *subjective*. He defined objective sensations as those that are 'socially organised' – in other words, they are more or less accessible to, and categorisable by, everyone. Subjective sensations are 'individually organised', and belong to just a single consciousness. He then attempted to unite these realms in a philosophical system called empiriomonism: the 'organisation of experience'.

This took him down a path towards perception and philosophy of mind, and well out of sight of the world beloved of those great materialist thinkers Engels and Marx. The material world. The world 'as it is'.

So much for philosophy. The philosophy matters – how much it matters will become clear over the next 400 or so pages – but this rather abstruse disagreement between old friends was also part and parcel of a widening political split between Lenin's faction and Bogdanov's. It seemed unbridgeable, though the writer Maxim Gorky was indefatigable in his efforts to bring the two old friends back together to sort out their differences. In 1908 Alexander Bogdanov was staying in Gorky's aristocratic mansion on Capri (a resort once favoured by Roman emperors). Lenin, living in Geneva, had a standing invitation to visit, and in April 1908 he was finally cajoled onto the boat. It took Lenin no time at all to get into an argument with his host, however: 'I know, Alexei Maximovich, that you're hoping to bring about my reconciliation with the Machists, though my letter has warned you that it's impossible. So please don't try!'[14]

It was probably Gorky who suggested a friendly chess game to clear the air. At any rate, he is there in the photograph taken by the film-maker Yuri Zhelyabuzhsky, looking on as Lenin, seated with his back to the Tyrrhenian Sea, steadily loses control of the game and his own temper. It astonished Gorky how angry and childish Lenin became.

Afterwards Bogdanov and the others went for a walk, and Lenin unburdened himself to his hostess, Gorky's common-law wife

Maria Andreyeva. Of Bogdanov and his set, Lenin said – more in sorrow than in anger – 'They are intelligent, talented people. They have done a great deal for the Party, they could do ten times more, but they won't go with us! They can't. Scores and hundreds like them are broken and crippled by this criminal system.'[15]

Lenin had disliked Bogdanov's philosophy for years, but he had held back from attacking a fellow Bolshevik in public. The appearance of *Essays on the Philosophy of Marxism* by Bogdanov and his allies in January 1908, however, seriously undermined Lenin's hold over his own party. Lenin's cautious policy of parliamentary participation was losing ground to Bogdanov's confrontational manifesto.

Lenin's response was a book – a fully worked-out version of the long letter he had sent to Bogdanov, rubbishing his philosophy. Written between February and October 1908 in the reading room of the British Museum, *Materialism and Empiriocriticism* was a rush job (Lenin even considered offering the printer a bribe of a hundred roubles to speed up his work) and advanced a brutally materialist argument. Materialism, Lenin said, is incompatible with any doubts about the comprehensibility of the world. He advanced a copy theory of knowledge, in which our sensations are 'copies, photographs, images, mirror reflections of things'. 'Matter', he explained, 'is a philosophical category which is given to man by his sensations, and which is copied, photographed and reflected by our sensations, while existing independently of them.' He goes on to assert – against evidence, reason, and every serious thinker since Plato – that what we see in our heads are *literally* copies of external objects.

Materialism and Empiriocriticism puzzled reviewers at the time and has been a running sore in Marxist studies ever since. It is not just shrill, not just offensive – it is, in many places, profoundly stupid.[16] It makes psychological sense, however. Lenin understood how thin the ice was beneath him, and his panic is palpable. His party was haemorrhaging badly; its membership fell from more than 40,000 in 1907, to a few hundred in 1910, fractured into cells that were themselves riddled with tsarist secret police.

But this was not how the document was remembered. It was Lenin's first full-blooded stab at philosophical analysis, and over the coming years it acquired the status of scripture. This is what makes Lenin's tract so frightening: the way the exigencies of that wobbly moment in Bolshevik history got frozen, not just into print, but into policy.

The meshwork of alliances binding Europe together at the beginning of the twentieth century was supposed to privilege diplomacy and prevent war.

In the four weeks following Archduke Franz Ferdinand's assassination on 28 June 1914, the unimaginable happened. The tsar mobilised 15 million men, mostly peasants. Conscripts were transported thousands of kilometres to the Eastern Front, or south to launch attacks on Turkey. By the summer of 1915, a million were dead, wounded or missing, and three-quarters of a million had been taken prisoner. Russian casualties in the First World War proved even more severe than those in Western Europe. When added to deaths during the civil war and ensuing famine, they total 16 million.

In late April 1915 a series of German offensives began that would lead to the so-called Great Retreat. The Russian Army was grotesquely under-equipped. Whole units were wiped out, retreats turned into routs, and hundreds of thousands of Russian soldiers – some without bullets for their rifles, some without rifles altogether – surrendered. Galicia was lost, then Warsaw, then the rest of Poland; by late August it appeared that Riga might be taken. Millions of refugees, starving or sick, got in the way of the army's retreat.

Among the governing classes, this catalogue of shortages, defeats and unpreparedness was greeted with dismay and anger.

They insisted the Duma be reconvened. It opened on 1 August, the anniversary of the outbreak of war, and straight away a coalition formed to press for social and political reforms.

But you didn't need to be a member of this 'Progressive Bloc' to regard the tsar's next move as the act of a madman. Nicholas announced he was quitting the capital and assuming personal command of the army. His strongest loyalists were probably the ones who greeted the news with the greatest horror: now the tsar would be saddled with the blame for any future defeats. In a desperate attempt to bring him to his senses, the Council of Ministers offered their resignations. The tsar, beside himself, accepted them. He closed the Duma for nearly half a year, and appointed reactionaries and unqualified yes-men in place of the loyalists he had fired. When they failed to come up to snuff, he fired them and appointed worse people in their place. Over the next year and a half Russia endured four different prime ministers and five ministers of the interior. Since government was impossible in such circumstances, in the summer of 1916 a 'temporary dictatorship' was set up to deal with the worsening food supply. But the peasantry couldn't be persuaded to sell more of their grain stock because the government didn't have the money to buy it. (What use had the peasants for money, anyway? Thanks to the war, there was no agricultural equipment available for them to buy.) On the front line, meanwhile, discipline was disintegrating. By 1917, nine out of ten officers in the field were reserves with next to no combat experience. More and more troops simply deserted.

Food riots in Petrograd in February 1917 turned into an anti-war protest, and then a revolution. Confined to the capital and lasting less than a week, the February Revolution was a spontaneous affair, and Nicholas appealed to his generals for loyal troops to put down the disorder. Instead, they advised that he abdicate.

Without the backing of the army, he had no choice. On 15 March 1917, the Romanov dynasty came to an end. The tsar was replaced by a Provisional Government made up of liberals and socialists. At the same time, socialists also formed the Petrograd Soviet, which ruled alongside the Provisional Government, in a 'dual power' arrangement.

The Bolsheviks, late to the party as usual, rushed to St Petersburg to lobby for the nation's scattered soviets – collective bodies of workers, soldiers and peasants who enjoyed popular support and wielded effective power, but lacked any legal status. The Bolsheviks were reasonably successful. Aside from their desire to end the war, the workers' demands were more practical than political: an eight-hour day, higher wages, respect at work. (Offensive and overbearing foremen and managers were beaten up. Sometimes they were daubed in red paint or tossed in a sack. A few were thrown into rivers.)

When Lenin returned from Switzerland in April, however, he rebuked the Bolsheviks' acting leader, Joseph Vissarionovich Stalin, for working so closely with the Provisional Government – and circumstances proved Lenin right, for the Provisional Government soon turned around and committed Russia to continuing the war. The protests that followed this declaration were such that both the foreign minister and war minister resigned – and the Bolsheviks, who claimed it was possible to conclude a just peace sooner rather than later, found their ranks swelling at last.

Disease was rife during the war. In 1917 Vladimir Vernadsky's niece, who lived with the family, died of tuberculosis. Vladimir himself contracted the disease. His doctor advised a diet of fermented mare's milk and a move south to speed his recovery. In June 1917 the remaining Vernadskys travelled south to the family's dacha in Shishaki, Ukraine. They made a sadly depleted party. Vladimir's son George had taken it into his head to join the army, against his father's advice. (Vladimir observed wearily that 'reason is often not the most powerful or important force in life'.)

From their new home, Vernadsky wrote to his old pupil Fersman, saying he felt caught between 'extremist Ukrainians' and 'extremist Russians'. He mistrusted the Bolsheviks, but as he wrote later, 'both the [Ukrainian] Volunteer Army and the Bolsheviks did a mass of unclean deeds, and in the final analysis one was not better than the other'.[17]

Once he was strong enough, Vernadsky went back to work: he left his family at their dacha and spent some weeks at the Academy's biological research station at Staroselsky, a village on the Dnieper River near the ancient city of Vyshgorod. Chance, Louis Pasteur once said, favours the prepared mind: Vernadsky arrived just in time to witness, at first hand, a curious phenomenon dubbed by locals 'the flowering of the waters'. At the right moment of the year, a warmish night would trigger an immense, overnight blossoming of algae in the local ponds and wells. Water that a few hours before had run clear was now clotted with life. The phenomenon is not particularly rare, but the sheer scale of this sudden growth brought Vernadsky up short: just how much did all this stuff *weigh*?

So much for luck: the rest was down to preparedness and hard work. Vernadsky knew that different proportions of different elements, reacting with each other under different conditions, give rise to different kinds of mineral. It dawned on him that it might be possible to look at living processes the same way – as a series of chemical reactions. Obviously this highly reductive approach wouldn't explain much about life, but it might reveal how living things harness the non-living stuff of the planet for their growth and expansion.

'There in the meadows I worked quickly,' he later recalled. 'I sorted out in my own mind the basic concepts of biogeochemistry, I clearly distinguished the biosphere from the other envelopes of the earth, and came to understand what the multiplication of living matter in the biosphere actually meant.'[18]

In the space of a month, Vernadsky had filled forty pages of graph paper with notes, and held in his hands the blueprint for a previously unheard-of science: ecology.

Following his month at Staroselsky, Vernadsky returned to Petrograd. His old friend and patron Sergei Oldenburg, Russia's foremost scholar of Buddhism and now minister of education in the Provisional Government, immediately appointed him his assistant minister.

They lasted a month, from the end of July to the end of August, but the awkward political arrangement in place at that time – with the capital jointly governed by the Provisional Government and the Petrograd Soviet – was never going to hold. Eventually, workers in the Red Guard took up arms and overwhelmed the government's guard at the Winter Palace. There was very little violence. On 7 November 1917 the Provisional Government was disbanded and authority was assumed by the All-Russian Congress of Soviets.

The next day Vladimir Lenin, the leader of the revolution, presented two bills: a peace decree, inviting Russia's enemies to enter into immediate negotiations; and a land reallocation act. All the vast estates of the imperial family, the church, the monasteries and the large landowners were to be expropriated without compensation, and the land distributed to the peasants.

In one crude stroke, the Bolsheviks achieved what Tsar Alexander II had failed to achieve: a true liberation of the serfs. The crudity of the bill was deliberate. The chaos it unleashed made it impossible for vested interests to game the system – because there wasn't one.

However, this then left the Bolsheviks with the business of putting the genie back in the bottle and reasserting control over the largest country on earth. That they even contemplated such a task says much for their optimism, and this optimism ran deeper than the euphoria of the moment. The Bolsheviks' philosophy preached optimism as a virtue, even a moral duty.

The Bolsheviks truly believed that the practice of government would fill the gaps in their political knowledge. Armed with Marx's science of government, they simply had to stay the course, hold on to power, stick by their beliefs, and wait for the world to deliver on its promise. In this, they were a kind of mirror-image of a very different, and much more recent sort of idealist – the market animists of Wall Street and the City of London who asserted, throughout the 1980s and 1990s, that a free market finds its own *natural* level, making big government redundant. Both sets of idealists hoodwinked their era into believing in a 'science' that turned out not to exist. Worse, both convinced themselves that they were being wholly 'scientific'.

Anyway, the circumstances for the Bolsheviks' brave experiment in practice-makes-perfect could not have been less promising. Why would Germany sue for peace now? Its high command was making hay of the total collapse of Russia's war effort, detaching the Ukraine from Russia and making it a German puppet state. Now the Ukraine was fast becoming a gathering ground for anti-Bolshevik Cossacks and anti-Bolshevik 'White' forces across the south.

At the peace negotiations at Brest-Litovsk, the Bolsheviks fully expected to have their hand strengthened any day, as copycat revolutions blazed across Western Europe. While they waited, they tried to drag out proceedings as long as possible. Leon Trotsky, president of the Petrograd Soviet and Lenin's number two, declared that Soviet Russia would make 'neither war nor peace' and left for Petrograd. The German Supreme Command threw up their hands in exasperation and resumed the fighting.

Lenin eventually managed to convince the Party leadership and the Congress of Soviets that only peace could save the revolution, and they had to accept terms. By then, however, German forces had advanced 250 kilometres further onto Russian soil, and their demands were far more severe.

The Treaty of Brest-Litovsk, signed on 3 March 1918 between

Russia and the Quadruple Alliance, was one of the harshest in modern history. Territorial losses amounted to 3.4 million square kilometres. Thirty-four per cent of the Russian Empire's population was lost or seceded. Over half of Russia's industrial capacity vanished overnight, along with invaluable natural resources. Russia also agreed to demobilise its armed forces. Eight months later, on 11 November, Germany's armistice with the Allied powers nullified the treaty, and the Soviet government eventually regained most of its lost territories. But the damage had been done. The cession of the Ukraine meant that Soviet Russia would have to obtain its grain from regions that could barely even feed themselves. The treaty left Russia with butchered national borders, a food crisis, and no friends. It hopelessly alienated the Bolsheviks' socialist allies, and a full-blown civil war seemed (indeed was) inevitable.

The Bolsheviks had control of government, but were hardly in a position to run the country. It wasn't the tiny party of Bolsheviks who had seized the nation's farms and factories. It was the people: without resources, or education, or training, or indeed much self-discipline. Throwing a foreman in the river is easy. How do you run a factory?

The Bolsheviks had won a great victory for the people; but the people were melting away. They regarded the Bolsheviks much as they had regarded the last lot of absentee landlords – with a mask of servility concealing a great quantity of guile. By the end of the Civil War, the total number of industrial workers in Russia had dropped to just over a million – a third of what it had been in 1917.

The Bolsheviks never meant to nationalise industry on a mass scale. Now, driven by necessity, they changed tack; by January 1919 nearly all large factories were owned by the state. Now came the harder job: finding people to manage them.

Communists, technical specialists and trade unionists rubbed up against each other, none too comfortably, in an attempt to manage the economy. The biologist Mikhail Novikov worked closely with state-owned heavy industry even while, as the rector of Moscow University, he was organising strikes. Vladimir Nikolayevich Ipatiev, a nobleman and monarchist, owned the holiday home in which Nicholas II and his family were murdered on 17 July 1918. Yet in 1920 he was picked to direct the Central Chemical Laboratory in Petrograd: Lenin called him 'the head of our chemical industry'.

Charming exceptions aside, the turn of the 1920s was a bad time for those who wore their learning on their sleeve. An open letter addressed to Lenin and published by *Pravda* gives a glimpse of the 'specialist baiting' that scourged the nation, and wasn't really quelled till 1921:

... these newly minted unconscientious Communists, made up of former lower middle class elements, village policemen, small-time civil servants, shopkeepers ... It is difficult to describe the full horror of the humiliations and sufferings caused at their hands. Constant, shameful denunciations and accusations, futile but extremely humiliating searches, threats of execution, requisitions and confiscations, meddling in the most intimate sides of personal life. (The head of a squad demanded that I, who am living in the school where I teach, absolutely must sleep in the same bed as my wife.)[19]

The All-Russian Union of Engineers spent a great deal of its time arranging for the release and safe-conduct of its members. Every once in a while someone in the uniform of the Cheka, the Bolsheviks' secret police, showed up at a Soviet technical bureau and handed over a rumpled, unshaven prisoner who turned out to be a senior employee.

The specialists did not take this sort of treatment lying down. In the Academy of Sciences – that venerable centre of learning founded in 1755 by Peter the Great himself – most members regarded the Bolshevik takeover as a national catastrophe. Vernadsky and Oldenburg had turned in their ministerial portfolios at once, and Vernadsky went further, involving himself in an effort to continue the Provisional Government's work underground, even though most of its members had been arrested.

Learning that Vernadsky was on an arrest list, the Academy voted to send him 'to the Southern part of the country because of his bad health'. The same day, he left Petrograd and headed back to the Ukraine; his life was saved.

Oldenburg and fellow academics formed a delegation to protest the Cheka's programme of arrests. They arrived at Communist Party headquarters, then established at the Smolny Convent on the edge of St Petersburg. The convent, famous for its elegant architecture and its dull blue dome with silver stars, was once a finishing school for young ladies. Now it resounded to the thud of muddy boots as thousands of soldiers, sailors and factory workers traipsed through wide hallways fitfully lit by a faltering electrical system.

In an upstairs room, often alone, and rarely showing his face to comrades on the ground floor, Vladimir Lenin sat working. Oldenburg knew Lenin already, as the brother of the student who had tried to assassinate Alexander III. Oldenburg, by now Permanent Secretary of the Academy, had a deal to put to Russia's new leader.

The Academy was quite small in 1917, boasting only forty-four full members and 220 employees. But it enjoyed an enviable international reputation. Besides, it was old, and for centuries had acted as a sort of expert arm of the Russian civil service. Oldenburg promised that Academy scholars would aid the Soviet regime by addressing issues of 'state construction'. In return, he wanted the Academy to receive financial and political support from the government. He was offering Lenin the chance to conduct business as usual – and since Lenin understood more than most the importance of specialists to the state, he agreed.

He was as good as his word. In 1918 the education commissar, Bogdanov's old friend Anatoly Lunacharsky, attempted a wholesale reform of the Academy. When Lenin's 26-year-old secretary Nikolai Gorbunov caught wind of this, he went straight to his boss, who warned Lunacharsky: 'If some brave fellow turns up in your establishment, jumps on the Academy and breaks a lot of china, then you will have to pay for it.'[20]

Lunacharsky shelved his plans, the Academy received some money, local authorities were stopped from using some of the Academy's property for housing, and Lunacharsky promised that the Academy's press would be left alone. Lenin told Lunacharsky to publicise the deal: 'The fact that they wish to help us is good. Tell the whole world that the Academy of Sciences has recognised the government.'[21]

In Bolshevik hands, revolutionary justice took on curious forms. Far from resembling Paris's notorious tribunal at the height of the terror, Bolshevik trials were characterised, at first, by their relative objectivity. They did not deliver a single death sentence in the first six months, and freed many declared enemies of the regime.

The idea was that the tribunals should teach Russians about the new state and how it worked. They were show trials, in fact, of an early, innocuous sort. They drew simple, easily understood lessons about what constituted a crime under the new regime. They were recorded, and re-staged as theatrical performances the length and breadth of Russia, with actors playing the parts of prosecutors and defendants. (It was risky work. On more than one occasion an actor playing the defendant had to flee lynch-mobs of angry peasants.)[22]

Alongside this benevolent but inadequate masquerade, however, there operated another form of justice. The Cheka, which had been set up to target counter-revolutionaries in the immediate aftermath of the October Revolution, was essentially unaccountable. It acted against opponents as it saw fit. Its head, Felix Edmundovich Dzerzhinsky, a Polish nobleman turned communist, proclaimed that he did not 'seek forms of revolutionary justice; we are not in need of justice. It is war now – face to face, a fight to the finish.'

Food shortages in the winter of 1917–18 prompted Lenin to demand the execution without trial of speculators and bandits. On 21 February, a decree, 'The Socialist Fatherland is in Danger', urged that all speculators, spies, hooligans and counter-revolutionaries should be shot on the spot. The Cheka obliged. When an attempt was made on Lenin's own life, a decree was issued on 5 September 1918 directing the Cheka to shoot opponents summarily: and they did. The Cheka shot well over 5,000 people in 1918 alone. Thousands of conservatives, liberals, Mensheviks and Socialist Revolutionaries were executed. Many technical specialists were among those arrested. The 'Red Terror' had begun.

The state had enemies aplenty in 1918 – it hardly needed to go looking for them. For a start there was the complexion Britain, France and the United States had put on Lenin's revolutionary success. It had been German assistance and money that had enabled him to return to Russia and seize power. When he then took Russia out of the war and ceded enormous territories to Germany, the Entente powers concluded that he was some sort of German agent. Nations that had fought Germany now felt an obligation to intervene against Lenin's new regime. Besides, large amounts of Allied war material were still in Russian hands. So, in early 1918, small parties of Allied soldiers occupied Russian ports.

Come May 1918, matters became still worse for the Bolsheviks. Under the tsar, 40,000 Czech and Slovak prisoners of war had been formed into a Czech Legion, to fight on the Allied side. With Russia out of the war, the Czech Legion found itself abandoned. After endless negotiations, the Legion agreed to leave Russia, travelling along the Trans-Siberian Railway to Vladivostok, then heading by ship for Europe. As they crossed Siberia, however, the soldiers of the Legion clashed with the local Bolsheviks, then mutinied, seizing control of the railway, and with it all access to the sparsely inhabited bulk of Russian territory. The Siberian wilderness became yet another zone where the Bolsheviks' opponents could organise resistance. In November 1918, Admiral Alexander Kolchak took charge of White forces in Siberia in a military coup. Using Allied supplies, and conscripting whoever he could find in that thinly populated region, Kolchak launched his offensive in March 1919.

It was a war calculated to alienate vast swathes of the country for a generation and more. Reds and Whites alike conscripted unwilling peasants in massive numbers. The peasants had no dog in this increasingly vicious fight and fled the ranks of both armies almost as quickly as they could be rounded up.

And it was a war designed to confound military historians, dividing Russia geographically and politically into a shifting patchwork of autonomous regions. The Red Army managed to hold on to both capitals, Petrograd and Moscow, but as the British spy and children's author Arthur Ransome would have it, 'the territory held by the Soviets was only a small part of Russia, consisting for the most part of districts in normal times either not self-supporting or barely capable of self-support. The revolution was cut off from the main sources of iron, cotton, oil, meat and bread.'[23]

The artist Yuri Annenkov recalled 'endless hungry lines, queues in front of empty "produce distributors", an epic era of rotten, frozen offal, mouldy bread crusts, and incredible substitutes'.[24] As bodies weakened, more than half the women living in large towns ceased to menstruate. Fatality rates spiralled. In the intense cold, drains and water pipes shattered, spreading typhoid, dysentery and even cholera. A malaria epidemic was under way even before the fighting began; now it got unimaginably worse. The greatest typhus epidemic in history took hold, with deaths running somewhere between 2 and 3 million. Typhus affected both sides equally – even a White general, Baron Wrangel, caught it. Refugees carried typhus-bearing lice around the country and into the cities on crammed trains – a disaster that prompted Lenin to remark, in his address to the Seventh Congress of Soviets: 'Either lice will defeat socialism or socialism will defeat lice.'

For the longest time, the lice held sway: in 1920 the population of Moscow nearly halved to one million. Petrograd's 700,000 represented less than a third of the pre-war population. Seven million homeless children wandered the nation's streets, or ended up in state orphanages, where up to half of them died.

Vladimir Vernadsky's apricot trees were grubbed up and stolen. The family's dacha at Shishak was sacked. Such misfortunes, serious enough in peacetime, now barely registered. In his diary Vernadsky recorded the state of his family's home in newly independent Ukraine without complaint.

In Ukraine's capital Kiev the streetcars weren't running, traffic was at a standstill and political chaos reigned. Between 1918 and 1920, more than a dozen governments tried to run the city. Once, a gang of army deserters who had already robbed all the apartments on the first three floors of the building stormed the Vernadskys' apartment. Vladimir turned and snapped at them: 'Can't you see I'm working?' Startled, the intruders left. Mind you, by then the family had nothing worth stealing.

Vernadsky at this time was one of the city's more important figures: founder and president of the Ukrainian Academy of Sciences, which, incredibly, survived all changes of political and military power in Kiev. But his position brought him no safety. He spent much of 1919 in hiding when a local paper mistook him for Bernatsky, another, less well-liked former minister of the provincial government. Between the spring and summer of 1919 the Bolsheviks held Kiev, taking well-born hostages, executing hundreds, and murdering Vernadsky's closest assistant.

Vernadsky's bolt-hole during his time in hiding was the biological research station at Staroselsky where he had previously witnessed 'the flowering of the waters'. Each week Theodosius Dobzhansky, Vernadsky's research assistant, set out on a seventeen-kilometre trek from Kiev, bringing his boss a knapsack full of mail and groceries. (A professor at a *rabfak* – a workers' university – Dobzhansky had papers that could get him past the Bolshevik checkpoints.) Vernadsky used his time in hiding well, studying and tabulating the chemical composition of different animals. By 1922 he had established that, of the ninety-two elements then known, over fifty were bound up in the history of living organisms. These elements comprised 99.6 per cent by weight of the whole earth's crust, leading Vernadsky to conclude that living organisms could reshape planets as surely as any purely physical force. It was the first step on an intellectual journey that would culminate in Vernadsky's concept of the 'nöosphere' – the idea that intelligence itself was yet another planet-changing component of earth's geological system, wreaking changes even greater and more significant than those achieved by life itself.

At the end of April 1919, Vernadsky, fearing for his Ukrainian Academy, travelled south to meet with the generals in charge of the region. Stricken with typhus, and caught between battle lines, he got stuck in the Crimea. At the same moment Mikhail Frunze, a professional revolutionary with no military training whatsoever, launched a devastating attack that routed White forces out of the south. The sheer scale of Russia's enormous interior did the rest. The Whites, exhausted and under-supplied, withdrew. The Reds recaptured Ukraine's major cities in late December and, in late March 1920, the remaining White forces were pushed into the Black Sea at Novorossiisk, leaving only the Crimean Peninsula as the Whites' final refuge.

Nursed back to health, Vernadsky found himself surrounded by old friends. They were teaching at the new Taurida University in Simferopol – according to Vernadsky 'one of the strongest academic schools in the country' – and virtually all of them were looking to emigrate. So too were his son George and his wife

Nina. They left with the British as the Red Army closed in, and Vladimir, too, sought British help in emigrating.

In the end, though, Vernadsky's sense of duty, and his extraordinary work ethic, kept him in place. As the Red Army overwhelmed the Crimea Vernadsky, who was already working at Taurida University, found a fiendishly simple way to protect his White friends: he issued them with student cards. The Bolsheviks caught on and questioned him, but in the nick of time the message came through that Nikolai Alexandrovich Semashko, the health commissar, had organised a hospital train to bring Vernadsky and his family, along with the wife and daughter of Academician Sergei Oldenburg, back to Moscow. Semashko was proud to count himself one of Vernadsky's students. (Vernadsky did not remember Semashko, but kept that to himself.) Between them, Oldenburg and Semashko saved Vernadsky's life.

The train arrived in Moscow in March 1921. The police did not altogether let go – Vernadsky found himself briefly under arrest yet again – but Oldenburg extricated him and took him to Petrograd, where they found grass growing on Nevsky Prospect. Vernadsky, happy nevertheless to find that 'thought is still alive' in the city, visited his old lab and met one of his students, Evgeny Flint, deep in an experiment – or so he thought.

It turned out that Flint was making moonshine to exchange on the black market for food.

In early November 1920 the Red Army smashed through the narrow land bridge that connected the Crimea to the mainland, and eliminated all remaining resistance in hardly more than a week.

The last major White army was gone, but the damage was done: peasants driven to violence by the theft of food and forced conscriptions began an insurgency to end the Red Army's policy of requisitioning grain, even when the peasants needed all of it to feed themselves. The Bolsheviks crushed this insurgency mercilessly. Responding to a massive rebellion in the Tambov region, they even used poison gas against peasants hiding out in the forests.

In the cities the food shortage was driving the urban population to its own acts of insurrection. The Kronstadt revolt erupted while the Tenth Party Congress was in session in nearby Petrograd. Kronstadt was an island fortress and naval base in the Gulf of Finland, about thirty kilometres west of the capital. After workers' strikes and scattered disturbances in factories and garrisons in other parts of the country, in March 1921 the sailors of the fortress rose up against the regime with the slogan 'Soviets Without Communists'. The rebels' newspaper declared:

The power of the police and the gendarme monarchy passed into the hands of the Communist usurpers, who, instead of giving the people freedom, instilled in them the instant fear of falling into the torture chambers of the Cheka, which in their horrors far exceed the gendarme administration of the tsarist regime.[25]

On 16 March the merciless Mikhail Tukhachevsky led about 50,000 Red Army troops across open ice, taking the fortress and filling the jails of Petrograd with prisoners. 'Months later,' a contemporary wrote, 'they were still being shot in small batches, a senseless and criminal agony.'

It was a pyrrhic victory, and the Bolsheviks knew it: they were now at loggerheads with the very class in whose name they had fought. Lenin's 'New Economic Policy' (NEP) was being discussed by the Tenth Congress even as the drama of Kronstadt was unfolding, and the policy was adopted less than a week after the fortress fell. 'War Communism' – the wholesale state takeover of industry and commerce – ended forthwith, and private individuals were permitted to own small businesses. Felix Dzerzhinsky's 'Red Terror' was brought to an effective end, too, replaced by a much more targeted persecution of opposition groups. Dzerzhinsky had realised that keeping tabs on those 'who might take part in active struggle against us' was much more effective than direct repression, and in September 1922 the state began compiling dossiers on the country's entire intelligentsia.[26]

On 14 July 1921 Vladimir Vernadsky was arrested. Telegrams to Lenin, health commissar Nikolai Semashko and education commissar Anatoly Lunacharsky from Alexander Karpinsky, president of the Academy of Sciences, almost certainly saved his life. After a two-hour interrogation, he was released. The next day, sickened and afraid, he left with his daughter Nina for the Kola peninsula and in May 1922 got passports to travel – via Prague, where his son George had settled as an emigrant – to Paris. They arrived on 8 July 1922.

Vernadsky had received an invitation to teach geochemistry at the Sorbonne. Marie Curie and her institute were also ready to fund him. Indeed, everything had been arranged for the family's permanent resettlement. Between August 1922 and September 1924 the Vernadskys lived in two small rooms on the rue Toullier, bang in the centre of the Latin Quarter, and for a while it seemed as though Vernadsky might be tempted to emigrate. He even searched for funds in the United States to establish a bio-geochemical laboratory there. He wrote to the British Association for the Advancement of Science, the US National Research Council and the Carnegie Institution in Washington, DC – all without success. Frank Golder, an old friend who had worked with the American Relief Administration, passed his proposal to Stanford University; again, he was turned down. In 1924 he tried to remain permanently in France, creating his new lab as part of the Natural History Museum in Paris – but sufficient funds could not be found.

And anyway, Vernadsky's closest comrades in the new Soviet Union were not inclined to let him go. With no carrots to offer, they applied the stick: Oldenburg and Fersman refused to renew the authorisation for his trip to Paris beyond the first year, and cut off his salary when he overran, so that the family had to move out to the suburbs, to Bourg la Reine.

Vernadsky's Paris acquaintances lobbied hard for him to remain in Paris. But Vernadsky turned sixty in 1923, and began to realise that the upheaval of emigration would be too much for him. 'If I were much younger,' he wrote to an old friend, living in exile in Prague, 'I would emigrate. My universal feelings are much stronger than my nationalist feelings. But now it is difficult and impossible, as one always needs to lose several years on securing a position.'[27]

By 1925, Oldenburg and the Academy were at last in a position to offer Vernadsky some positive reasons to return. A promise of freedom, and funding to pursue his own lines of research, combined with the introduction of Lenin's New Economic Policy, suggested that Bolshevism was a spent force.

So Vernadsky returned to the USSR – or tried to. He had extraordinary problems getting a visa and making his way back home. Still lacking his Academy salary, he worked his way across Europe, taking whatever speaking engagements he could find. Friends and colleagues lent him money for warm clothes for the journey. In the spring of 1926 Vernadsky arrived back in Leningrad.

The state of the city astounded and appalled him. For one thing, many of his acquaintances were dead.

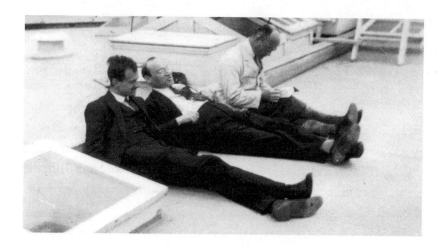

3 : Entrepreneurs

*Communism cannot be built without a fund of knowledge,
technology, culture, but they are in the possession of
bourgeois specialists. Among them the majority do not
approve of the Soviet regime, but without them we cannot
build communism.*[1]

Vladimir Lenin to Anatoly Lunacharsky

The famine of 1921–2 began with the dry winter of 1920–1. The
1921 spring sowing was a predictable failure, and the autumn
harvest not much better.

[By 1922] there were abandoned homes in the communes by the score,
the roofs and wooden parts taken off for fuel, and the walls of mud
and straw falling into decay. Everywhere we found emaciated starving
children, with stomachs distended from eating melon rinds, cabbage
leaves and anything that could be found, things which filled the
stomach but did not nourish . . .[2]

Aid came from private organisations like the Quakers and the
Save the Children Fund, and indirectly from the US government
through the American Relief Administration under Herbert

Paul Dirac (left) and guests of the Sixth National Congress of the
Russian Association of Physicists sail down the Volga – a grandiose
gesture, never repeated.

Hoover. The Norwegian explorer Fridtjof Nansen won a Nobel Peace Prize for his efforts in helping famine victims. Despite all this, around 5 million people died.

It was an era of heroic achievements, of groundbreaking novels by Babel, Bely and Bulgakov, Platonov and Zamyatin, iconoclastic music by Prokofiev and Shostakovich, paintings by Malevich, poetry by Blok and Mayakovsky, films by Vertov and Eisenstein. In theatres, Meyerhold and Stanislavsky held sway. But life was cheap. The novelist Maxim Gorky recalled a peasant confessing to a city gent that he had killed a native Bashkir and stolen his cow: 'Will he be prosecuted for the cow?' Frank Golder, an observer for the American Relief Administration, recalled how one professor called on a Crimean government representative. He informed her that professors in the Crimea were dying of hunger. 'What of it?' she replied. 'Let them die!'

They did: in Petrograd, seven of the Academy of Sciences' forty-four members died of starvation. From his spacious quarters at the Academy's residence on Vasilevsky Island, the physiologist Ivan Pavlov, then Russia's only living Nobel laureate, witnessed the catastrophe for himself. Two academicians living in his building died of cold and hunger; others stood by helplessly as the authorities billetted complete strangers in their apartments. Pavlov's Nobel Prize money was confiscated. So were the gold medals awarded to him by St Petersburg University and the Academy of Sciences. He scavenged for firewood and fed his family from a garden he tended himself at the Institute of Experimental Medicine.

By 1918 Pavlov's groundbreaking research work was grinding to a halt for lack of assistants, experimental dogs and the food to feed them. He couldn't even get decent lecture demonstrations out of his lab results, let alone publishable work. As winter approached he had 'no candles, no kerosene, and electricity is provided for only a limited number of hours. Bad, very bad. When will there be a turn for the better?'[3]

Arriving in Petrograd in the spring of 1922, Frank Golder visited the permanent secretary of Russia's Academy of Sciences, Sergei Oldenburg. 'It is so pitiful and so heart breaking that it completely upsets me,' Golder wrote to his friend Ephraim Adams. Oldenburg, now in his late fifties, was bedridden. He was barely able to reach the scraps of toast on his table, let alone chew them. He had spent the past four years trying to support his own family, the orphan children of his brother, his daughter-in-law and his grandchildren. The children had bread every day: for the rest of the family it was a weekly treat.

In Moscow, too, conditions were desperate, and ingenuity was essential if any of that city's scientific life was to survive. The Moscow Practical Institute, a former trade school, announced the opening of three departments, in biology, agriculture and economics, to harness the fast-developing biological sciences to serve the needs of the people. All that large building lacked was equipment – so they stole it. One of the teachers, Nikolai Vladimirovich Timofeev-Ressovsky, used to dress up in his military uniform, get into a two-horse buggy, and ride off to whatever establishment had what they needed. People handed over to this curt 'Red Cross official' quite extraordinary treasures: new microscopes, binocular loupes, microtomes,[4] thermostats, crates stuffed with chemical glassware. By 1923 the Practical Institute was better equipped than the biological laboratories of the university.

Most intellectuals had welcomed the revolutions of 1917. Some enthusiastically embraced Lenin's October coup. Others endured. The physics community paid hardly any mind to the revolution until winter brought their work near to ruin. (Meeting after dark proved so dangerous, the discipline's various committees and boards only ever gathered on Sundays or other holidays.) In February 1919 the education commissariat provided emergency funding so that over 100 physicists could meet to celebrate Russia's most famous scientific son, Dmitry Mendeleev, and his periodic table of elements. They met in a 'melancholy Petrograd of iced-over buildings without food or heat or the promise of reimbursement of expenses'. The Third Petrograd Pedagogical Institute offered accommodation in its dormitories and meals in its dining hall, but 'it is necessary to bring with you some foodstuffs, e.g., sweet crusts, sugar, tea, jam, sausage, meat, and so on'.[5]

Out of the congress was born the Russian Association of Physicists, the brainchild of Abram Fedorovich Ioffe, an expert in electromagnetism. Ioffe found the association a two-storey brick building in a park on the outskirts of Petrograd. On Sundays, scientists gathered to clean the place up. They mended and scrubbed and hauled furniture from room to room. Electrical gear, pipes, laboratory equipment and even furniture were in short supply. The Commissariat of Education, with few resources, showed willing by cannibalising the inventory of the few museums. Chemicals, instruments, books and journals and marble slabs were lifted from the Agricultural Institute. The Winter Palace donated items 'without artistic value': an oak table, lamps, a clock, curtains, a piano, padded chairs, a rug.

There was no heating. Even the main building of the Polytechnical Institute went unheated. Practical demonstrations and examinations were held in a small room with a brick stove in the centre and an exhaust pipe poking through a window. Experiments were conducted in dense smoke.

The 1922 famine was an entirely predictable disaster. For years, while botanists in the universities and the Academy were studying wild species, European-educated agricultural experts had struggled without support or patronage, studying crop varieties and trying to modernise Russian agriculture.

The Bureau of Applied Botany, established in 1894 in St Petersburg, was supposed to bring botanical expertise in line with the needs of agriculture. Its first director, Alexander Batalin, conducted a huge amount of research into cultivated crops. But there was a limit to what he could do: he was, after all, the grandly titled Bureau's only member of staff, and had to run the Imperial Botanical Gardens in Petersburg at the same time. Ivan Borodin took over in 1899 but he too juggled too many posts to do the Bureau justice. Robert Eduardovich Regel became director in 1905 and got the place running at last, beginning its first collection of cultivated plants. He too struggled with scant resources. To a keen graduate applying for an internship he wrote: 'We expect you in Petersburg in the near future. PS. We have a spare microscope. It would be good if you could bring your magnifying glass with you.'[6]

The intern was Nikolai Ivanovich Vavilov, the grandson of an indentured peasant. Nikolai was one of four children, all of whom became scientists. Lydia, a microbiologist, died young, succumbing to the smallpox she had rushed to treat during an outbreak in Voronezh, in the south-west of the country. Alexandra was a physician. Sergei, a physicist, would in time become Permanent Secretary of the Academy of Sciences under Stalin. And Nikolai? He was destined to become Stalin's leading agronomist, a champion of genetics, and arguably the most celebrated scientific martyr of the twentieth century.

Nikolai Vavilov's interest in botany, and in collecting, began early, as he played with his brother Sergei in the ponds and birch forests of their home in Ivashkovo, 100 kilometres west of Moscow. The two also developed a keen interest in chemistry, on one occasion setting off an explosion that permanently damaged the vision in Nikolai's left eye.

In 1906 Nikolai Vavilov enrolled at the Petrovsky Agricultural Institute and studied under Dmitry Pryanishnikov – a fearless man who, much later, would expend his dying energies trying to prise his former student out of death row.

In 1911 Vavilov's thesis on the ecology of field slugs, a serious pest, won a science fair competition at the Moscow Polytechnic Museum. Robert Regel spotted the young man's talent straight away, and his year's internship at the Bureau evolved into a regular and eager correspondence as Vavilov embarked on a strenuous programme of foreign study and exotic field trips.

Returning from an expedition to the Pamir mountains, the fabled 'roof of the world', Vavilov took up a professorship at Saratov University, arriving there a month after the October Revolution. Saratov had once boasted a music conservatory and the country's first free art museum, but during the Civil War it was reduced to little more than an army camp. Drunken soldiers mugged the unwary. Anyone with money fled, and the city was crammed with hungry refugees from the countryside. Typhus and cholera ran amok. Spanish flu, measles, diphtheria and dysentery went untreated for lack of medicines. Soldiers fresh from the front slept in the streets. A great fire rendered one-eighth of the population homeless.

Vavilov was married by this time – none too happily. His wife Ekaterina elected to stay in Moscow, while Vavilov, unable to find an apartment, slept on the floor of his laboratory. His plight had its comic compensations. His department was staffed almost exclusively by earnest young women, who promptly fell in love with him. Young, dashing and energetic, Vavilov took Elena Ivanovna Barulina, a Saratov native, as his lover. He spent the rest of his life with her.

Between 1918 and 1920 Vavilov's correspondence with Regel intensified. Regel, born a German, knew he had no future in Soviet Russia and was grooming Vavilov to be his successor. Vavilov did not want to leave Saratov, but Regel's death by typhus made a move inevitable; the opportunity, along with Vavilov's own sense of obligation, was irresistible. At the end of October 1920, Vavilov caught a train to Petrograd to take over the running of the Bureau of Applied Botany.

The Bureau took up six apartments in 44 Herzen Street, in the centre of Petrograd. 'There are millions of troubles,' Vavilov wrote, a week into his stay. 'We are fighting against the cold at home, and for furniture, flats and food . . . I must confess that it is quite a problem to arrange a new laboratory and an experimental station as well as to settle sixty employees. I am accumulating patience and persistence.'[7]

His twenty-strong team of researchers from Saratov arrived on 5 March 1921. One remembered 'a picture of almost complete destruction, as if there had been an enemy invasion . . . The rooms were freezing, the central heating pipes had burst, the bulk of the seeds had been eaten by starving people, there was dust and dirt everywhere . . .'[8]

'Already thrice I have cabled the Experimental Department about our catastrophic situation with funds,' Vavilov complained. 'We are unable to remunerate employees, hire day-workers, or pay for horses, and in general it is essentially impossible to work . . .'[9]

Supplies were virtually unobtainable, but there were empty buildings everywhere. Vavilov moved the Bureau around the corner to the former tsarist Ministry of Agriculture buildings at 13 St Isaac's Square. He set up an experimental farm in Pushkin, sixty-five kilometres from Petrograd in the grounds of the tsars' summer palace of Tsarskoye Selo. (Its headquarters were a replica of an English country house sent, piece by piece, by Queen Victoria to her godson the ex-Grand Duke Boris Vladimirovich.)

There was no food to feed them, but specialists flocked to the Bureau. According to one enthusiastic arrival, Vavilov had created 'a Babylon . . . a fairytale island of freedom'. Gavryl Zaitsev, an expert in cotton, wrote to his wife, 'He reminds me very much of Mozart, who according to Pushkin's Salieri, was an "idle reveller". But nobody can do as much as or equally well as he does.'[10]

Russian society under the tsars had been frighteningly simple. There were very few institutions, and private societies and associations were, if not banned outright, then looked on with suspicion. From the top, if you wanted to get anything done, you generated paperwork. A colossal, and colossally inefficient, bureaucracy stood in for the ordinary business of delegating duties to other organisations.

If you were somewhere further down the pecking order, and you wanted to get something done, you did your level best to avoid paperwork altogether. Instead, you cultivated people. You looked for patrons to support you, and in your turn you did your best for your clients. Whole empires have been built on less, and no one was particularly surprised to find the old pre-revolutionary system of 'friendship circles' surviving and persisting into the Soviet era.[11] State ministers were the new patrons. It was a lot easier, if you could get to them, to deal with Bolsheviks of high calibre – with Lenin, Alexei Rykov (head of the Council of People's Commissars), Anatoly Lunacharsky (education commissar), and Nikolai Semashko (health commissar) – than with the local authorities, who were by and large ill-educated, and tended to regard any 'bourgeois specialist' with suspicion.

For their part, the commissars remembered pre-revolutionary times and knew the part they were supposed to play. Lunacharsky, in particular, built his entire commissariat out of friendly non-communists, the wives of comrades, and the cash-strapped granddaughters of famous artists.

Some patrons who had operated under the tsars continued in post even after the revolution. Maxim Gorky (a pen name: he had been christened Alexei Maximovich Peshkov) was one of those rich Russians without whose assistance Lenin himself could scarcely have survived the revolution.

'Highly contradictory, intentionally ambiguous, and often extremely false,' according to one account, Gorky was also a phenomenal patron. Entire institutions owed their existence and survival to him; so too did countless students and writers, who received stipends and even accommodation in Gorky's home. Gorky's strange late career as an apologist for Stalin ruined his posthumous reputation, but as the poet Anna Akhmatova recalled in the 1960s, 'It is fashionable to curse Gorky now. But after all, without him we would have all starved to death.'[12]

'There is practically no food and it can be said without exaggeration that there will soon be famine in Petrograd,' Gorky wrote to his friend H. G. Wells. 'I just can't imagine how our scholars are going to survive.'[13] The lack of food was critical but scholars faced many other challenges. Their laboratories were being confiscated, their libraries and homes overrun by refugees from the countryside. If you went off to conduct fieldwork, you ran a serious risk of coming home to find it stripped of books, letters and even furniture.

Gorky set up Committees for the Improvement of Living Conditions in cities all over Russia – and Wells, on a visit, found these 'salvage establishments' offering food rations, baths, barber shops, tailoring, cobbling 'and the like conveniences'. Oldenburg and Pavlov, looking 'careworn and unprosperous', both buttonholed Wells on a visit to the Petrograd 'scholars' house' and begged him for new scientific publications. On his return, Wells recruited London's Royal Society in an effort to salve Russia's book famine.

At the height of the famine proper, the Council of People's Commissars unified Gorky's committees under a central commission, TsEKUBU, and over the next five years TsEKUBU's list of scholars grew until it included nearly all of Russia's scientific and literary intelligentsia. By 1926 TsEKUBU boasted six sanatoriums, a dormitory in Moscow, pensioners' homes in Moscow and Leningrad, and a House of Scholars in Moscow with a library and a cafeteria. Such was the power of Gorky's well-run patronage circle – and this and like efforts put the fear of God into the authorities.

The Civil War had dragged rich and poor alike down to abject poverty. Now, under Lenin's NEP, some sections of the population were growing more prosperous, and those who knew how to look after their friends, while fostering the interest and support of patrons, did best of all. The intelligentsia rose quickly from the poverty of the Civil War years to become, by the mid-1920s, a sort of new Soviet bourgeoisie. Specialists employed by the government earned very high salaries. Professors, while quick to complain of ill treatment, enjoyed housing privileges and assured places in higher education for their children.

By the spring of 1922 Lenin was beginning to rue his government's dependence on these old and frankly nepotistic networks of specialists. 'If we take Moscow,' he declared, 'with its 4,700 Communists in responsible positions, and if we take that huge bureaucratic machine, that gigantic heap, we must ask: who is directing whom? I doubt very much whether it can truthfully be said that the Communists are directing that heap. To tell the truth, they are not directing, they are being directed.'[14]

There didn't seem to be any way out of the bind. Articles like the one in *Pravda* by Valerian Pletnev in September 1922 imagined the rise of a new generation of 'social engineers' capable of 'uniting all fields on a grand scale' but, as Lenin scrawled sarcastically across his copy, which of Pletnev's proletarian 'social engineers' was going to build the locomotives?[15]

Lenin decided that if the revolution was not to dissolve in a meshwork of informal friendship circles, drastic action was required.

One obvious way out of the bind was to take the loss of skilled personnel on the chin, and expel those who threatened the Bolsheviks' hegemony. The trick would be to throw out representatives of those disciplines the state could afford to do without on a daily basis. Gorky, for all his support of scientists, was better known as a patron of the arts, and as early as October 1920 Lenin was trying to persuade his old friend to leave. 'If you don't go,' he said, as though it were a joke, 'we'll send you away.' By the summer of 1921, he had grown insistent. There was no chance for Gorky to work in Russia: 'In Europe, in a good sanatorium, you can get treatment [Gorky had tuberculosis] and do three times as much work. Truly. We don't have treatment or work – all we have is bother. Time-wasting bother.'[16]

Gorky was slow to take the hint, and his barrage of personal appeals on behalf of imprisoned intellectuals finally broke Lenin's patience. '[Don't] waste yourself on the whining of decaying intellectuals,' he wrote. 'The intellectual forces of the workers and peasants are growing and getting stronger in their fight to overthrow the bourgeoisie and their accomplices, the educated classes, the lackeys of capital, who consider themselves the brains of the nation. In fact they are not its brains, but its shit.'[17]

The arrest in mid-August 1921 of Gorky's brainchild, TsEKUBU's famine relief committee (for orchestrating 'counterrevolutionary intrigues'), marked the final break between the two old friends. In some disgrace, Gorky left Russia for Italy. He stayed abroad until 1928, aghast at the news of his friends' arrests and the way Nina Krupskaya, Lenin's wife, was removing philosophers from Plato to Tolstoy from the libraries.

The obliteration of pre-revolutionary culture from the universities was part of an attempt to bring on, as fast as possible, a new revolutionary generation. In 1918 the Council of People's Commissars had decreed that all citizens sixteen years old and over had the right to a higher education without regard to sex, nationality or social origin. Working people's faculties, *rabfaki*, had been set up in universities and colleges to prepare working-class students for higher education. But the *rabfak* system had achieved very little. By the early 1920s, hardly any university students had a working-class background. So the government took control of the universities: in 1921, a new charter gave the government control over the appointment of deans and rectors. This attack on the autonomy of the universities couldn't have been better designed to rile the liberal establishment, many of whom remembered the hash made of university education under

Nicholas II. In 1922, the staff of Moscow University went on strike.

The strike played directly into the hands of the government. At Lenin's urging, non-Marxist philosophers and social scientists were stripped of their right to teach, publish and organise academic societies. In 1922, 161 of the most eminent were driven into exile. The Bolsheviks did not need philosophers. They needed engineers. In the early morning hours of 17 August 1922, Soviet security operatives fanned across Moscow and Petrograd, jarring hundreds of intellectuals and their families awake. In the Butyrka and Shpalernaia Street prisons, an assortment of the country's most prominent philosophers, historians and litterateurs shared jail cells before boarding the steamships *Oberbürgermeister Haken* and *Preussen* for their one-way voyage to the German town of Stettin (now Polish Szczecin).[18] Meanwhile scientific figures considered vital to the regime were being feted in grandiloquent ceremonies of reconciliation like the solemn (and expensive) Congress of Scientific Workers of 1923.

There was, finally, one other way Lenin's government could reconcile its reliance on old specialists with its need to make the universities hotbeds of revolution. It could take the specialists out of the universities altogether, and ensconce them in their own well-appointed institutes.

And this, of course, was music to many a specialist's ear. Was this not precisely the policy Germany had followed, in creating the Kaiser Wilhelm Institutes? Was this not what France had done in Paris, with the Institut Pasteur? Was the humble-sounding John Innes Horticultural Institute in Merton Park, England, not leading the world in experimental biology?

When it came to new buildings, regular funding and a routine that did not involve teaching undergraduates, scientists and Soviet officials found themselves singing from the same hymn sheet. Almost every pre-revolutionary institution had managed to survive under the Bolsheviks, finding support from one governmental agency or another. Now a host of new institutes was created. Over forty were created during the Civil War – more than half of the seventy-odd that were running by 1922. Most were very small, poorly housed and poorly supplied. But this was about to change.

Vernadsky's friend Alexander Fersman, writing in January 1922, caught the mood of enthusiasm among the surviving specialist elite. 'The time of amateurish work has gone,' he declared,

the time of individual endeavours, scattered unsystematic scientific literature, chaotic methods has gone. The heroic time of science is in the past, now we need to build science differently, [we] need laboratories, research institutes, expeditions, conferences, world congresses, [we] need to coordinate and unite.[19]

4 : Workers

Why are there mountains of books written on thermal energy, furnaces, boilers, steam machines, electricity, anthracite, white coal, and electrification, but none on the energy of the worker?[1]
  Alexei Gastev

Alexei Kapitonovich Gastev was born on 26 September 1882 in Suzdal, a small town in the middle of Russia. The townsfolk were mostly craftspeople – cobblers, tailors, hatters and painters – and Alexei grew up playing in the large cherry orchards which lay a short distance from the centre of the town.

The child in the cherry orchard would become the most popular, reprinted and influential workers' poet of the early Soviet era, his works universally considered some of the finest and strangest blooms of the Russian futurism. A writer less like Chekhov it would be hard to imagine, and not everyone was impressed. One comrade, coming away from Gastev's recitation of his poem 'We Grow from Iron', complained, 'That worker doesn't speak so much as he yells, like it's coming out of a trumpet.'

Making the New Soviet Man one reflex at a time: innovative biometric studies at Alexei Gastev's Central Institute of Labour.

Gastev, like his literary peers, held Nikolai Chernyshevsky's utopian novel *What Is to Be Done?* of 1863 in high esteem. (Lenin called it 'the greatest and most talented representation of socialism before Marx'.) Chernyshevsky conjured up a vision of the ideal twentieth-century person – rational, self-disciplined, loves work for its own sake, and is selflessly committed to society and the greater good – and enlivened this rather dull-sounding formula with a plot about free love and references to bodily mortification, as when one minor political activist in the novel, Rakhmetov, sleeps on nails to harden his political resolve.

The other great influence on Gastev was Walt Whitman. The New York poet had been introduced to Russian readers through Kornei Chukovsky's translations in the literary journal *Vesy*. In fact futurist poetry generally, in its freedom, roughness and *chutzpah*, owes a lot to Whitman. But it was Gastev, and Gastev alone, who really understood how to marry Whitman's style with the poetry of machines, factories and urban life. And this hardly comes as a surprise, since Gastev was the only worthwhile poet to have spent years working on the shop floor.

Gastev's love of tools and machinery began early. Around 1898, while attending Moscow Teachers' College, he devoted himself to metalwork and carpentry. Expelled just before his finals in early 1902 for arranging a student demonstration, Gastev devoted himself entirely to political agitation in Moscow. (He had joined the Social Democratic Workers' Party in 1900, at the age of eighteen.) In December 1902 he was arrested, charged with disseminating illegal propaganda, and sentenced to three years' exile in the province of Vologda, 500 kilometres to the north.

It was not a severe sentence, and Gastev took it in his stride, becoming the leader of the exile colony. He even arranged a demonstration there against the police's ill-treatment before skipping Russia entirely for France.

In June 1904 Gastev contacted the Bolshevik apparatus in Paris and took a job as a fitter in the Citroën plant. He spent his evenings attending courses organised by Russian émigrés, and learned to speak fluent French. By the end of the year he was in Geneva, where he published at his own expense a short story in Russian entitled 'The Accursed Question'.

The main character, Vasily, is a man in his thirties who has renounced his personal life and dedicated himself to working for the good of society. But Vasily has a weakness, an 'animal passion' which fills him with self-loathing and becomes 'the most accursed, the most tragic question of his life'. Spring comes, and his repressed instincts boil over. He visits a prostitute, regrets it, and vows that 'this would never happen to him again, that he would never again be torn between his passion and his conscience'.

Biology, in Gastev's story, reduces man to an animal. The only way to 'let the human being within him triumph' is self-denial. Gastev soon found more elegant avenues of self-expression, and his poetry is full of vigour and excitement and a sort of horripilating enthusiasm for everything mechanical. It is noticeable, however, that he never again invented a single individual to write about. All terror and excitement in his poetry is collective: heightened, and at the same time sanitised. (The historian Kendall Bailes dubbed Gastev's mentality 'a kind of Puritan ethic in secular garb' – a summary hard to better.[2])

Gastev returned to Russia in the early spring of 1905 and drifted from job to job: on an assembly line, as a lathe operator, and as the driver of a tram.

... with the motors whirring mightily you cut through the crisp air saturated with the fragrance of fresh greenery. Slowly, quietly, as if on velvet, you drift onto the Stroganov Bridge and rein in the car on the slope. A stop. And after it, ignoring the protests of the over-dressed passengers and disregarding safety precautions, I switch on both motors at once; with a terrible lurch and bursts of sparks, I take off as if I had been stung by a bee down the twisting lane of Kamennoostrovsky Prospekt.[3]

A professional revolutionary his whole adult life, Gastev now began to drift away from the Social Democratic Workers' Party. A practical man, what mattered to him most were improved working conditions, better wages, health insurance and the like, and these issues, it seemed to him, were better settled on the factory floor, by the workers themselves, than abstracted into policy and discussed by bureaucrats in a capital that might lie half a world away. Who better to realise the revolution than the workers in whose name the revolution was being fought?

Gastev found himself more and more drawn to the Syndicalist movement (*'syndicat'* is French for trade union), which was then reaching a high point of popularity and influence. Gastev's dream was to put the entire economy into the hands of skilled trades unions, and across Europe many socialists were drawing the same conclusion: that it was the business of workers themselves to take over factories, using the weapon of a general strike. (Mass industrial actions in Holland, Russia and Italy had been making headlines since 1904. The socialists had no such victories to brag about. There had been a revolution in Russia, in 1905, but they had missed it.) The International Congress of Syndicalists, meeting in Paris in April 1907, bemoaned Marxism's 'decomposition', while the Italian journalist Arturo Labriola put the syndicalist case succinctly: five minutes of 'direct action' in the streets were worth many years of parliamentary debate.[4]

In 1910, Gastev returned to Paris. He wrote: 'I want to try not only to observe, not only to learn, but actively to work there.' He found the city flooded with fellow self-exiles. In the Seine department alone, one in twelve residents was Russian. He found jobs as a metal-worker and a taxi driver, waking at five in the morning, returning home for supper around eight. He spent his evenings at lectures or gatherings of the Metal-Workers' Section, the Bureau of Labour and the Workers' Club. The bohemian existence led by many other Russian residents – visits to second-hand bookstores, cafés and museums, premieres of plays by Anatoly Lunacharsky, the doyen of the Paris scene – passed Gastev by.

In the spring of 1913 he returned to Petersburg and after six months he was arrested again. This time he was sent to Siberia: to Narym on the Ob River, 400 kilometres north-west of Tomsk. Again, he took exile in his stride, baking bread, teaching, repairing machinery and writing. 'Express: a Siberian Fantasy' was completed here: a visionary prose poem about the industrial future of Siberia.

In a curious Russo-American pidgin all its own, 'Express' describes a journey on an enormous, futuristic express train, *Panorama*, through Siberia's 'ungreened land' – now a dense network of factories, tunnels, roads, canals, stations and towns. Buildings tower into the sky and plunge kilometres deep under the earth. Most utopias are as ossified as statues. Gastev's future Siberia is a contested place, as 'Americanoid trusts' struggle for dominance with Siberian peasant socialists.

Panorama crosses Siberia, from the Urals to Irkutsk, and then on a new line up to the Bering Straits and America. The steppe and tundra have been made the breadbasket of the world. Monstrous machines work the land. Human workers live in rows of houses laid out in a straight line for hundreds of kilometres. One after another the man-made wonders pass. Kurgan, 'Kitchen of the World'. The 'City of Steel', Novo-Nikolaevsk. Krasnojarsk, the city of research, has a seismograph that can accurately predict earthquakes.

Irkutsk, city of commerce: moving sidewalks carry passengers from level to level. Energiya: all the volcanoes have been capped with steel and asbestos to collect geothermal energy. Bering: a palace has been built of amber harvested from the sea. There are plans to melt the North Pole, so the Arctic may be fit for gardening. In front of the Bering tunnel stands a giant lighthouse, the tallest building on earth, made of concrete, metal, paper and non-melting ice. 'Onward! Through the dangerous swamps, to the end, to the farthest, farthest end!'

Gastev's fascination with machinery dominates – but already we can see what became his abiding obsession in later life, when he gave up poetry to train the Soviet worker for the modern factory. Through an alchemical wedding to the machine, men become Man. The train itself longs 'To drown man in metal, melt all the souls and out of them create one great one'.[5] There are even hints of a cybernetic future here, and a world run not by people, but by mainframes.

Alexei Gastev returned to Petrograd in April 1918. He was now thirty-four. He tried to make a go of a career in trade unionism; he was elected secretary of the All-Russian Metal-Workers' Union on its foundation on 27 June. But he found that the Bolsheviks, having taken control, were less inclined than ever to give real political power to the workers. The Bolshevik trade union boss Solomon Lozovsky was dogmatic: 'We must emphasise clearly and firmly that the workers ... must not believe that the factories belong to them.' The trade unions existed to organise the workers. What the factories actually did was the preserve of government, and was to be coordinated on a national scale. Gastev attempted to protest, which is probably why he was not re-elected.

Gastev's disappointment was tempered somewhat by meeting Sofia Abramovna Grinbald, a young secretary in Lenin's office. They married, and in 1919 they travelled south to the Ukraine, where Gastev devoted his time to writing. It proved a profitable occupation. His literary reputation grew quickly: a genuine worker who wrote poetry that poets could admire, in language that embodied everyone's idea of the technologically advanced future.[6]

Whistles

When the morning whistles resound over the workers' suburbs, it is
  not at all a summons to slavery. It is the song of the future.
There was a time when we worked in poor shops and started our
  work at different hours of the morning.
And now, at eight in the morning, the whistles sound for a million
  men.
A million workers seize the hammers at the same moment.
Our first blows thunder in accord.
What is it that the whistles sing?
It is the morning hymn to unity.[7]

Lunacharsky called Gastev 'our most gifted proletarian writer'.
The poet Nikolai Aseev dubbed him 'the Ovid of miners and
metalworkers'. Gastev's young contemporary Mikhail Svetlov
once remarked, 'Alexei Gastev? That's not just a poet, that's a
phenomenon.' Gastev's *Poetry of the Factory Floor* (1918) was the
first literary work published by Proletkult, the Bolshevik cultural
organisation. It was already in its sixth edition by 1926. His works
were regularly read and staged in Proletkult theatres in Petrograd
and Moscow, and agit-prop trains[8] at the front during the Civil
War had 'The Tower' and 'We Have Usurped the World' on their
repertoire.

Gastev affected to have little time for the Russian language,
with all its endless circumlocutions. Whenever he could, he
used foreign loan-words. This makes him surprisingly easy to
read for non-Russian speakers, who can still get an idea of his
poetry from all the numbers, musical terms, and industrial and
military jargon. 'I grant you that, so far, we lack an international
language,' he explained:

Still, there are international gestures, and international psychological
formulae which millions know how to use. It is this which imparts
to proletarian psychology a striking anonymity – one that lets us
classify a single proletarian unit as A, B, or C, or 325, 0.75, 0 and so
on. This psychological flattening and streamlining is the secret behind

the enormous spontaneity of proletarian thinking ... In the future this process will make individual thought impossible. It will, by imperceptible stages, be transformed into an objective and universal class psychology: a system of on-switches and off-switches and short circuits.[9]

Gastev's *Poetry of the Factory Floor* was probably the inspiration – certainly a huge influence – upon one of the great works of satire to come out of the iconoclastic 1920s, Yevgeny Zamyatin's funny and devastating novel *We*.

A prominent Old Bolshevik, Yevgeny Ivanovich Zamyatin was born in the Tambov region, 300-odd kilometres south of Moscow, in 1884. His father was a Russian Orthodox priest, his mother a musician. He had joined the Bolsheviks while studying naval engineering in St Petersburg. In 1916 he had been sent to Britain to oversee the construction of icebreakers, and lived for a while in Newcastle upon Tyne.[10] The protagonists of *We* have numbers rather than names, and these were taken directly from the specifications of Zamyatin's favourite ship, the *Saint Alexander Nevsky*.

In *We* Zamyatin's hero-inventor D-503 describes with lyrical enthusiasm how work is organised in the One State. He is, by the lights of his far-future society, a literary-minded fellow, and lets us know, by repeated literary references, how influential Gastevian poetry has been in the shaping of his utopia: 'We all (and perhaps all of you also) have read as schoolchildren that greatest of all the monuments of ancient literature which have come down to us: *Timetables of all the Railroads*. But place even that classic side by side with *The Tables of Hourly Commandments* and you will see, side by side, graphite and diamond.' But his humanity, though deeply buried, and regimented nearly out of all existence, plays him false in the end. At first his malfunctioning existence is revealed in moments of high comedy. But as his love for the delicious E-330 grows, farce turns to tragedy, and ultimately to horror, as the One State rolls out its final solution for these sorts of emotional breakdowns – wholesale mechanisation, inside and out.

The door of the auditorium on the corner was gaping and out of it issued a slowly, ponderously trampling column of fifty men. However, men is not at all the word: they did not have legs but some sort of heavy forged wheels turned by an invisible drive; not men, these, but some sort of humanoid tractors.[11]

Zamyatin composed *We* in the early years of the Bolshevik experiment, and this makes it a dangerous book to pontificate over. Is it a send-up of the production-line methodology espoused by that prophet of industrialisation, Frank Winslow Taylor? Of course it is. Is it a savage parody of the regimentation that awaited Soviet Russia under Stalin? Probably not, though it's very hard for a modern reader not to read it that way. Is it an attack on technically narrow-minded trade unionists like Gastev, or patrician old specialists and their patrons – men like Felix Dzerzhinsky? He was probably tilting at both. Zamyatin's delirious, scatter-gun determination to send up virtually every Bolshevik pretension is rather lost, because subsequent events made it prophetic. So too did the author's fate. In 1921, *We* became the first work banned by the Soviet censorship board. In 1924 Zamyatin arranged for *We* to be smuggled to London for publication and in 1927, an unauthorised Russian edition (translated *back* into Russian from Polish!) made Zamaytin's further career in Stalin's Russia impossible. He was finally given permission to leave in the country in the autumn of 1931.

Alexei and Sofia Gastev's idyll in Kharkov came to an end towards the end of 1919, when they were forced to flee Denikin's advancing White army. In early 1920 they came to Moscow, and Alexei found work as a technical supervisor at the engineering company Elektrosila. It was here that he first 'plunged into an analysis of the work of the automaton as the most perfect machine' – a truly *We*-ish formulation to describe his growing interest in the ergonomics of factory labour. Inspired by the work of Frank Taylor, Gastev wondered whether the workings of machines held lessons for how humans performed simple tasks.

Gastev's work at this time was both practical *and* poetic. His last poetic work, 'A Packet of Orders', was written deliberately so as to sound as if it is being fed into a machine. In his note to performers, Gastev explains that 'A Packet of Orders' is supposed to be read in an entirely flat manner: 'No intonation, no pathos, not pseudo-classical elation and no stressed emotive places in the reading. Words and phrases follow one another with even speed.'

Order 2

Chronometer, report to duty.
To the machines.
Rise.
Pause.
Charge of attention.
Supply.
Switch on.
Self feed.
Stop.[12]

Gastev found practical expression for these ideas in a pilot project conducted with five or six friends in two small rooms in the Hotel Elite in Central Moscow – now the site of the (rather good) Budapest Restaurant. Like everyone else, they spent most of their time scrabbling around for equipment. Pencils and paper were in short supply, and for a while they had to do without tables and chairs. In a letter dated 21 October 1920 Gastev complained, 'I have one colleague, a man who has made extremely valuable contributions, whose shoes literally lack soles, and none of my fellow workers has a room.'[13]

This did not stop Gastev giving his time-and-motion experiment a grand title. As director of the 'Institute of Labour', Gastev contacted Lenin in person and Lenin (who by now was sticking Taylorist charts on his walls and jotting ideas for his own book on scientific management) granted it funding. The Central Institute of Labour was born.

If Gastev's efforts to become a sort of Soviet Taylor seem ripe for parody (and Zamyatin certainly thought so), it is worth noting what Gastev himself said about the mechanisation of factories. For Gastev, who had actually worked in these places all his life, the point of mechanisation was to clear out the chaos, dirt and danger of the old sweatshops – places where, if you didn't get your fingers ripped off, your labour still came to nothing. As far as Gastev was concerned, the modern factory actually *dignified* labour, and he had very little time for people who assumed that machines were there to replace people.

People are waiting around like Oblomovs for a machine to come along like some sort of saviour – a machine that will eliminate the need for skills or so-called heavy labour ... An enormous number of literary popularisers of Scientific Management look upon the machine as Chekhov's characters gazed upon the 'diamond sky' in *Three Sisters*.[14]

Lenin was never entirely convinced on this point. He eventually embraced Taylorism and its promise of enormous gains in productivity, but he worried that the new assembly-line practices his party was introducing would simply extract the last ounce of sweat from the worker. As early as 1912, he had neatly skewered Taylor's supposed 'advances', when recounting the improvements visited upon one American factory:

A mechanic's operations were filmed in the course of a whole day. After studying the mechanic's movements the efficiency experts provided him with a bench high enough to enable him to avoid losing time in bending down. He was given a boy to assist him ... Within a few days the mechanic performed the work of assembling the given type of machine in one-fourth of the time it had taken before!

What an enormous gain in labour productivity! ... But the worker's pay is not increased fourfold, but only half as much again, at the very most, and only for a short period at that. As soon as the workers get used to the new system their pay is cut to the former level. The capitalist obtains an enormous profit, but the workers toil four times as hard as before and wear down their nerves and muscles four times as fast as before.[15]

These doubts never really vanished. When it came to running the country, however, scarce physical and human resources forced his hand. Practicality soon overrode every other consideration. His wife Nadya Krupskaya's appraisal of Taylorism was brutally practical: 'The division of functions and the introduction of written instructions allow for the placement of less qualified people in any given job.'[16]

This goes a long way to explaining why Gastev's frankly eccentric Central Institute of Labour received such enthusiastic official backing.

Not that this backing made much difference – at least, not at first. The institute did not have stopwatches, training aids, fuel for heating, or, indeed, food. Its staff tried to simulate factory conditions in their experiments, but they had to mock up their machinery out of wood.

Gastev proposed to study, measure and improve all the basic motions involved in physical labour. 'We begin with the simplest, the most elementary movements,' he explained, 'and proceed to the mechanisation of man himself.'[17] One of Gastev's admirers – Tatyana Popova, a Moscow graduate from a textile-making family – wrote to her husband in 1924: 'The Central Institute of Labor is a new Institute . . . Everything is done in a new manner, not in the way it was done by the bourgeoisie. The Institute is striving to introduce science into production. The interests of the director are those of a metalworker, therefore the Institute studies mostly the work of a metalworker and his two main procedures: chiseling and filing.'[18]

A class at the Central Institute of Labour was a sort of drill practice. Pupils stood before their benches in set positions –

places were marked out for their feet. They rehearsed separate elements of each task, then combined them in a finished performance. This was the regimentation of human beings taken into the realm of choreography, and Gastev, for all his modernist toothbrush moustache and crew-cut hair, wasn't blind to the artistic potential of his creations. Take, for example, his Social Engineering Machine – a giant structure of pulleys, cogs and weights of no fathomable use whatsoever. It was, by Gastev's own admission, his 'last work of art', a sort of mascot for the whole venture, and his aim was to install them all over the USSR.[19]

As funding improved, the institute's classrooms started to resemble modern muscle gyms. New-fangled machines trained workers in the movements they needed to operate their tools. On the walls, huge charts dealt with every detail of the work process: what and how much you had to eat to perform different work; how to swing a hammer; how to build a wall.

Going by the sheer popularity of the classes, and the speed of the institute's expansion, the classes must have been quite enjoyable. The problems started when, having completed your training, you went off to work in a real factory. The Central Institute's classes, for all their memory training sessions and educational films, frequently left you with next to no understanding of *why* you were performing the actions the factory demanded of you. The institute's well-ordered instruction shops taught you how to work, but they taught you nothing about factory life. If you were a peasant, arriving in the big city and looking for work and to better yourself, attending one of Gastev's courses could indeed get you a job – but in truth you'd still be stumbling around in a rather lost fashion, and a factory floor, even a modern one, is an unforgiving place.

In January 1921 Leon Trotsky, then at the height of his political influence, convened a conference to discuss these issues: the First All-Russian Initiating Conference on the Scientific Organisation of Labour. Gastev's Central Institute of Labour was a fascinating experiment, but the conference revealed a need for change. By the time the institute completed its biometrical investigations of manual labour, Russian industry would be so changed by mechanisation that the institute's results would mean nothing, and Gastev had a fight on his hands to convince the authorities and his critics that his institute was still relevant to the development of the economy.[20]

In photographs, Gastev's mean-looking *pince-nez* and his uncanny resemblance to his hero Frederick W. Taylor suggest an extremely reserved man. One contemporary described him as 'a tightly wound steel spring' and a hard taskmaster. But he was also quick-witted, never at a loss for words, and a good and considerate friend. An evening out with Alexei Gastev was likely to be an entertaining one: athletes and circus artists fascinated him, particularly jugglers, acrobats and magicians, whose feats of precision and agility made an art of physical training.

Gastev had, by his own account, left his own artistic practice behind him. He once wrote that he had turned to poetry only when other avenues of expression were cut off, and when the 'revolution broke out [it] presented an opportunity to work directly as an organiser and creator of something new'. Sure enough, his 'practical' writings weren't much less surreal than his verses. He wrote in *Pravda* in July 1922: 'In the human organism there is a motor, there are springs, the finest regulators, there are even manometers. All this we need to study and use. There should be a special science – biomechanics. This science does not have to be narrow, about "work" only; it should border on sports, where movements are powerful, precise, and at the same time as light as air and as artistic as mechanisms.'[21]

He was not at all an isolated figure. For example, there were numerous attempts at this time to finds ways to record and categorise dance movements. Valentin Parnac, who introduced jazz bands and jazz dance to Russia, developed a system of dance notation. Nikolai Bernstein's 'chronophotographs' used to hang alongside pictures of dance performances in exhibitions.

Bernstein's efforts are particularly interesting because the 26-year-old joined Gastev in 1922 and worked with him for three years, before moving biometrics out of the workshops of the Central Institute of Labour and into the physiological laboratories of Moscow, where he worked in collaboration with Tatyana Popova, his sister-in-law. Their experiments on the 'biodynamics of piano touch' were notoriously noisy: now and again a siren would howl, while a pianist, interrupted by shouted commands, hammered on the same two keys of a piano, all to tease out the 'scientific organisation of musical labour'.[22]

Nikolai Alexandrovich Bernstein, a lifelong Moscow-dweller, came from a family of celebrated intellectuals (his father was a psychiatrist, an uncle a mathematician) and he cultivated all manner of interests and hobbies. He played piano, drew, assembled radios, and constructed models of steam engines and bridges. Bernstein worked out ways of mechanically registering movements and analysing those records mathematically. It was quite an achievement: movement is usually too quick for the human eye to grasp with any accuracy, and wasn't really accessible to science until the advent of film. Even then, a visual record of points in motion is extremely hard to interpret. It took Bernstein several years to come up with the necessary mathematics.[23]

He developed a high-speed camera called the 'kymo-cyclograph'. The shutter, a round plate with holes in it, rotated before the camera lens, so that the photographic plate would record multiple images, each exposed a fraction of a second after its neighbour. The shutter span so fast, Bernstein had to use a tuning fork, blowing air through the holes and comparing the pitch of the fork and the whistle, to work out the machine's exact shutter speed.[24]

Bernstein saw all manner of applications for his experimental equipment. He worked with musicologists and mathematicians, and collaborated for a while with psychologists Alexander Luria and Lev Vygotsky. To begin with, he was happy enough with Gastev's machine metaphors. In his first scientific paper on the subject, 'The Biomechanics of the Worker's Stroke', he wrote:

The laws of mechanics are the same everywhere, no matter if they concern a steam locomotive, a lathe, or a human machine. Therefore, we do not have to derive some new, special mechanical laws. We must only compile a description and a characteristic of this living machine in the same way as we would do it for an automobile or a loom.[25]

In the course of his work, however, Bernstein discovered something that undermined Gastev's doctrine. He discovered that a physical movement – say, 'the hand movement with the hammer' – had to be treated and examined as one continuous action: 'one could not alter a single detail without the entire construction changing in a quite regular way'.

With practice you can hit the same sort of nail into the same sort of block of wood, at the same point, and with the same force, again and again and again. But film those actions with a kymocyclograph and study the results: no two strikes are alike! The result remains the same, but how you *get* the result varies each time.

His conclusion that 'the movement responds as a living being' meant Gastev's project – to break human movements down into mechanical components – simply didn't make sense.

By 1925 Bernstein and Popova found themselves increasingly out of sympathy with Gastev's project, and left the Central Institute of Labour. Bernstein spent the next ten years jumping from post to post at a dizzying number of institutions: the Moscow Institute of Psychology, the State Institute of Labour Preservation, the State Institute of Musical Sciences, the Scientific Research Bureau of Prosthetic Appliances, the All-Union Institute of Experimental Medicine, the Central Institute of Physical Culture . . .

The more he worked on his discovery, the more radical it turned out to be. It suggested that behind the simplest action lay a high-level command – a command that couldn't possibly arise from simple reflexes.

Put it this way: why is it that if you sign your name in sand with your foot, your signature bears a remarkable likeness to the signature you make with your hand, holding a pen, while sat at a desk? And if you think about it, how is it that a signature written while sitting at this chair and this desk is so similar to a signature written in *that* chair, at *that* desk? Conventional wisdom had it that handwriting style is a muscular habit. But if that were so 'then every new posture of a subject would require the establishment of a whole new system of muscular regulations' – an obvious nonsense.[26]

So Bernstein began to assemble a wholly new and revolutionary model of physiology: one which compared the nervous system not to a telephone switchboard, as Ivan Pavlov did, but to something Bernstein frankly struggled to find a name for. A servo-mechanism? A feedback device?[27]

And so, a full decade before the American mathematician Norbert Weiner coined the term, Nikolai Bernstein invented cybernetics – and embarked on a course that would set him and his colleagues at loggerheads with Ivan Petrovich Pavlov, the most revered scientist in all Russia.

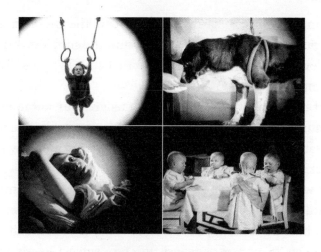

5 : Exploring the mind

*A human being is not a satchel stuffed with reflexes; neither
is it a hotel for reflexes that happen to drop in.*[1]
  Lev Vygotsky, 1925

Ivan Petrovich Pavlov was born in 1849. Nine years later he fell
off a wall. He recovered slowly, and spent much of the next two
years with his godfather, the abbot of a monastery near his home
town of Ryazan, 200 kilometres south-east of Moscow. The
monastery's simple, regular life and efficient bureaucracy had a
clear influence over Pavlov's later administration of his famous
laboratories.

Pavlov left Ryazan for good at twenty-one, abandoning a
religious career to join the faculty of physics and mathematics
of the Military Medical Academy of St Petersburg. While he was
there, student demonstrators harried, attacked and eventually
destroyed the career of his adored mentor Ilya Tsion, a gentle man
who was also an arch-conservative. This 'wild episode' left Pavlov
alienated from his liberal colleagues and shaped his bleak view of
the Russian character. 'To say anything against the general mood

Vsevolod Pudovkin's film *Mechanics of the Brain: The Behaviour
of Animals and Man* popularised Ivan Pavlov's physiology as a
'materialist' science.

was impossible,' he recalled, speaking to the Women's Medical Institute in the spring of 1918. 'You were dragged down and all but labelled a spy.'[2]

Before the war, Pavlov led a life of great comfort: he was, after all, a member of the Academy of Sciences and a Nobel Prize winner. He enjoyed a spacious apartment with his wife and four children. He collected books and paintings. He spent summers at a dacha in Finland, where he swam, rode his bicycle, and socialised with other moneyed and educated holidaymakers.

Even Ivan Pavlov, a monarchist with two sons at the front, considered the tsar a dead duck, especially following the destruction of the Russian fleet off the coast of Korea in 1905. He scorned Nicholas II as an 'idiot' and 'a degenerate'. He welcomed 1917, and came to call himself a liberal in solidarity with his colleagues, though his actual views were rather more conservative.

Pavlov was a choleric, pugnacious man, not above rapping students about the head with a beaker when they exasperated him. As far as he was concerned, 'the only virtue, the only joy, the only attraction and passion, is the achievement of truth'. His wife Serafima, a gentle, profoundly religious woman, also tended to bang on about how worthwhile it was, supporting her godless husband in his pursuit of science. (If she talked about anything else Ivan would fall into a sulk and then stalk out of the room.)[3]

Pavlov spent the 1890s, his most productive decade, studying the physiology of digestion at the Imperial Institute of Experimental Medicine in St Petersburg. 'Pavlov was not an eloquent orator,' a former pupil recalled,

but his lectures were amazingly interesting . . . He spoke vividly and gesticulated expressively . . . It was a rule at his lectures that any student could raise his hand at any time and ask a question . . . It grieved him to hear none; in such cases he considered his lecture inadequate and would plead: 'Questions, questions, I don't hear questions.'[4]

Pavlov spent a long and illustrious career working out the role of the nervous system in bodily functions such as circulation and digestion. He deliberately omitted from his work any consideration of thought. He banned the word 'learn' from his laboratory and fined assistants who used any mental terms. By the turn of the century, however, thought was becoming increasingly hard for Pavlov simply to ignore.

The problem was straightforward. Beginning around the 1830s, a new experimental discipline had arisen. Psychology was supposed to explain the relationship between the senses, consciousness and normal behaviour. It was the science of the mind, and it was supposed to be an independent field. But in the course of half a century it had failed to explain anything very much, and had left a vacuum into which other disciplines were being drawn.

In 1874 the German philosopher Franz Brentano opened his major work *Psychology from an Empirical Standpoint* with a complaint about the unscientific clutter of 'psychologies' with which he had to deal. It was time, he said, to fuse them into one discipline. This was what his book was for. It failed.

William James's *Principles of Psychology* of 1890 was altogether more iconoclastic: a self-mocking anatomy of a discipline which James doubted very much existed at all. He had arrived in Germany in 1867 as a young man, eager to be in at the birth of the new science, only to find his colleagues heaping up psychophysical measurements on top of each other in the ever more desperate hope that they would cohere to form a natural science of the mind. It was an approach, wrote James, that 'could hardly have arisen in a country whose natives could be bored'.[5]

James had reached the conclusion that there was no 'science of the mind' – only the hope of one. The much touted 'new psychology', James wrote, was:

a string of raw facts; a little gossip and wrangle about opinions; a little classification and generalisation on the mere descriptive level; a strong prejudice that we have states of mind, and that our brain conditions them: but not a single law in the sense in which physics shows us laws, not a single proposition from which any consequence can causally be deduced.[6]

Leon Trotsky, who had first-hand knowledge of psychoanalysis from his time in Vienna, wondered whether we could indeed arrive at knowledge of the sources of human poetry by studying the saliva of dogs,[7] and nursed the idea that while Ivan Pavlov – Russia's only living Nobelist and a national treasure – was approaching mind from *underneath*, by studying the smallest units (reflexes) from which mind was built, the father of psycho-analysis, Sigmund Freud, was approaching mind from the top down, scraping away at consciousness to reveal the structures beneath:

Both Pavlov and Freud think that the bottom of the 'soul' is physiology. But Pavlov, like a diver, descends to the bottom and laboriously investigates the well from there upwards, while Freud stands over the well and with a penetrating gaze tries to pierce its ever-shifting and troubled waters and to make out or guess the shape of things down below.[8]

Given time and encouragement enough, might these great pioneers not meet in the middle, and create a truly scientific psychology?

Well – *no.* The problem has turned out to have as many dimensions as there are writers to write about them, but the chief problem in Pavlov's day was practical. Physiology is the study of how living systems function, and this turns out to be virtually impossible unless you – literally, physically – cut mind out of the equation. The most famous and disturbing description of this problem came from the English physiologist Charles Sherrington, who tried to map the nerve circuits that controlled the scratching movement of a dog's leg when its back is rubbed.

Sherrington found himself embroiled in the most grotesque series of vivisectional horrors, creating 'puppet animals', mutilated living preparations, in order to solve quite simple physiological problems.

Several Russians who made important contributions to physiology found themselves banging their heads against the same wall.[9] Ivan Sechenov was the first Russian physiologist of European standing, while at home he acquired an additional reputation as a radical and a nihilist when, in 1863, he published 'Reflexes of the Brain', a speculative essay arguing that all human behaviours are woven together out of reflex actions. The problem was how to map these reflexes in any meaningful way. By 1869, Sechenov was ready to quit. He lived to 1905 but spent his time studying gas absorption in blood, rarely returning to the grand but failed big ideas of his youth.

Vladimir Bekhterev enrolled at the Military Medical Academy in St Petersburg in 1873, three years after Ivan Pavlov, and graduated as a medical doctor in 1878. After a series of gruesome experiments that demonstrated how plastic the nervous system was, and how good it was at compensating for damage, Bekhterev threw up his hands and decided he did not actually *need* to map physical structures at all. Instead, he just stretched the idea of a 'nervous reflex' to cover whatever it was he wanted to talk about.[10] Free from real, headache-inducing physiology, Bekhterev toyed with all kinds of approaches to mind, from parapsychology to Marxism. This sort of dilettantism is usually a bad idea, but for as long as the field of psychology refused to cohere, bringing every idea under one roof at least got people talking with each other. In 1907 Bekhterev founded his Psychoneurological Institute – an extraordinarily diverse, diffuse and confusing place that Bekhterev liked to think of as 'a university of the human sciences'. (It was also one of the first co-educational colleges in Russia.) Its staff included the Darwinian zoopsychologist Vladimir Vagner, an educational psychologist specialising in personality studies, a liberal sociologist, a historian of culture, and a religious philosopher. Disagreements were welcomed, controversies thrashed out with courtesy and wit. It was a good place to work. But it produced no coherent psychology.

Then there was Pavlov.

Ivan Pavlov was a meticulous and brilliant surgeon who hated the sight of blood and took great care over his experimental animals. He had developed a technique of using surgical 'windows' or fistulas to study the functions of various organs, so that he didn't have to resort to vivisection. This meant that he was confronted, day in day out, with the well-known phenomenon of 'psychic secretion' – the way dogs salivate not just when food is placed in their mouths but also when food is put before them – even when they simply hear the sounds of food being prepared.

He was perfectly well aware of this phenomenon, of course, but it was the kind of observation he rigorously excluded from his work, for the very good reason that his work could say nothing about it.

He was fifty-four, and on the verge of winning a Nobel Prize for decades of patient investigation of the nerves that regulate secretions in the digestive system, when he changed tack, startling a congress of physiologists in 1903 with his 'conversion from purely physiological questions to the field of phenomena that are usually called psychic'.[11]

Quite why he stuck his neck out so far, so late in his career, had less to do with ambition and more to do with his desire not to revisit matters he considered closed. The great advance of Pavlov and his generation had been to explain bodily functions in terms of nerve signals. Prior to their arrival, physiology was still steeped in the humoral tradition, which held that the body's workings were steered by the comings and goings of chemicals in the blood. Pavlov and his peers had taken great delight in demolishing this medieval idea.

In 1902, however, English physiologists William Bayliss and Ernest Starling discovered a hormone which appeared to control the release of juices secreted by the pancreas. According to Bayliss and Starling, the pancreas was not directed by nerve signals after all, but by a 'humoral agent'. A whole string of 'neo-humoral' findings followed, threatening Pavlov's tidy 'nervism'.

So, rather than review ten years' worth of experimental results, Pavlov chose to move into a fresh, less contested, less crowded area: psychology. And, in doing so, he attempted to become what Trotsky had made of him: a down-to-earth materialist whose dogged good sense would explain uncomfortably complex behaviours in simple, reductive ways.

Pavlov tried to explain mind in terms of reflexes. Simple 'hard-wired' reflexes were familiar to physiology students the world over. Pavlov concentrated instead on those reflexes that are acquired as one interacts with the world. He called these 'conditional reflexes', first because they depended upon external conditions, and second because he was still not entirely convinced of their existence. (The functional ambiguity of the word 'conditional' is the same in Russian as it is in English.)

A conditional reflex is acquired by repetition. In the laboratory, for example, a buzzer, light or mild electric shock (never, funnily enough, the handbell of popular myth) excites a dog's 'hearing analyser' – a more-or-less hypothetical structure somewhere in the dog's cerebral cortex. Then the dog is fed, and this excites a different locus, the 'feeding centre'. Between the two points a temporary nervous link is established. After sufficient repetitions, the link is strong enough to set off a salivary reflex all on its own – but only for so long as feeding follows the bell. If food does not follow, then the conditional reflex is 'inhibited' and the salivatory reflex dies away.

'Conditional reflexes' were a way Pavlov could say things about mind while still generating repeatable, demonstrable effects in the laboratory. It was a fudge, enabling him to shoehorn his research into psychology while churning out dissertation topics that his junior staff could handle.

Not everyone was convinced. Though they had been friends for years, and had served on the same dissertation defence committee since 1895, Pavlov had a major falling-out with Vladimir Bekhterev when Bekhterev – eight years younger, and as a practising doctor more experienced in these matters – began to criticise Pavlov's change of direction. The men argued bitterly about the functions of the brain. They stopped talking to one another and even refused to shake hands. Six years after Bekhterev's death in 1927, Pavlov admitted to a friend: 'Only now I feel how much I miss the arguments with Bekhterev.'[12]

Pavlov suffered his share of the tragedies visited on the country by its civil war. His son Viktor contracted typhus and died in Kiev while on a mysterious mission to White territory. During the Red Terror of 1918–20 the Bolshevik secret police repeatedly searched the Pavlovs' home and briefly detained both Ivan and his eldest son, Vladimir. A third son, Vsevolod, was also a source of constant anxiety: declaring his intention to 'fulfil my citizenly duty to the Homeland', he had joined the White resistance. He did not return home until 1928, and ever afterwards the family remained on tenterhooks, fearful of his further arrest.

Throughout the 1920s the secret police followed Ivan Pavlov closely, and informers repeated his conversations in the lab, on the street and even with his family. The pressure took its toll. Pavlov's temper grew worse and worse. He was even known to strike his subordinates – though if you had the bottle to shout back at him, you could usually bring him to his senses.

It was a wonder he stayed. As early as June 1920 he was writing to the highest governmental agency, the Council of People's Commissars, for permission to 'begin a correspondence (even a controlled one) with my foreign scientific friends and colleagues about finding a place for me outside of my homeland'. He felt isolated and, worse, he was broke: he held three academic posts and together the salaries weren't enough to support his family.

I am compelled to work as a gardener in the appropriate season, which at my age is not always easy, and even to constantly play the role of a servant, of an assistant to my wife in the kitchen and maintenance of the apartment's cleanliness, which, taken together, occupies a large, and even the best, part of the day. And despite this my wife and I eat very poorly, in both amount and quality; for years we have not seen white bread, for weeks and months at a time we have no milk or meat, feeding largely upon black bread of poor quality, on millet, also of poor quality, and so forth, which naturally is causing us gradually to waste away and lose our strength. And this after a half-century of intense scientific work crowned by valuable scientific results recognised by the entire scientific world.[13]

Pavlov was laying it on a bit thick. He had august connections all over Europe and had he wanted to leave the USSR, no one at that time could have stopped him. He wanted signs of support at home much more than he wanted, or needed, permission to leave. He was advancing an argument already tried by the Academy of Sciences and other institutions: fund us, or we emigrate.

While he waited for an answer, Pavlov tested the ground by asking Anatoly Lunacharsky if he could have his medals back. They were returned to him straight away.

Contemplating Pavlov's letter, Lenin found himself in a bind:

It doesn't make any sense to let Pavlov go abroad . . . being a truthful person, in the event someone starts that sort of conversation, he won't be able to stop himself speaking out against the Soviet regime and Communism in Russia. At the same time, this scientist constitutes such a big cultural asset, our keeping him in Russia by force, in conditions of material want, is unthinkable.[14]

Not long afterwards, Lenin's hand was forced. In November 1920 the Swedish Red Cross wrote to him, suggesting that the Soviet government express its gratitude for all the aid given to Petrograd's hospitals by allowing Pavlov to emigrate.

Around the same time, a report landed on Lenin's desk, revealing that Pavlov's sob stories had drifted hardly at all from reality: his lab really was closed down and frozen through; of his staff of twenty-five, only two die-hards remained; and a hundred valuable lab dogs had died of poisoning after eating refuse from a local artificial bread factory.

There was no point procrastinating. On 24 January 1921 the Council of People's Commissars issued a decree 'On Conditions Facilitating the Scientific Work of Academician I. P. Pavlov and His Co-workers', tasking a special commission (headed, naturally enough, by Maxim Gorky) with creating 'the most propitious conditions' for Pavlov's scientific work. A 'luxurious edition' of Pavlov's works was announced, the proceeds from its sale to go to his laboratory. His living quarters, as well as his main laboratory, were to be furnished 'with all possible conveniences'. Pavlov balked at the offer of a double food ration until the authorities agreed to extend the offer to his staff.

Pavlov's wife Serafima summed up what this all meant: 'Give Ivan Petrovich everything that he wants, but under no circumstances allow him to go abroad.'[15]

Pavlov's attempt to explain higher nervous functions in terms of reflexes was fraught with difficulty. 'Psychology almost entirely overshadows physiology,' he complained, as he struggled to devise lab experiments that could handle mental operations. But that, of course, is the problem: laboratory work is all about spotting the repeatable effect, while mental operations are, by their very nature, notoriously changeable. (It wouldn't be much of a brain that couldn't change its mind.)

Pavlov was reaching the end of his useful working life and like many an ageing scientist, before and since, he inclined to philosophy. He had no worked-out methods for studying people and no data at all on human reflexes, but he was quite happy to extrapolate from dogs to people. Pavlov tried to prove that all levels of animal behaviour are made up of chains of conditioned reflexes. If a dog salivates to a light that precedes a bell that precedes feeding, then it follows that when a man goes off to the shops, he is responding to a longer, more complicated, but still recognisable chain of conditioned stimuli.

That anyway was the idea, but it didn't work. Pavlov's dogs hardly ever salivated to the light that preceded the buzzer, and not a single dog ever salivated to the electric shock that preceded the light that preceded the buzzer. Chains of reflexes proved impossible to establish. By then, however, Pavlov, once the great and honest experimentalist, had ceased to care about his own data.

Pavlov was not the first great scientist to publish, in his dotage, what we might charitably dub 'twilight' works. In 1935, readers of 'General Types of the Higher Nervous Activity of Animals and Men' had to wait until the last couple of pages to read anything about people, and then what Pavlov had to say was wholly speculative. Such very late work ought to have been charitably forgotten, but the more vapid the old man's handwaving became, the more his writings came to resemble what was fast becoming a scriptural document of the Bolshevik project – Lenin's dismal *Materialism and Empiriocriticism.*

As far as the Bolsheviks understood it, science was *supposed* to be easy to understand. Science was difficult only because capitalism had shattered it into elitist, self-serving specialties. As these specialties fused, they were bound to become more straightforward, until, with the advent of Marx's one science, everything became explainable in the terms of everything else.

The Bolsheviks considered the brutal simplicity of Lenin's first philosophical work to be a strength, not a weakness – and now Pavlov's late studies were revealing by experiment how right Lenin had been in saying that the material brain reflected, without flaw, the world as it was!

To Bolshevik eyes, Pavlov wasn't old, or tired or out of ideas.

In his last years, and through his last works, Pavlov was proving them right.

Even into the 1930s, it was still possible to disagree with Pavlov. Nikolai Bernstein had been disagreeing with Pavlov since 1924, when he gave a talk at Gastev's Central Institute of Labour, arguing that Pavlov's theory could not explain human skills. Skills are acquired and practised with a purpose in mind. They are not dumbly reactive. They are *intentional* – an aspect Pavlov's theories did not address.

Bernstein's disagreement with Pavlov rumbled on until Pavlov's death in 1936. He never came under any pressure to change or muffle his views, and even wrote a book, *Contemporary Studies on the Physiology of the Neural Process*, which was strongly critical of Pavlov's theory. The book was never published in his lifetime, but the reason had nothing to do with politics. Pavlov died just a couple of months before the book was due to be published, and Bernstein withdrew it rather than be seen to attack a recently deceased colleague. This act of personal decency cost the field of psychology dear. By the time Bernstein got back to taking Pavlov's theory apart, after the Second World War, the political climate was far more hostile: it resulted in his forced early retirement.

Bernstein and Bekhterev and other sincere critics of Pavlov were able to make their case, but Pavlov's adoption by Bolshevik leaders and commentators had a huge influence on the way their field was organised, and what kinds of research received funding and support.

The Shchukina Institute of Psychology of Moscow Imperial University (later called the Moscow Institute of Psychology) was officially opened in 1914 by its founder and director, Georgy Chelpanov. Chelpanov resisted Marx's psychological ideas, and had very little time for materialist explanations of mind. This brought him no trouble until 1923, when the government forced hundreds of 'idealist' intellectuals into exile. Almost immediately Chelpanov, who had managed to preserve the institute's building, equipment and staff through years of ruin and famine, came under attack by an ambitious colleague and former student, Konstantin Nikolaevich Kornilov.

Politically, Kornilov's background could not have suited the times better. He had come from nothing. His father, a provincial bookkeeper, had died, probably from drink, when Konstantin was six, so that he never acquired more than a very basic primary education. At nineteen, he was a provincial schoolteacher. In his mid-forties, accomplished and undoubtedly talented, he fitted the role of director of a Soviet institution far better than an élite professor like Chelpanov.

Chelpanov tried to argue back but Kornilov had mastered the ever more rigorous and precise rhetoric of the commissariats on whom everyone relied for funding. 'For fruitful development,' he declared, 'this discipline needs a materialist and Marxist approach.' This speech, delivered in Moscow at the First Russian Congress on Psychoneurology in 1923, was published in the main official newspaper, *Izvestiia*. Chelpanov could only respond in brochures paid for out of his own pocket. Chelpanov complained that if psychology was reduced to the study of reflexes, it would disappear, absorbed by physiology – and he wasn't wrong.

In December 1922, the Association of Research Institutes of Moscow University closed the Moscow Institute of Psychology and made its staff redundant. Under cover of administrative reform, Chelpanov was not returned as director.[16] That job fell to Kornilov.

Alexander Luria, later a world-renowned psychologist, was one of Kornilov's new intake:

It was suggested that our institute should transform the whole of psychology, abandon Chelpanov's old idealism and create a new, materialist science, or a Marxist psychology, as Kornilov used to say. Meanwhile the transformation of psychology took two different forms: first, renaming; second, moving. As far as I remember, we renamed perception as receiving reaction signal, memory as storing and reproducing reactions, and emotions as emotional reactions. In one word, everywhere we could, we used the term 'reaction' and sincerely believed that we were doing important and serious business. At the same time, we shifted furniture from one laboratory to the other, and I very well remember that, while dragging tables on the staircase, I was totally convinced that in this way we would reorganise the work and lay out a foundation for Soviet psychology.[17]

In fact, very little changed. Kornilov's rhetoric was always more show than substance, and once he was in post he more or less copied Chelpanov's tolerant and eclectic approach to the field. So long as his junior staff used Pavlovian terminology, they were free to do what they wanted.

For Alexander Luria, this was all the licence he needed to explore any number of exotic ideas.

Luria came from a high-achieving medical family. His father, a gastroenterologist, taught medicine at the University of Kazan. His mother was a dentist and his sister eventually became a psychiatrist. In the chaos of the Civil War Luria, barely twenty, was already holding down a research position in one institution, doing graduate work in another, attending medical school part-time, and running tests on mentally ill patients (including Dostoyevsky's granddaughter). He also started a journal, organised a commune for adolescents, published his own study of psychoanalysis and established the Kazan Psychoanalytical Society (complete with letterheaded paper in Russian and German). Sigmund Freud thought he was dealing with a tenured professor. Addressing Luria as 'Dear Mr President', he granted permission for the society to translate a minor essay into Russian, and accepted it into the International Psychoanalytic Association.[18]

'I wanted a psychology that was relevant, that would give some substance to our discussions about building a new life,' Luria wrote in his memoir.[19] Invited to Kornilov's Moscow Institute of Psychology in 1923, he set about developing a new physiological method for recording motor responses in psychologically disturbed patients. In one test, subjects were asked to respond to a word by a verbal association while simultaneously squeezing a small rubber bulb connected to a paper-and-ink recorder. Luria compared the (psychoanalytically assessed) verbal responses with the (physiologically assessed) motor reaction curves.

In another test, subjects were hypnotised and told they would not be able to say certain words. Wakened, they were tested, and measurements were taken of the amount of trembling and aphasia they experienced when they were asked to repeat forbidden words.

The results were promising, but the emotions Luria was able to elicit with his volunteer subjects were weak beer: he needed subjects who were altogether more charged-up. How about students facing a purge? (Rounds of expulsions were quite usual, for academic as well as ideological failings.) From there he went to actual or suspected criminals, studying them after their arrest, during interrogation and on the eve of trial. Fear, aggression and despair were real enough among these subjects, and Luria noticed something interesting: people who turned out to be innocent didn't react with any especial force to words associated with the crimes for which they had been accused.

Over several years, Luria and colleagues studied more than fifty subjects, asking them to respond to words both innocent and connected with the crime. Eventually they were able to pick actual criminals out of a field of suspects. Though the idea didn't find any practical application for another fifty years (when Luria dug out his old papers and sent them to the Institute of Criminology) it is fair to say that Luria had just invented the lie detector.

The year after Alexander Luria's arrival at the Moscow Institute, virtually its entire staff decamped briefly to Petrograd to attend the Second Russian Psychoneurological Congress, and launch a campaign for Marxist psychology.

The state's very visible public backing for Ivan Pavlov had not yet cast a pall, and what 'Marxist psychology' should look like was still a matter of open debate. Should psychology become a laboratory science, or should it, like psychoanalysis, become a much more social, cultural project? Luria neatly summed up the conundrum that faced (and still faces) the field: 'One way will lead to a psychology which is scientific but artificial; the other will lead to a psychology which is natural but cannot be scientific, remaining in the end an art.'[20]

Into this debate stepped a newcomer: Lev Semyonovich Vygotsky. We do not know who invited him to Petrograd, or whether his controversial subject – the relationship between reflexes and thinking – was his own idea, or set as some kind of audition piece. We do not even have a transcript of his talk, though the impression he made was so vivid, so powerful, we can get the gist from other people's letters and memoirs.

Vygotsky's point was that while it was absurd to try to explain consciousness in terms of simple reflexes, that wasn't any reason to ignore the existence of consciousness. After all, what was thought but (according to the theories of the day) inhibited speech? What people said was as much a scientific datum as what they did. There was no reason why study of the mind should be any less scientific than any other study of human behaviour. Vygotsky argued that any psychology that ignores consciousness is not 'scientific': it is purblind.

'Although he failed to convince everyone of the correctness of his view,' Luria recalled, 'it was clear that this man from a small provincial town in western Russia was an intellectual force who would have to be listened to.'[21]

Once the speech was over, Luria edged to the front of the crowd gathering around Vygotsky, who appeared stunned by the fuss he had caused. He was still clutching his notes, and Luria angled himself to see what Vygotsky had written. The paper was blank. A mere prop. In that instant, Luria knew he was in the presence of genius.[22]

Vygotsky was born late in 1896 into a highly educated family in the city of Gomel (in Belorussia), in western Russia's pale of settlement for Jews. Vygotsky wanted to study philosophy and literature, but graduates of these subjects became teachers and Vygotsky, being a Jew, was banned from the profession. Law and medicine were the only options open to him.

Vygotsky managed to get one of the very few medical places reserved for Jews at Moscow University, took classes in history and philosophy at the Shaniavsky, and flung himself into the city's literary life. He fell in with the 'formalist' school – young critics who were interested in the way language relates to meaning. Vygotsky submitted his study of *Hamlet* as his thesis in 1915.[23]

Vygotsky's central theme was something that had troubled Karl Marx when he contemplated the Greek classics: how can the beautiful works of a slave-owning society still be beautiful in a capitalist one, and show every sign of remaining just as beautiful under socialism? What aesthetic qualities persevere through such changes?

Vygotsky was an enthusiastic Marxist. He had been studying Marx, Engels and Plekhanov long before the revolution, and by the time he was invited onto the junior scientific staff of the Moscow Institute of Psychology in 1924, he knew more about Marxism than almost anyone outside the Kremlin.

Hamlet, the Greek classics and other great works, Vygotsky argued, are timeless because they manipulate our sympathies and expectations at a structural, psychological level. They are not static 'opinion pieces'.

Later, Vygotsky began to piece together exactly what this 'psychological structure' might be that responded so strongly to beauty. Art was not a mere decorative cloth draped over human nature. It wasn't even a vessel for human nature. It *was* human nature. Art offered glimpses of the inner lives of other people – the kinds of glimpses from which an individual's sense of self is made. 'Knowing others' is the basis for 'knowing myself'. More than that: 'The social dimension of consciousness is primary in time and in fact. The individual dimension of consciousness is derivative and secondary.'[24]

Social exchanges, particularly between parents and children, gave children the tools with which to hold internal conversations with themselves. It was these internal conversations that generated a sense of self. The self did not exist until it was spoken and acted into being.

Vygotsky's best known example is that of a child pointing out objects to its mother. At first the child tries to grasp a distant object, and fails. That unsuccessful grasping manoeuvre, that gesture 'in itself', becomes a gesture 'for others', because the mother deliberately interprets it that way. The child is the last person to realise that it is gesturing for others. When it does, it can finally internalise that gesture into a thought. So 'all higher mental functions are internalised social relationships'.[25]

Vygotsky's fascination with the way children learn made him one of the era's leading experts in child psychology. At that time there existed a sort of portmanteau discipline called 'pedology' (a term coined by the American psychologist G. Stanley Hall), which embraced early learning, education and training, and developmental psychology. The Party had always taken its mental health responsibilities seriously, and even as the Civil War was decimating the country, it had promoted the study of child development and education. In 1922, there were over twenty pedagogical and psychological research institutions in Moscow alone. The commissariats of education, health, railways and heavy industry all had dedicated pedological departments. And they kept on multiplying: the Moscow Testing Association, the All-Russian Psychotechnical Association, the Soviet Pedological Association . . .

Vygotsky's reputation in this burgeoning field was second to none. When he ran consultation sessions, not only co-workers at the Institute, but also teachers, doctors, psychologists and students from all over the city turned up in the auditorium. In the summer they threw open the windows so people who could not fit inside the building could hear what was going on. Lev Zankov, a student of Vygotsky, remembered:

Many observers were amazed how Vygotsky conversed with the child while examining him. Vygotsky was always able to establish an atmosphere of trust in his rapport with the children, he always talked with them as though they were equals, always paid attention to their answers. In turn, the children opened up to him in a way they never did with other examiners.[26]

Consultations with individual children were valuable, but much more could be learned from children in groups. So Vygotsky, Luria and others found themselves drawn into the running of the White Nursery, founded in 1923 by Vera Shmidt, a talented psychologist and wife of the Bolshevik explorer, planner and bureaucrat Otto Shmidt.

Vera worked in collaboration with the great, unsung psycho-analytic pioneer Sabina Spielrein, an idealistic intellectual who had been Jung's client, then his lover, then Freud's protégée and Piaget's analyst, before returning to Moscow at the age of thirty-eight, armed with years of experience and learning and new ideas about child development. Lev Vygotsky and Alexander Luria both worked at the nursery, and it is clear enough from their writings that they were hugely influenced by Spielrein's research on how children acquire language, though neither man gave her any published credit.[27]

The school's nickname was the children's idea: to its founders, the White Nursery was known as the Solidarity International Experimental Home, and it was meant to develop a form of early education that could address the needs of the country's millions of homeless orphans.[28] The school, which shared with the Psychoanalytic Institute a magnificent art nouveau mansion on Malaya Nikitskaya Street, was typical of progressive projects at that time: it adopted much of Jean Piaget's thinking, and was not so different in spirit to A. S. Neill's experimental school at Summerhill, which opened in England in the same year. But there was an important difference: unlike Piaget, workers at the White Nursery believed that a child relied on the company of others to develop its language and social skills. People were not 'hard-wired', but responded to the society in which they grew up. It was up to socialist psychologists and teachers to discover what environment brought forth the healthiest, happiest personalities.

Spielrein conducted experiments at the school to find alternatives to regimentation and discipline in the teaching of children. Activities included drawing, collage, modelling and educational games. Teachers prepared reports, diagrams and graphs detailing each child's development. There were no punishments and no excessive displays of affection. Acts earned neither praise nor blame: children were left with their dignity intact. A little girl who enjoyed smearing herself with excrement was calmly washed and changed, again and again, and was eventually given paints to play with. The girl happily switched her interest to paints and brushes. The approach impressed the nursery's many and distinguished foreign visitors, including the radical Austrian psychoanalyst Wilhelm Reich.

The nursery was the most celebrated of a whole slew of Bolshevik social experiments, designed to overturn centuries of incarceration, regimentation and punishment, and put in their place rational, scientific and humane forms of social engineering. The principle of 'no punishment' was not unique to the nursery. The Education Commissariat recommended that schools do away with all punishments, examinations and homework. The terms 'guilt', 'crime' and 'punishment' were removed from the first Soviet criminal code of 1919 because they obscured the social causes of crime.

These ideas were all very idealistic, very rational, but they proved impossible to implement in a society reeling from revolution, civil war and famine. Two years after opening, the nursery ran out of funds and had to close, the mascot of a social revolution that never happened. As projects of progressive education were abandoned, one by one, even the movement's patron, Lenin's wife Nadya Krupskaya, was forced to admit her 'leftist mistakes'.

Vera Shmidt put the best face she could on the nursery's eventual closure in 1925, telling Reich that she herself had shut up shop because 'the requisite conditions for that type of work were not yet available'. And it is true to say she couldn't get the properly trained analysts on whom the nursery's work depended.

By many measures, the White Nursery was a failure. It never got around to caring for any actual orphans. Instead, it looked after the children of Communist Party bigwigs – including Joseph Stalin's son Vasily. Intellectually, however, the White Nursery had been a resounding success. It gave researchers such as Luria and Vygotsky vital data in their quest to refashion psychology.

The Historical Meaning of the Crisis in Psychology was written between 1926 and 1927, while Vygotsky was bedridden with an attack of tuberculosis. 'I have been here a week already,' he wrote to a colleague from hospital, 'in large rooms for six severely ill patients. Noise, shouting, no table, and so on. The beds are set next to each other without any space between them, as in a barracks. On top of this I feel physically in agony, morally crushed, and depressed.'[29]

Vygotsky, who had spent the Civil War nursing his mother during two separate episodes of tuberculosis, as well as his two brothers (one with tuberculosis and the other with typhoid fever – both died before the end of 1918), suffered constant recurrent bouts of TB throughout his life, requiring operations, painful treatments, and regular periods in overcrowded hospitals and sanatoriums. Even as he and his colleagues compiled their book, doctors were measuring his remaining life in months.

The *Crisis* was yet another brave attempt to arrive at a single definition for psychology.

Psychoanalysis, Gestalt psychology, the Würzburg school: every school of psychology was battling every other, and to listen to them you would think that each had discovered some universal principle explaining all the phenomena of the mind. It was time to accept that the field *was* fractured, and would be for some time to come.

Rather than attempt to break down all these fiercely defended positions, Luria, Vygotsky and their co-author Leontiev went back to basics, wrote up the experimental data they had gathered from Spielrein and Shmidt's nursery, and then from Vygotsky's own educational laboratory, and attempted to recast psychology as a set of questions about human development. They turned psychology into a practical programme, bringing what they knew to bear on medicine, education and industry.

Vygotsky's own view of the project was modest: 'What has the new psychology succeeded in yielding? Not much so far . . . some methodological premises, the outline of a science, its plan . . . but what is most important: Marxist psychology has an objectively and historically justified will toward the future.'[30]

Vygotsky had by now grown rather dismissive of Pavlov. 'A human being is not a satchel stuffed with reflexes,' he had written in 1925; 'neither is it a hotel for reflexes that happen to drop in.' But he had not reckoned on how appealing this vision was.

Pavlov's reflex theory of higher nervous activity had one huge advantage over Vygotsky's richer, more nuanced psychology: it was a science of prediction. It said that people's behaviour was controllable. In this, it resembled the Bolshevik grail of one simple self-consistent science much more closely than Vygotsky's frank admission of muddle.

As the Pavlovian juggernaut gathered pace in scientific journals and administrative meetings, competing disciplines found their funding cut, their theories attacked, and their institutions either abandoned or simply banned. Psychoanalysis was the first to go. Freudianism, like Marxism, claimed to unify both scientific and literary ideas of self. Freud, however, nursed an essentially tragic idea of human destiny: humans weren't just the pawns of history; they were the pawns of their own animal natures. Socialist programmes of improvement were pointless. Civilisation as it advanced only increased repression to ever more pathological levels.

By the mid-1920s, then, Freudianism's philosophy was being derided, while Freudian practitioners were tolerated as 'specialists', using Freudian techniques like spanners to treat the mentally ill. They were not considered important – and in a country with very few psychiatrists or mental hospitals, they probably weren't. Freudian therapy isn't much good against extreme pathology.

In January 1927 Trotsky published his article 'Culture and Socialism', in which he made his most enthusiastic appeal for a sympathetic understanding of Freud's work. To receive such support from the man who at that very moment was being defeated by Stalin in the struggle for control of the leadership of the Communist Party was the kiss of death for Soviet psychoanalysis.[31]

In April, opposition in both Party and literature forced Luria to resign as secretary of the Russian Psychoanalytic Society. In 1928, soon after Trotsky's arrest and expulsion from the country, the society president, Moshe Wulff, resigned and emigrated. Sabina Spielrein retreated to Rostov-on-Don. In January 1930 Freud's theories were bitterly attacked by A. B. Zalkind, a psychologist who had once been the Freudians' most vocal apologist, and the Russian Psychoanalytic Society was disbanded. The last work by Freud to be published in Russian translation appeared in 1930. Pedology itself was banned six years later; industrial psychology and applied psychology fell to the same sweep of the administrative scythe.

Something horrible was happening to the field, and it was happening so quickly, to so many disciplines, that Alexander Luria, for one, couldn't quite take it seriously. In 1932 he published a summary of his early Freudian monographs in a book titled *The Nature of Human Conflicts*, and blithely sent a copy of his summary off to Ivan Pavlov.

The next day Pavlov blazed into Luria's office, pulled out the monograph, tore it in half, and tossed it on the floor. Luria was officially prohibited from teaching, researching, or publishing anything to do with psychoanalysis ever again.

Konstantin Kornilov, in his battle with Chelpanov over control of the Moscow Institute of Psychology, had first mooted the idea of 'Marxist psychology'. Now the idea came back to bite him. In 1930, at a special meeting, ambitious juniors accused Kornilov of not being loyal to Marxism, and threw him out.

By then, most of the institute's significant figures (including Bernstein and Vygotsky) had already left. The three authors of the *Crisis* headed for the city of Kharkov, looking for work in whatever disciplines still seemed open to them: in child development, medicine ... On 26 June 1933 Alexander Luria wrote to Lana Linchina, his wife-to-be:

I am completing my studies of aphasia patients and trying to convince them that the brother of the father is not the same as the father of the brother ... Currently we're coming across lots of very interesting material: cases of agnosia, agraphia, postnatal psychoses with aphasia ... we are drowning in an abundance of the rarest cases. I am thoroughly enjoying medicine: I am spending time with Vygotsky to study pathophysiology, and, of course, thinking about you.[32]

In his final years, Vygotsky's interest in human development broadened to encompass the way society influences and shapes the individual. After all, human beings transmit huge bodies of knowledge across generations not biologically, but culturally. It follows, then, that people's potential is constrained by the level of sophistication of their culture. This prompted him to look for parallels between a child's thinking and the reasoning power of so-called primitive societies.

Luria and Vygotsky planned a study of the 'intellectual functions' of adults from 'a non-technological, non-literate, traditional society'. In some respects the 1930s were a good time to conduct such studies because the state's impressive literacy campaigns were being extended into the remotest corners of the Soviet Union, where people up till then had had no exposure to formal schooling. Together Luria and Vygotsky settled on Uzbekistan and Kirghizia in Central Asia for their most ambitious studies. Ill-health kept Vygotsky at home, so Luria became his corresponding researcher.

Changes were coming thick and fast to the tenant cotton farmers of Uzbekistan. Schooling was being introduced, agriculture was being collectivised, and efforts were being made to emancipate women. Luria studied five groups of people: illiterate women unaffected by these changes, illiterate peasants in remote villages not yet gathered into collectives, women who had had some training in teaching nursery-age children, experienced collective farm workers, and literate women who had had a few years' teacher-training.

Luria and Vygotsky's aim was to see how different levels of culture affected the way individuals perceived the world. One of the first experiments the research team carried out involved showing people at different stages of social development some classic visual illusions. Could very basic-seeming perceptual assumptions be affected by one's upbringing? Indeed they could: Luria was so excited by the preliminary results that he telegraphed Vygotsky to say, 'The Uzbeks have no illusions!'[33]

The assumption behind this project – that cultures can be

ranked in a line from 'primitive' to 'sophisticated' – would never pass today. But Luria's findings are fascinating and foundational. When 'non-modern' subjects were presented with a set of objects, they tended to group them according to practical considerations ('all these objects are needed for chopping'). Other, to us very obvious, characteristics were completely ignored. Nobody bothered to gather together all the yellow objects. Indeed, for this group, grouping by abstract characteristics like colour seemed 'stupid'.

These people made 'excellent judgments about facts of direct concern to them' and exhibited 'much worldly intelligence', but in theoretical thinking they were limited. Told that in the far North, where there is snow, all bears are white, and that Novaya Zemlya is in the far north, they answered the question, 'What colour are the bears in Novaya Zemlya?' with comments like 'I've never been in the North and never seen bears,' or 'There are different kinds of bears. If one is born red, he will stay that way.'[34]

Groups who had had even the slightest training, however, responded very differently. They handled simple logical problems with ease, and they used abstract categories far more frequently than practical ones.

Luria concluded that 'basic changes in the organisation of thinking can occur in a relatively short time when there are sufficiently sharp changes in social–historical circumstances, such as those that occurred following the 1917 Revolution'.[35]

So far, so politically correct: Luria and Vygotsky had demonstrated the efficacy of Soviet education, and the improvement potential of even the least educated Soviet citizen. On top of that, they had garnered clear evidence of the malleability of mind, and the power social improvements wielded over an individual's life-chances. Karl Marx had promised that human natures would change for the better under socialism: here was the evidence.

When Luria's train pulled into Moscow station, agents of the NKVD were waiting for him. They had intercepted his telegram to Vygotsky, and read it to mean that the Uzbeks had no illusions about Soviet authority in the area. This farcical misunderstanding seems to have been cleared up fairly quickly. But it set the mood for what was to come. As news of their expedition spread, critics accused Luria of insulting the national minorities of Soviet Asia, whom he had, apparently, depicted as an inferior race unable to behave reasonably. A 1934 review of the Uzbek expedition in the journal *Literature and the Proletarian Revolution* claimed that:

Instead of showing the process of development and the cultural growth of the workers in Uzbekistan, they search for justifications for their 'cultural psychological theory' and 'find' identical forms of thought in the adult Uzbek woman and a five-year-old child, dangling before us under the banner of science ideas which are harmful to the cause of the national cultural construction of Uzbekistan.

Such malicious misreadings were all too common. Every kind of non-Pavlovian psychology was becoming politically unacceptable. In her memoir, Luria's daughter Elena Alexandrovna recalled her father saying, 'I was accused of all mortal sins right down to racism, and I was forced to leave the Institute of Psychology.'[36]

Vygotsky's confidence was utterly undermined. In a letter to Luria he wrote:

I am still beset with thousands of petty chores. The fruitlessness of what I do greatly distresses me. My scientific thinking is going off into the realm of fantasy, and I cannot think things through in a realistic way to the end. Nothing is going right: I am doing the wrong things, writing the wrong things, saying the wrong things.[37]

Vygotsky took it very hard that he and Luria were unable to continue their Uzbek research, and witnesses said it was this that broke his spirit at last, so that he gave in to his illness. In May 1934 he fell ill and was brought home from work suffering a throat haemorrhage. He died a few weeks later, on 10 June.

Luria retrained as a physician. He spent the late 1930s working in the Institute of Neurosurgery under Nikolai Burdenko.

The Crisis in Psychology had peaked and passed, and Pavlov – or rather, a narrow, Bolshevik idea of Pavlov – now towered over its wreckage.

6 : Understanding evolution

*Whereas Westerners were inclined to go in through the
traditional front door, our Soviet colleagues seemed at
times to break in through the back door or even to come
up through the floor.*[1]
Leslie Dunn

The Augustinian friar Gregor Mendel was no stranger to big
questions. In his scientific writings – and in particular in the
introduction to *Experiments on Plant Hybridisation* (1865) – he
asks how species manage to evolve into new forms, when their
young look so much like their parents. Equally puzzling to him
was the business of intraspecies variation. How can a black
chicken give birth to a white chicken, which then gives birth to a
black chicken (and where are all the grey chickens)?

Mendel asked good questions. More impressive, from a professional standpoint, was his willingness to narrow his field of
enquiry, to accept with good grace that some of his questions
were beyond his ability to unpick.

Professor Zange (Bernhard Goetzke) hugs his only friend in
Salamandra (1928), a film based on the life and work of Austrian
Lamarckist biologist Paul Kammerer.

Rather than give in to handwaving, Mendel contrived some deceptively simple experiments in plant hybridisation – the crossing of related but distinct varieties of the same species – and through them attempted to reveal the smallest, simplest unit of heredity that could be passed from one generation to the next.

Crossing green peas with yellow peas over several generations, Mendel discovered that particles of heredity combined and recombined in statistically coherent ways. You could see the mathematics playing itself out in the colour of peas of each successive generation: yellow peas were produced three times more often than green peas.

Mendel's explanation for this three-to-one ratio is simple and ingenious. Hereditary particles must come in two forms: dominant and recessive. During fertilisation, particles from each parent combine. If two dominants combine, the child will express a dominant characteristic (yellowness, in the case of Mendel's peas). If a dominant and a recessive combine, once again the dominant trait will be expressed. A recessive trait only gets expressed if two recessive particles combine. This means that dominant characteristics will be expressed three times as often as recessive ones.

$$\begin{array}{lclcl}
\textbf{dominant} & + & \textbf{dominant} & = & \textbf{dominant} \\
\textbf{dominant} & + & \text{recessive} & = & \textbf{dominant} \\
\text{recessive} & + & \textbf{dominant} & = & \textbf{dominant} \\
\text{recessive} & + & \text{recessive} & = & \text{recessive}
\end{array}$$

Mendel's findings brilliantly fulfilled his ambitions. Mendel had proved the existence of units of heredity. We call them 'genes'.

What Mendel had *not* done, of course, was reveal any link between heritability and evolution.

Living things in the real world almost never behave like the peas in Mendel's experiment. (Even Mendel's peas didn't behave that simply: the good friar wasn't above massaging his statistics.) Most important characteristics aren't like light switches: they don't simply appear and disappear. They blend, from generation to generation. In fact, heredity is almost always a matter of blending, of slow and subtle change. We see it in the mirror, in the way we combine the qualities of our parents; and we see it in our children, and how they embody elements of ourselves and our mates in uncertain and unpredictable ways.

Mendel had his enthusiasts – men who believed he had unlocked the secrets of evolution with one seed packet. But sceptics like Ernst Haeckel in Germany and Kliment Timiryazev in Russia were more reasonable, arguing that one random variation couldn't possibly spread through a whole population in any simple way. That would be like squirting a drop of paint into the waves and expecting the whole ocean to change colour. One variation, however advantageous, is going to get diluted, generation by generation, until it vanishes entirely. The mathematics of the day supported these criticisms. Developing new techniques in statistics to prove his point, Francis Galton, Charles Darwin's first cousin, calculated that successive generations descended from even the most gifted individual will naturally regress to the species' norm.[2]

This left genetics out on an intellectual limb for years. Theodosius Dobzhansky, Vernadsky's young assistant, who would go on to be one of the most celebrated geneticists in the United States, was once told by one of his professors that genetics was a 'passing fad'; why was he wasting his time on something so 'intellectually perverse'?

What alternative model was there for evolution? Haeckel and Timiryazev hypothesised that changes in species were driven by changes in the environment. Whole populations, being exposed to these changes, would adapt to them as a group. This seemed much more likely than to imagine that one lucky sport of nature spread its inheritance through an entire population. This theory is called Lamarckism, after the extremely talented Jean-Baptiste Lamarck (1744–1829), the man who coined the very word 'biology'.

Lamarck's argument, first made in 1809, is compelling. Imagine an animal that relies on speed. The faster it is in pursuing its prey, the more it gets to eat. The more it eats, the healthier it is, and the more offspring it will have. The need for speed in capturing prey won't just apply to one individual. It will apply to all its fellows, too. An entire generation, forced to scamper about for a living, will all acquire a honed physique. According to Lamarck, this athletic generation will then pass their muscular strength to their children. What they worked for, their children are born with. It is a hard-hearted biologist who denies the elegance of Lamarck's idea about the heritability of acquired characteristics. Charles Darwin himself thought there was something to it.

The Bolshevik state invested a great deal of time and money in making science a subject of popular interest. Their effort was sincere: they truly meant the Soviet Union to be the world's first scientifically run state, so public education about science was essential. And in weaning the populace off the old gods of the Orthodox religion, monarchy and property, they had quickly found that popular education, in the form of lectures, magazine articles and books, was much more effective than any number of antireligious festivals. (Indeed, those godless carnivals, timed to coincide with major church festivals, had frightened and angered many working people.)

The cult of science lay at the foundation of Soviet rule, and Soviet rulers considered it their duty to keep science constantly before the people.[3] The irony – that 'Soviet Rule has bestowed on science all the authority of which it deprived religion'[4] – was not lost on Western commentators.

Soviet institutions devoted to science – and over a hundred 'people's universities' were already operating by 1919! – took their public duties seriously. Moscow's Timiryazev Biological Institute, founded and led by the biologist Boris Mikhailovich

Zavadovsky, churned out impressive quantities of Marxist popular science. Zavadovsky, a key propagandist in the Militant League of the Godless, was particularly keen to spread evolutionary theory and other Darwinian ideas, and – this is significant – he and almost all his peers adhered to Lamarck's theories.

They cleaved to the idea that acquired characteristics could be inherited and did so, for the most part, out of ordinary scientific conservatism. They had grown up with Lamarck; his ideas seemed to fit the facts, and they saw no pressing reason to rethink them. New experimental data appeared constantly, but they never indicated a clear defeat of Lamarck. Set against it, meanwhile, were all those experiments that seemed to substantiate Lamarck's views on heredity. You can darken the wings of moths by feeding them metallic salts characteristic of smoke deposits in industrial areas. You can alter the breeding frequency of silkworms by changing the temperature of their surroundings and by transplanting ovaries. You can induce mutations in *Drosophila* by turning up the heat. (Naturally, Lamarck's opponents claimed these experiments for themselves and gave their results quite different interpretations.)

Sincere as it was, though, the commitment of Marxist scientists to Lamarck also had a political component. Marx had claimed that socialism could reform and improve people's physical and mental well-being within a single generation (at least, that was how Lenin and Bogdanov had read him), and the inheritance of acquired characteristics seemed the only possible mechanism by which this miracle might be achieved. How else – except by improving social conditions and hoping for the best – could doctors ever hope to treat and cure humanity's numerous systemic and inherited illnesses: their cancers and weak hearts, their diabetes and arthritis? S. P. Fyodorov's brochure *Surgery at the Crossroads* neatly encapsulates the problem faced by physicians at this time:

We can best treat diseases whose causes we understand. These, however, are diseases that are triggered by infections, parasites, traumas, and relatively crude pathological changes. Our constitutional and functional diseases are the burden of modern surgery. Their challenge is far more difficult than anything we have faced before, because not only are their causes unknown; their basic natures and processes are equally mysterious.[5]

Being impatient, Bolshevik physicians in particular looked to Lamarck's inheritance of acquired characteristics to cure these diseases in a single generation.

In 1926, a Society of Materialist Biologists was formed within the Communist Academy, a Marxist research institute created in 1918 with the idea that it would ultimately surpass and replace the Academy of Sciences. Later events would set the society's president – the doctrinaire Marxist philosopher Isaak Izrailevich Prezent – and its most active members – Izrail Agol and Solomon Levit – at each other's throats. To begin with, though, these men imagined themselves allies, pursuing the inheritance of acquired characteristics through experimental studies.

Agol and Levit were of that younger generation of Soviet scientists who desperately needed the support of Bolshevik institutions, since all the important posts within the academic science system were already filled by older colleagues. Solomon Grigorievich Levit in particular owed everything he had to the revolution. He was born in 1894 to the poorest family in the small town of Vilkomir, in what is now Lithuania. His grandmother went out of the way to buy stale bread so they would eat less. His invalid father worked as a night watchman. The only member of his family to receive an education, Levit worked his way through public school and high school by coaching and tutoring. As a teenager, he had joined the Bund, the most left-leaning of the Jewish parties in the Jewish Pale, and later he joined the Bolsheviks.

Levit was his generation's most articulate defender of Lamarck's theories, and he had a frank dread of what it might mean for medicine if genes turned out to be real, physical structures.

'Chromosomal' genetics depressed and worried him. The term had come into use in 1902 when a German, Theodore Boveri, and an American, Walter Sutton, had suggested that Mendel's 'factors of inheritance' (what we would call genes) were located in large thread-like protein structures in the cell, called 'chromosomes' because they absorbed dyes so easily. This meant that specific stretches of the chromosome could be identified visually and tracked across generations.

The trouble with imagining genes as solid, physical entities, strung along a protein thread, is that there can only be so many of them. It follows that the number of variations they generate must, in turn, be finite. Levit feared that if chromosomal genetics was real, then all evolution could ever do was shuffle an ageless, pre-existing deck of genetic cards.

Levit objected to this idea. As a Marxist, and therefore an atheist, he wanted an explanation of where these supposedly ageless and changeless genes came from. Since 1925, when the Scopes Monkey Trial had made an international laughing stock of the state of Tennessee,[6] Christian fundamentalists had been hitting the headlines, and it was not hard to imagine them claiming these timeless and adamantine particles as God's handiwork.

Besides, the idea – that genes were a finite set of cards, shuffled endlessly – was simply not borne out by the facts. The fossil record hardly suggested that Nature had run out of ideas. Quite the contrary, in fact.

For Levit, a sincere Bolshevik and therefore a man focused on brass tacks, on work and the welfare of the people, his severest objection was practical. The idea that an unborn child had a fixed genetic endowment, which nothing could alter, suggested a mechanistic sort of predestination – a sort of scientific Calvinism.[7] It would render many new and exciting medical ideas completely useless. If people's genes were fixed, then heritable diseases and weaknesses were locked in, and nothing doctors or hygienists or other health workers might do would be of any benefit. Genetics, Levit declared, 'smacks of desperate pessimism and impotence. If, indeed, pathology is determined by one's genes, and these develop solely under the influence of "internal forces," independent of the environment, what will become of human efforts to change these pathological forms?'[8]

If your family carries the gene for cancer, and nothing in the environment can change what that gene does, what's the point in looking for a cancer cure?

Levit refused to give in to such fatalism. Better, surely, to embrace Lamarck.

At the University of Vienna, meanwhile, Paul Kammerer, a well-established scientist working under Hans Prizbram, head of the city's Institute for Experimental Biology, was conducting fascinating experiments to demonstrate the inheritance of acquired characteristics in amphibians. In a series of trials stretching back more than a decade, Kammerer had been raising the yellow-spotted newt *Salamandra maculosa* on differently coloured soils. Newts raised on black soil gradually lost their yellow spots. Those raised on yellow soil gained larger and larger yellow spots. In another experiment, Kammerer experimented with the cave-dwelling newt *Proteus anguinus*. *Proteus* is totally blind; its rudimentary eyes are buried deep beneath the skin. Exposing blind newts to ordinary light produced a black pigment over the eye and sight never developed. Raising *Proteus* under red light, however, produced newts with large, perfectly developed eyes. Experiments like these convinced Kammerer that what parents learned was somehow being passed on to their children.

Kammerer, who was something of a showman, drew big conclusions from his work. In 1924 he wrote a visionary book, *The Inheritance of Acquired Characteristics*, about how his theories could be harnessed to the betterment of human society. Improve the material conditions of life, and people would give birth to a generation of children superior in every way to their parents! Lamarckism made us 'captains of the future'. Mendelian genetics offered only a shuffling of existing characteristics: it made us 'slaves of the past'.

Paul Kammerer's work took him all over the world, and he received an enthusiastic reception from the *New York Times* when he toured the United States in 1923. In Russia, Kammerer's claims were met with even greater keenness and interest. His huge treatise on general biology appeared in translation in 1925. Two years later Efim Smirnov discussed Pavlov's and Kammerer's experiments at length in a major survey of 'the problem of the inheritance of acquired characteristics', and two different publishers released Kammerer's book *The Enigma of Heredity* in translation. And Russia's enthusiasm for Kammerer extended beyond the page: in 1926 the Communist Academy invited Kammerer to head a special laboratory, affiliated to Pavlov's institute, to look into the inheritance of acquired characteristics. Kammerer accepted the offer. He had already met young Moscow enthusiasts of his work, and discussed the possibilities for research in the Soviet Union with Anatoly Lunacharsky, the education commissar.

(Later, discussing the controversy between genetics and Lamarckism, Lunacharsky confessed: 'I am not sufficiently qualified in biology to be able to say right now and for sure whether one side or the other in this dispute is absolutely right. But it is hard to discount worldwide sympathy for advocates of the idea that the living organism depends directly on the environment for its hereditary characteristics.'[9])

Kammerer's equipment had already been shipped to Moscow, and he was literally just a couple of days from boarding a train there himself when controversy and disaster overwhelmed him, in the guise of a paper in *Nature* by the American Gladwyn K. Noble – a paper that presented incontrovertible evidence that his most celebrated experiment was a fraud.

The experimental animals at the heart of this controversy were toads. Kammerer's latest experiments took advantage of a peculiarity of the species *Alytes obstetricus*, which, unlike other toad species, mates on dry land.

Toads that mate in water have slippery skins and, prior to the mating season, the male toad grows small spines along the edges of his palms and fingers of his hands to give him some extra grip. They are quite distinctive because, being heavily pigmented with melanin, they are darker than the surrounding tissue. Kammerer had run a breeding programme, forcing several generations of *Alytes obstetricus* to breed in water, and he reckoned he had produced a generation of toads whose males sported pigmented spines along their hands.

The problem was in demonstrating his achievement. All he could do was preserve his animals and cart them around Europe, as carefully as he could, from laboratory to laboratory – and this is what he did. In Cambridge and in London he delivered enthusiastic, competent lectures about his research, while the scientific community, led by the British geneticist William Bateson, conducted a sceptical study of his jarred specimens. No one doubted Kammerer's technical skill or his integrity. Breeding amphibians in captivity was not for the faint-hearted (Bateson, who spent fourteen years battling Kammerer's claims, marvelled that he had managed to breed *Alytes* at all). Neither Bateson nor anyone else doubted Kammerer's dedication to his work.[10]

The longer Kammerer travelled, the faster his specimens began to deteriorate. Phenomena that were subtle in the first place now became virtually impossible to study. It was clear that the toads would only withstand one more inspection – so it had better be a good one.

In 1926 Gladwyn Noble, from the American Museum of Natural History, went to Vienna. With Kammerer's head of department, Hans Prizbram, he made a detailed examination of Kammerer's celebrated toad – and discovered that the black pigmentation in its 'nuptial pads' was not melanin, or anything like it. It was Indian ink.

The fraud was so crude, several conspiracy theories have been thrown up to explain it. Was a rival out to discredit Paul Kammerer? Had Kammerer's imminent departure for the Soviet Union triggered a reactionary smear campaign? Or had an enthusiastic but hapless junior 'touched up' his boss's ever-more-putrid specimen?

We will never be sure. As soon as Noble's paper was published, on 7 August, Kammerer shut up shop. He wrote to the Soviet authorities and withdrew from his Moscow postings. The deception was not his, he wrote, but his position had been made untenable. On 23 September 1926, he set off along a mountain path in the Alps, found a quiet spot, drew a revolver, and shot himself in the head.[11]

Kammerer's death only heightened the scandal surrounding his name. Those who knew him were not altogether surprised at his fate. Lunacharsky put the matter delicately: 'On the other hand, there were quiet rumours about the other, dark sides of Kammerer's life, its social and family aspects, etc.'[12]

Kammerer's manic depression had been tripping him up for years. A string of affairs had disordered his personal life quite severely before he was ever invited to Russia. His first marriage had broken up, and his second lasted hardly any time at all. Even as Noble's paper was going to press, Kammerer's lover, Grete Wiesenthal, had pulled out of her promise to accompany him to Russia, and this disappointment, along with the humiliation occasioned by Noble's paper, was probably what tipped him over the edge.

The response of Marxist biologists and their patrons to the news was a strange mix of horrified sympathy and political haymaking. Marx had promised that science, as it developed and coalesced under socialism, would improve the human lot. Lamarckism fitted seamlessly into this ideological scheme. If

Lamarck's great champion had been exposed as a fraud, then it could not possibly be the science that was wrong!

Izrail Agol's sympathetic obituaries of Paul Kammerer capture this kind of shrill, defensive thinking perfectly. He was actually quite reserved about Kammerer's scientific work. He did not believe that Kammerer had altogether vindicated Lamarck's ideas. However, Kammerer's 'consistent, monistically materialist position' had driven mysticism out of biology, and clearly this had led directly to his persecution in the West. The future of science lay in the Soviet Union: 'Where else but in the land of the victorious proletariat could [Kammerer] find comradely sympathy and support for quiet, objective, scientific research?'[13]

Agol compared the conditions of his homeland to the ones that pertained in America, where (as the Scopes trial had reminded everyone) there was a law against teaching evolution in Tennessee – a state which Agol charmingly if inaccurately called 'one of the most enlightened areas of the United States'.

Lunacharsky was aware of the hash Kammerer had made of his private life, but this didn't stop him from turning Kammerer into a political martyr. You only have to look at his peculiar vanity project, the feature film *Salamandra*, which came out in late 1928. By then, research had moved on, and Lamarck's ideas were being rejected by the very scientists who had supported his invitation to Moscow. But this story was too good to be derailed by a few facts.

In the film, a biologist at a university in Central Europe (we'll call him Kammerer for convenience) is working with salamanders. By changing their surroundings, he has changed the colour of succeeding generations. A priest, learning of this, realises that this discovery will spell an end to the power of the Church (*why?*) and hatches a conspiracy with a young prince whom he has had appointed as Kammerer's assistant. Kammerer is persuaded to announce his discovery at a formal university meeting. The night before the big day, however, the priest and the prince enter Kammerer's laboratory, open the jar in which the specimen salamander is kept, and inject it with ink. During Kammerer's presentation, someone takes out the salamander, and dips it into a jar of water – and all the colour runs out. An immense uproar follows, and Kammerer is kicked out of the university.

Next we see the scholar, down and out, begging for loose change with an experimental monkey which has followed him into the streets. A former student (played by Lunacharsky's wife) takes a train to Moscow, and obtains an interview with Lunacharsky (playing himself), who straight away sets about saving the life of this victim of bourgeois persecution. The last scene shows Kammerer and his old student riding east on a train bearing a large streamer. It reads, 'To the land of liberty'.

Salamandra was hugely popular. It was still in the cinemas when Arthur Koestler, Kammerer's best-known English-language biographer, visited Moscow six years later. It was dreadful tosh, of course. Alexander Sergeevich Serebrovsky, a Marxist biologist with impeccable Party connections,[14] thought the whole effort shameful: 'Comrade Lunacharsky . . . arranges class elements about this problem: in support of the inheritance of acquired characters are the revolutionary intelligentsia, the People's Commissariat of Education and others, while against it are clerics, bankers, fascists, and counterfeiters.' This fairy tale especially irritated Serebrovsky since he had been for years the leading Marxist *critic* of Lamarckism in biology. Since when, he wanted to know, did being a Marxist make you a Lamarckist?

Serebrovsky considered this a rhetorical question. It turned out, however, to have a frighteningly precise answer. At what point did being a Marxist make you a Lamarckist? The fateful date is 1923.

Ivan Pavlov, too, subscribed to Lamarck's theory that acquired characteristics were inherited. In this he was no different to many other physiologists and psychologists of his generation.

Being a physiologist with an interest in psychology, however, he did something to test his allegiance, and oversaw experiments to show how acquired *behaviours* might be passed from generation to generation.

Between 1921 and 1923 one of Pavlov's co-workers, Nikolai Studentsov, trained a mouse to run to a feeding rack only after a buzzer had sounded. The mouse was an unresponsive pupil – it took 298 tries for its conditioned reflex to be established. Studentsov then trained the mouse's offspring. They established the reflex much faster than their parents (after 114 repetitions); their children, in turn, were even faster (29 repetitions), the third generation faster still (11 repetitions), and the fourth were running to the feeding rack at the appointed moment almost immediately (6 repetitions). These astonishing results became the talk of the life-sciences community, and soon enough reached the ears of Nikolai Konstantinovich Koltsov, Russia's leading light in the new science of genetics.

Koltsov paid Pavlov a visit. Studentsov's experiments, he explained, were badly flawed. Studentsov had had no previous experience in training mice, so chances were that 'it wasn't the mice which had learned, but the experimenter'.

Pavlov wasn't ready to give up on such spectacular results. That summer, as he lectured across Britain and the United States, he regaled audiences with news of Studentsov's success. An article in the American journal *Science*, presenting Pavlov's report, quoted him as saying, 'I consider it probable that after some time a new generation of mice will run in response to the bell to the feeding place without preceding lessons.'[15]

'Had that article not borne Pavlov's signature, we would have simply ignored it,' Koltsov later remarked. But Pavlov's opinions mattered a great deal – a rash declaration from him was quite capable of derailing the young science of genetics. Koltsov orchestrated a concerted newspaper campaign against Pavlov's assertions.

It proved unnecessary. Pavlov had, after all, been listening. Returning to Russia, he had assigned one of his most senior researchers, Evgeny Ganike, to design a much more robust experiment – one that entirely automated the training process. When Koltsov visited Pavlov once again, in 1925, he found the old man in a penitent mood: Pavlov showed him Ganike's automated experiment, and cheerfully admitted that it had disproved Studentsov's findings. 'Now I'm just working with dogs, I don't want to work with mice any more!'

A published opinion in *Pravda* in 1927 sealed the matter: the idea that conditioned reflexes could be inherited vanished from the physiology journals, and Pavlov himself organised a laboratory to study the genetics of behaviour. On his order, three sculptures were erected in front of the building: of René Descartes, Ivan Mikhailovich Sechenov – and Gregor Mendel.

Neither Koltsov nor Pavlov had any idea just how much damage Pavlov's mouse experiments had done to the cause of Russian genetics. Neither man had behaved badly. Indeed, Pavlov's honest attempts to prove the inheritance of behaviours, the failure of those experiments, and the way he had handled that failure, were by any measure exemplary.

Nonetheless, the damage was done. For four years, the Russian public lived with the notion that Ivan Pavlov, their most celebrated scientist, had *proved* that acquired characteristics could be inherited.

A sincere and well-connected Bolshevik, Alexander Serebrovsky led a campaign against Lamarck's ideas from inside that bastion of Bolshevik learning and research, the Communist Academy.

At the same time – and despite the men's serious political differences – he was being drawn into the intellectual orbit of his old tutor Nikolai Koltsov. Koltsov was politically suspect: he came from a well-to-do family (his mother was related by blood to the theatre director Konstantin Stanislavsky) and had already been thrown in jail and threatened with state execution. Nonetheless, Serebrovsky recognised the genius of the man who had put Pavlov straight about his mice, and he worked energetically in support of Koltsov's independent research institution, the Institute of Experimental Biology.

Koltsov was a liberal. Like many, he had walked out of Moscow University in 1911. Within the year he had created an outstanding zoology programme and a research laboratory at the Shaniavsky. Serebrovsky and Mikhail Zavadovsky attended. (Mikhail was the older brother of Boris Zavadovsky, the Darwinian propagandist.) Koltsov's other base was the Beztuzhev Advanced Courses for Women, better known as Moscow Women's University; he set up a laboratory there, too.

In Europe at the turn of the century biology was still a descriptive science. Koltsov's first published work was on the development of the pelvis in frogs. But he reckoned purely descriptive approaches had 'exhausted their research program and their vitality'. Study abroad and at marine biological stations at Naples, Rostov and the Russian station at Villefranche persuaded him to explore new, experimental approaches to his subject: genetics, cytology, protozoology, hydrobiology, physicochemical biology, endocrinology, experimental embryology – even animal psychology.[16]

In 1913 Koltsov became co-editor of the newly launched popular science journal *Priroda* [*Nature*] and three years later, in late 1916, he managed to secure a large grant from the will of Russian railway magnate G. M. Mark to create an Institute of Experimental Biology. It was housed in Moscow's merchant quarter – a splendid building donated by the city *duma* – and Koltsov filled it with his favourite co-workers from the Shaniavsky, including Serebrovsky.

Koltsov's experience of persuading private donors to part with their money served him well in his dealings with the Bolsheviks. Patrons were soon found within the health commissariat, the agriculture commissariat, and KEPS, the Commission for Studying the Natural Productive Forces of Russia – brainchild of Vladimir Vernadsky and his colleague Alexander Fersman.

Even such illustrious backing could not insulate the Institute of Experimental Biology from the hardships of revolution and civil war. Koltsov kept the lab open around the clock, and students had to earn their places by lecturing, carpentry or repair work. They soldered, mended, made whatever they could, before settling to study.

They learned to determine species, did live culturing. Everyone had his own cultured amoebas, flagellates. Every stage of division and multiplication had to be fixed, compared, drawn. Then the same with sponges. All this was done independently. They dissected all kinds of bugs and beetles, observed regenerations and transplantation in guppies and tritons. Everyone had his own research, made discoveries, gasped, made mistakes, asked questions, and felt like a real researcher.[17]

It was a happy and productive working environment, and the political divide between Serebrovksy's revolutionary generation and Koltsov's bourgeois one could safely be left at the door until 1920, when the Cheka announced they had exposed an organisation called the Tactical Centre. This clandestine organisation had been formed, they said, to create a conspiratorial alternative government for Moscow in preparation for the city's occupation by White forces under General Denikin. What the Cheka actually uncovered has never been entirely clear. No one has yet found evidence of a structure capable of fomenting an armed uprising in Moscow. More likely, it was just a loose coalition of grumblers.

The discovery of the 'Tactical Centre' conspiracy gave the Cheka *carte blanche* to interrogate their most highly placed opponents. Even the Secretary of the Academy of Sciences, Sergei Oldenburg, was chucked – albeit briefly – into jail. On 19 August 1920 Koltsov was arrested. Twenty-four people were executed by firing squad that day but Koltsov, having being held for thirty-eight hours without food, was eventually released. (It is likely that Maxim Gorky managed his rescue: prior to the revolution Gorky had met him several times in Italy, where they were both living.)

Koltsov made a good show of shrugging off this terrifying experience. In the first issue of the *Proceedings of the Institute of Experimental Biology* he added a laconic footnote to a paper describing the effects of malnutrition. Speaking of his own case he described just how much weight he had lost in jail. The incident made him much more circumspect, however: the wit and political innuendo of his earlier writing is missing after the mid-1920s.

Serebrovsky's relationship with Koltsov cooled in the years following the Tactical Affair, though this made little professional difference. Serebrovsky had already relocated to the institute's field station in Anikovo, on the banks of the Moscow River, to the west of the city. Here he spent his time breeding chickens.

Serebrovsky was Koltsov's expert in genetics. This was a field Koltsov was keen for his institute to occupy, since studying the genetics of agriculturally important subjects – wheat, poultry and cattle – attracted government funding. Genetics had another advantage, in that it was largely a paper science, requiring little in the way of imported laboratory equipment. Serebrovsky's work on the genetics of poultry proved so important – a foundation of so much future genetics – that it was translated and reprinted all over the world well into the 1970s. More immediately, during the Civil War, Serebrovsky's experimental subjects could be converted into welcome dinners for the institute's staff. (This was no joke: Koltsov saved several associates from starvation by transferring them to his experimental stations outside Moscow.)

Meanwhile Koltsov was bringing on another talent. Sergei Sergeevich Chetverikov was rather less interested in politics, and rather more of Koltsov's own social stripe (indeed, the two men were distantly related). His parents were both from wealthy textile families, It was 1895, and Sergei was around fifteen years old when he began to collect butterflies. His father, quietly appalled by his son's ambition to be a zoologist, bundled him off to Dresden to study engineering, but it was hopeless; within a year he relented and Sergei joined the department of comparative anatomy of Moscow University.

Chetverikov would prove to possess extraordinary theoretical talent, but it wasn't this, or even his distant blood relation to his boss, that first landed him a place in Koltsov's institute; it was his ability to look after insects without killing them.

The same year he had launched Serebrovsky's poultry-breeding career, Koltsov had received a copy of Thomas Hunt Morgan's book *The Physical Basis of Heredity* from friends in Germany. This was the undisputed bible of the new field of genetics, detailing the pioneering experimental work of Morgan and his team at Columbia University in the United States. For a while, Koltsov's copy of Morgan's book was the only one in Russia. It travelled between Moscow and St Petersburg, and was then taken out of its binding so that Koltsov's students could translate individual chapters.

In their tiny cupboard of a room at Columbia, Morgan, Alfred Sturtevant, Calvin Bridges and Hermann Muller had begun trying to unpick the genome of a living creature, the fruit fly *Drosophila melanogaster*.[18] To help them, they had looked for ways to induce unmistakable, visible mutations in their stock of fruit flies – mutations they could follow from generation to generation. One promising avenue seemed to be to expose the flies to X-rays. Koltsov duly set a researcher, Dmitry Romashov, the task of repeating this work. Alas, Romashov had no more experience of insects than anyone else at the institute, and the experiments were a flop. Chetverikov, who had worked with Koltsov already on the Beztuzhev courses, and organised an insect room there, was invited to join the institute in 1921 to stem the slaughter of the fruit flies.

Chetverikov was a meticulous collector and a talented zoologist, whose interests extended far beyond the practical work of the institute. Through his mathematician brother Nikolai he had become interested in biometrics, applying statistics and careful measurements to the study of whole populations. His first notable piece of writing, *Volny Zhizni* (*Waves of Life*, 1905), is an account of the fluctuating population size of various butterflies and moths observed near his family's dacha:

It can be said without any exaggeration that the fauna is not permanent for a minute . . . Hence anyone who has more or less carefully studied some fauna of a single location knows that no two years are the same: that which last year was rare or even absent is met this year in abundance, and conversely that which last year struck the eye with each step now demands careful search.[19]

Waves of Life teased out how changes in the size of a population affected the amount of variation it displayed. This was a first step towards reconciling genetics and evolution, and according to Chetverikov the paper 'produced a sensation in Russian readership circles'.

Once at the Institute of Experimental Biology, Chetverikov revived this project, gathering together several of Koltsov's best students, including his wife, Anna Ivanovna, Nikolai Timofeev-Ressovsky (the man who had used a Red Cross uniform to blag scientific equipment for his teaching establishment), and Elena Alexandrovna Fiedler, a former assistant of Vladimir Vernadsky.

In August 1922, the Institute of Experimental Biology received a windfall: Hermann Muller, co-author of *The Physical Basis of Heredity*, visited Moscow. A radical socialist and a committed materialist who had fallen out with Thomas Hunt Morgan over whether a gene was an actual *thing*,[20] Muller was every inch a scientist in the Soviet mould. And he brought gifts: more than a hundred strains of *Drosophila melanogaster*, full of the genetic markers the Morgan lab had used to demonstrate the chromosomal theory of inheritance.

The visit was not without precedent. From its inception, genetics was an international discipline, dependent upon distributing to other researchers material developed in one's own laboratory. Whether it was the fruit fly, the evening primrose, maize, mice or wheat, the exchange of new mutants, varieties and stocks was an essential part of doing genetics. For Russian researchers, receiving such a significant collection from an overseas ally was an unprecedented coup. The collection went first to Moscow's premier geneticist, Alexander Serebrovsky at the Anikovo station. But Koltsov had no trouble persuading him that Chetverikov was better placed to look after the collection, and arrangements were made so that everyone in the institute would get the chance to work on *Drosophila* in addition to their own work.

From 1922 to 1924, most of the genetics work at Koltsov's institute was directed at mastering the theory of genetics as it had emerged from Morgan's 'Fly Room'. Chetverikov and his students studied the hard way. He scoured Western scientific journals for articles on genetics and handed them out to his seminar members to translate as best they could – word by word from the dictionary, if they had to.

As a consequence, Chetverikov's group got the gist of problems, but found it hard to penetrate the mathematical detail. Ignorance served them well, because buried in that detail were a whole host of assumptions about the nature of genetics, some of which turned out to be false. Chetverikov and his colleagues worked away at fusing genetics, evolutionary theory and biometric studies of the sort that informed *Waves of Life*. They did not know that this was supposed to be impossible. They did not know that naturalists and experimental geneticists were supposed to be at loggerheads. They did not know that evolution and genetics were incompatible. They blithely assumed that what they learned in the laboratory would in some way link up with what they observed in nature. And it did.

Around 1925 Chetverikov organised his seminar rather more formally, though the name the group chose for themselves, the Screeching Society, says something about their sense of humour.[21] The seminar was deliberately kept small, and prospective members could be admitted only by unanimous vote.

The seminar group set out to answer one of the great imponderable questions of genetics: how do species remain so distinct from each other, while at the same time evolving and dividing into new forms? To this end, it conducted the first study of a wild population of *Drosophila*, and laid the groundwork for the development of population genetics. In the autumn of 1925, Chetverikov completed a paper describing their achievement, and what it meant for biology.

Many genetic variations appear in the laboratory setting. The veins of a *Drosophila*'s wings, for instance, occasionally turn out deformed, and these deformations are then passed down through generations of flies. It doesn't do them any good. Indeed, mutations observed in the laboratory are almost always harmful. Natural populations, on the other hand, appear remarkably uniform. Their wings are not deformed. Their eyes are not too close together, and their ears do not stick out. They look so alike, they might have sprung from the same egg.

In the wild, harmful mutations show up very rarely, since 'in the severe struggle for survival, which reigns in nature, the majority of these less viable mutations, originating among normal individuals, must perish very quickly, usually not leaving any descendants'.[22]

In nature, then, among individuals old enough to breed, there is an incredibly small gap between the fittest and the least fit. Indeed, this gap is so small, it is hard to see how natural selection ever gets a look-in. The fittest individuals don't survive longer or reproduce any more often than the less fit. A large population of very alike individuals is incredibly stable: it will keep churning out versions of itself till the last trump. So what makes a species change?

Mutations (which are usually harmful) are always recessive. For this reason, the pioneering British biologist William Bateson believed all that recessives are accidents – mere genetic waste. Chetverikov looked at recessive mutations differently; for him, they were a mechanism for *storing* wild and unusual genetic ideas.

In a large population the chances of two adults mating who share the same recessive mutant gene are vanishingly small. When the going is good, then, populations absorb recessive mutations like sponges. But when the population crashes, or becomes fragmented, then individuals are much more likely to mate with someone genetically related to them. Then the chance that recessive genes will get expressed rises astronomically. Most of these recessives will prove harmful, but a few will confer an advantage.

Take a large population of, say, finches, laying finch eggs that hatch into finches virtually identical to their parents. Once that population gets fragmented, it will speciate into many families of finch, each adapting to its own unique environment.[23]

Published in the *Journal of Experimental Biology* under the title 'On certain aspects of the evolutionary process from the viewpoint of modern genetics', Chetverikov's paper is today considered one of first and strongest bonds established between genetics and evolutionary theory.

In January 1923, pioneering German neurologists Oskar and Cécile Vogt travelled to Moscow to participate in the First All-Russian Congress for Psychoneurology.

The couple's lecture, distilling twenty-five years' study of the cellular structure of the cerebral cortex, left a deep impression on specialists in Moscow. It offered them a unique view of a new and fast-developing science. Oskar explained how he and his wife would study thin slices of brain tissue through a microscope, and bit by bit, slice by slice, assemble a model of the structure of the brain, down to the level of the individual cell. This work required good vision and keen insight: 'It is like flying over a landscape, when one sees a number of towns; only the talented investigator of architecture can rapidly spot characteristics (like peculiar buildings) to identify individual towns.'[24]

Their lecture was also peculiarly, upsettingly topical: Lenin, father of the new nation, was even then dying, of repeated insults to the brain.

Lenin suffered his first stroke on 26 May 1922, and several more followed over a two-year period, along with prolonged epileptic seizures. An international medical team headed by the Breslau neurologist Otfried Foerster was struggling to save his life. During the Vogts' visit, the team invited Oskar to visit Lenin's bedside. There was no advice he could give, and Lenin died on 21 January 1924, aged fifty-three. Here the story would have ended, had some bright spark in the halls of government not decided, against the family's wishes, to preserve the leader's brain.

This was not, at the time, such an unusual thing to do. The practice had begun in the mid-1850s in Germany, where Rudolph Wagner studied the brain of the physicist and mathematician Carl Friedrich Gauss (and noted its 'remarkably convoluted' appearance). In 1876, French scientists founded the Mutual Autopsy Society of Paris, each member offering brains and other organs up to post-mortem dissection by his fellows. Pavlov's rival Bekhterev founded a Pantheon of Brains in 1927 – an institution maintained to this day. Bekhterev's own brain is part of the collection.

Foerster was asked who might best study Lenin's brain in detail; he suggested Oskar Vogt, and on New Year's Eve 1924, Vogt received the invitation. It was a signal honour for an entrepreneur who had never held any academic position beyond ones he had invented for himself. (On 2 February 1925 the Academy of Sciences elected him a corresponding member to bolster his CV.)

At the end of February 1925 Oskar's wife Cécile arrived in Moscow, bearing with her all the equipment required for their investigation of Lenin's brain.[25] The Lenin Institute accommodated the Vogts' laboratories in Moscow's historic Dmitrovka district, and between 1925 and 1927 sections of Lenin's brain were stained for microscopic investigation.

Oskar delivered his first major report on 19 November 1929. 'In layer three of the brain cortex, in many cortical areas, especially in the deeper regions of this layer, I found pyramidal cells of a size I have never before observed and in a number that I have never before observed,' Vogt declared. What did this signify? No one had a clue, but Vogt was generous: 'Our brain anatomical results show Lenin to have been a mental athlete,' he announced, delivering one of the most frequently cited bullshit statements in the history of science.

Away from the microscope, Oskar Vogt devoted much of his time to planning a permanent institute to study the brain. The Moscow Institute for Brain Research (incorporating Bekhterev's Pantheon of Brains) was officially founded in November 1928. Modelled on Vogt's Berlin institute, it reflected and consolidated the two countries' excellent post-war record of scientific co-operation.[26]

Vogt's institute was more than just a piece of intellectual empire-building. The plan was that it should research the genetic basis of all kinds of mental disorders and diseases. The Vogts were puzzled by the way certain inherited neurologic disorders varied tremendously in frequency and severity. They assumed that these variations had a genetic origin, rather like the mutations they had collected among bumble bees and beetles during their long walking holidays: 'Just as the body hair and colour spots of these insects are subject to variation,' Vogt argued, 'so also is the brain of man . . .'[27]

Vogt was friends with the health commissar Nikolai Semashko and Nikolai Gorbunov, secretary of the Council of People's Commissars, a former chemical engineer and the Bolsheviks' leading patron of the sciences. He also visited Koltsov's Institute of Experimental Biology, looking for a young associate he could take back to Berlin to jump-start genetic research into the brain. It did not take long before Nikolai Timofeev-Ressovsky and Elena Alexanderovna came to his attention. By now a married couple, Nikolai and Elena had found a single mutation in the fruit fly species *Drosophila funebris* that produced all manner of deformations in a vein in the fly's wings. That a single kind of mutation could produce many variant wing-types caught Vogt's attention. He discussed the matter with Koltsov and Semashko: would the young couple (they were still in their mid-twenties) be interested in a long stay in Berlin?

Nikolai Vladimirovich Timofeev-Ressovsky had been born into a family of impoverished nobility, on a modest estate which bordered the Ressa River.[28] His father, Vladimir Viktorovich Timofeev-Ressovsky, was a railway engineer. The family, if not especially wealthy, had an extraordinary pedigree. In their family tree there were Cossacks, including Stepan Rasin, the Russian Robin Hood; the descendants of Rurik, ninth-century ancestor of the tsars; admirals of the Russian Navy; the anarchist Prince Peter Kropotkin (himself a brilliant biologist); and many military officers and intellectuals. Family legend had it that one ancestor set off to explore the North Pole, contrived to wind up in an African prison, escaped, stole a ship in Turkey and sailed it back to Sevastopol.

Timofeev-Ressovsky was studying biology at Moscow University when the revolution erupted. He entered the war as an infantryman in a Cossack unit and ended the war a sergeant-major in the cavalry of the Red Army (and first bass in the Moscow military chorus). His wartime progress was anything but orthodox.

In 1918 he was captured by a band of anarchist 'Greens'. The Greens were a non-aligned force defending Russia against the Germans. They had no love for Red Army NCOs, but were quickly charmed by Timofeev-Ressovsky's stories of Prince Kropotkin, who was his grandmother's cousin. ('He gave us raspberry jam, which, by the way, was a gift from Lenin . . .')

He fought for the Greens a while, was hit on the head with the flat of a sword, woke up abandoned and set off to rejoin the Reds. He participated in the Red Army's attack on the White Army of General Denikin in the Russian south, before typhoid brought an end to his hectic career. In 1922 he returned to Moscow. 'I think, nevertheless, that all in all the life was merry,' he later recalled. 'Very few hungry, very few frozen. Rather, people were young, healthy, and vigorous.'[29]

Nikolai Koltsov accepted Timofeev-Ressovsky into his institute in spite of his lack of higher qualifications (he had only a high school gold medal), and soon enough one of his students, Elena Alexanderovna Fiedler, caught the young man's eye.

Elena was a long-time associate of Koltsov's; she had been working with him since attending the Shaniavksy, where Koltsov ran a research laboratory. Once, on a field trip led by Koltsov's student Mikhail Zavadovsky to the nature preserve in Askania-Nova, the road to Moscow was cut off by fighting (civil war was raging throughout the Ukraine) and the expedition broke up. Reaching Kiev, Elena fell under the orbit of Vladimir Vernadsky, and was one of those who had helped him map the chemical compounds of which all living things are composed. It was Elena who passed on Vernadsky's 'biogeochemical' ideas to Timofeev-Ressovsky, sparking his interest in the biology of entire populations. Elena and Nikolai's co-written papers on the subject formed the prologue to a lifelong partnership.

Genetic research at the Institute of Experimental Biology had begun with Serebrovsky's studies of poultry. Nonetheless, there was no doubt that the most useful subject for genetic study was the fruit fly: an animal so unprepossessing and apparently useless that it was attacked, rebuked, mocked and held up as an example of science removed from reality well into the 1950s. Nikolai Timofeev-Ressovsky once delivered a passionate speech in praise of the three-millimetre-long fly. His student Nikolai Luchnik remembered the gist:

Irreplaceable subject! Reproduces quickly. Has many offspring. Clear hereditary signs. Can't confuse a mutation with a normal individual. Red eyes, white eyes. All serious laboratories around the world work with the *Drosophila*. Ignoramuses like to talk about the fact that *Drosophila* has no economic significance. But no one is trying to develop a strain of dairy and meat fruit flies. *Drosophila* is needed to study the laws of heredity. The laws are the same for elephants and fruit flies. You'll get the same results with elephants. But a generation of fruit flies grows in two weeks.[30]

Drosophila were ideal laboratory animals, but they weren't much to look at. Hearing that their institute was to receive an official visit from health commissar Nikolai Semashko, Timofeev-Ressovsky and his colleagues were concerned. Their senior colleague Serebrovsky was studying chickens at Anikovo Genetic Station. At least chickens looked useful. Why should Semashko take the time and trouble to visit a research station in Zvenigorod, fifty kilometres out of the city, just to peer at a bunch of petri dishes?

Timofeev-Ressovsky decided to kidnap him. The stunt – a bunch of emaciated geneticists surrounding the staff car of a bemused commissar with makeshift clubs – paid off. Semashko, who must have had a sense of humour, saw to it after his inspection that the Institute of Experimental Biology was given an even grander home: a three-storeyed mansion at Vorontsovo Pole (now Obukha Street) with a garden and a yard where field experiments could be performed. Of course it is also possible that when Vogt offered to take Timofeev-Ressovsky out of the country, Semashko breathed a sigh of relief.

In any event, an interview was arranged. Vogt and Timofeev-Ressovsky hit it off almost instantly. Their similarly colourful family backgrounds may have helped. Vogt, half-Dane, half-German, came from a line that included liberal Lutheran ministers, sea captains and at least one pirate. The Timofeev-Ressovskys' trip to Berlin was registered in March 1925. It faced predictable bureaucratic hurdles. In Moscow the authorities thought the trip signalled the defection of a politically unreliable scientist. In Berlin the authorities suspected the Timofeev-Ressovskys of being spies or communist propagandists. Thanks to intense lobbying by Koltsov and Vogt, however, at the end of June the Timofeev-Ressovskys and their two-year-old son Dmitry were on a train to Berlin.

Nikolai was just twenty-five years old, Elena twenty-seven. 'When I went for work to Germany I was proud of it,' Timofeev-Ressovsky told his biographer, the novelist Danyl Granin, many years later, 'proud that this time the Germans came to us, instead of us to them.'

7 : Shaping humanity

*Eugenics has before it a high ideal which also gives meaning
to life and is worthy of sacrifices; the creation, through
conscious work by many generations, of a human being
of a higher type, a powerful ruler of nature and creator of
life. Eugenics is the religion of the future and it awaits its
prophets.*[1]
  Nikolai Koltsov, 1922

One of the nineteenth century's last independent 'gentlemen
scientists', the Englishman Francis Galton, has several claims to
fame. He was Charles Darwin's first cousin, a pioneer of statistics,
and first conceived the use of fingerprints in forensics. He was
the proud inventor of underwater reading glasses, an egg-timer-
based speedometer for cyclists, and a self-tipping top hat.

He was also a proponent of eugenics. A contemporary of
Gregor Mendel (the two were born in the same year, 1822),
Galton was firmly convinced that heredity cannot change during
one's lifetime. 'Will our children be born with more virtuous
dispositions', he asked, 'if we ourselves have acquired virtuous

'The acquisition of a profound knowledge of brain anatomy and
physiology is one of the most important tasks of the new century':
the Moscow Institute for Brain Research.

habits?' His answer, delivered in an 1865 magazine article entitled 'Hereditary Character and Talent', was a resounding 'no'.

Galton's first book on human heredity, *Hereditary Genius* (1869), was published two years after the first volume of Karl Marx's *Das Kapital*. Both books are about the betterment of the human race, but their philosophies are diametrically opposed. Marx supposed the environment was everything. Galton assumed the same for heredity. We already know how to improve whole populations, he argued. We do it by manipulating bloodlines. Ask any farmer. Ask any pig breeder.

If a twentieth part of the cost and pains were spent in measures for the improvement of the human race that is spent on the improvement of the breed of horses and cattle, what a galaxy of genius might we not create! We might introduce prophets and high priests of civilisation into the world, as surely as we can propagate idiots by mating cretins.

'This', Galton argued, 'is precisely the aim of Eugenics.'[2]

Livestock breeders mate choice specimens to improve their stock. For the same reason, they prevent runts and weaklings from breeding, usually by slaughtering them. Human beings are animals, and are just as capable of being shaped through a planned breeding programme as wheat, or chickens or dogs. The question is, what would a human breeding programme look like? Would it be a positive programme, using health education to promote couplings that produce genetically healthy offspring? Or would it be negative, discouraging or preventing pairings that would otherwise spread disease or disfunction? Was some sort of combination of both possible? How prescriptive would it be? Would it work by persuasion, or by compulsion?

Galton considered eugenics in a mostly positive light. Though he had some minatory things to say about the breeding habits of the feckless, he was, at heart, a sucker for genius. He focused his research on the bloodlines of remarkable families, and wondered how eugenic policies might foster future talent.

The study of degeneracy fell to a New York social reformer, Richard Louis Dugdale. Dugdale became interested in the matter when, during an 1874 inspection of conditions at a jail in New York State, he learned that six of the prisoners there were related. Using prison records, relief rolls and court documents, he traced the Jukes family tree back six generations to a Dutch colonist named Max. More than half of over 700 people related to this family by blood or marriage were criminals, prostitutes or destitute. Dugdale concluded that degeneracy, like genius, runs in families.

His response to this finding was actually quite measured. (It helped that he was a follower of Lamarck.) 'The licentious parent makes an example which greatly aids in fixing habits of debauchery in the child. The correction', he wrote, 'is change of the environment . . . Where the environment changes in youth, the characteristics of heredity may be measurably altered.'[3]

Later writers were not so circumspect. An Indiana reformatory promptly launched a eugenic sterilisation effort. It began in 1899 as a voluntary experiment with prisoners, but rapidly expanded. In 1907, Indiana enacted the USA's first compulsory sterilisation statute. California followed suit in 1909, and in 1917 its statute was extended to cover any patient at a state mental health institution 'who is afflicted with mental disease that may have been inherited and is likely to be transmitted to descendants . . .'

The desire to restrict population growth was quite alien to the Russian tradition. At the First International Eugenics Congress, held in 1912 in London, Prince Peter Kropotkin delivered a passionate diatribe against the idea. Who were unfit, he asked: workers, or moneyed idlers? Working-class women who suckled their own children, or society ladies who farmed the duties of a mother onto others? Those who produced degenerates in slums, or those who produced degenerates in palaces? 'Before recommending the sterilisation of the feeble-minded, the unsuccessful, the epileptic, was it not their duty to study the social roots and causes of these diseases?'[4]

Subsequent events only strengthened Russian hostility to negative eugenics. In the years between 1917 and 1920, Moscow lost half its population, Petrograd a staggering 71 per cent. Russia needed not fewer births, but many more.

Eugenics found its first Soviet home at a State Museum of Social Hygiene created by the Commissariat of Health in January 1919. Koltsov led the museum's consultative group on 'the biological question', covering 'general biology, physiology, anthropology, and racial hygiene'. (That last term no doubt has set the reader's alarm bells ringing, but the term was a common one: it referred to studying the relationship between disease and ethnicity. What it most certainly did *not* imply was any attempt to enforce apartheid between Russia's many peoples.)

Eugenic studies were an important part of a major health initiative orchestrated by health commissar Nikolai Semashko, which travelled under the broad label 'social hygiene'. This was basically preventative medicine: a health system that would not just heal the sick, but which would prevent illness from happening. Under Semashko's sponsorship, enthusiastic physicians used statistics, surveys and questionnaires to learn about health and disease in Soviet society. They also evaluated living and working conditions, arguing that social interventions would bring about revolutionary changes in the quality of health. At its height in 1927, the list of topics studied in the State Museum for Social Hygiene included nutrition, housing, demography, urbanisation, migration, occupational hygiene, alcoholism, drug addiction, sexual hygiene, prostitution, sanitary education . . .

It was the job of Koltsov's Institute of Experimental Biology to back this effort up with solid biological data and the latest genetic theory. The institute conducted eugenic studies alongside its genetic work. The distinction between the two activities was not particularly important in those early days. Besides, the government wanted to see practical work conducted for the betterment of the people, and eugenics was an obvious practical application of the new genetic science.

Eugenics also needed a lobbying group that would explain these activities both to the public and to the authorities themselves. Koltsov created the Russian Eugenics Society for this purpose in 1922. Because there was no money for another, separate scientific association, the society doubled as a clearing house for the latest eugenic information through its mouthpiece, the *Russian Eugenics Journal*. Koltsov was editor-in-chief, and himself wrote genealogical studies of Darwin, Galton and Gorky – the supreme Bolshevik example of a self-made man (Gorky's relatives were all drunks, savants, petty criminals and blaggers). In 'Betterment of the Human Race', presented at the inaugural meeting of the Society in 1921 and published in 1922, Koltsov exuberantly explored the implications of eugenics. In particular, he described a vivid thought experiment illustrating the power of selective breeding.

Civilisation, he argued, was vanishingly far from controlling people sufficiently to make practical eugenics possible. 'To do so we should either be carried by our imagination to long ago to times when mighty lords ruled their subjects as slaves, or unleash our fantasy and imagine for a moment that the idea of the famous British writer H. G. Wells came to pass, and dwellers of the planet Mars, who possess the greatest knowledge and technology, which vastly surpasses ours, landed on Earth . . .'[5]

Koltsov imagined the invasion of a group of super-intelligent Martians who treat us the way we treat our livestock and pets. People would be domesticated, just as dogs have been made from wolves. Rebellious types would be eradicated; docile breeders would give rise to an obedient workforce; those with fine motor control would be bred to create craftsmen, the most beautiful bred for show, and so on.

'In as little time as a century', he wrote, there would be 'endless individual races of domesticated people as sharply distinct from one another as a pug or a lapdog is from a Great Dane or St Bernard.'

At the time of publication, Koltsov's *jeu d'esprit* was received in the humorous spirit in which it was intended. But his brand of eugenics was still controversial. Underlying his studies of the bloodlines of remarkable families was an unspoken political assumption, that the most precious element of the population was its intelligentsia. Koltsov's argument against contraception reflects his view – common at the time but no less offensive to the Bolsheviks – that low-quality genes reside in the lower classes. Koltsov feared that it was the intelligentsia and the aspirational working classes who would respond best to a contraception campaign, while sluggish peasants and workers would carry on over-breeding.

This undiscussed and unexamined assumption exasperated Boris Zavadovsky, who fulminated against

the attempts of bourgeois biologists to draw conclusions from the laws of genetics and Mendelism, which are clearly aimed with a sharp point against the power of the proletariat . . . forcing even Marxist biologists to distance themselves from genetics . . . what's good for us is death to a Koltsov.[6]

Even setting aside Zavadovsky's gift for hysterical overstatement, this is rough stuff, but it stemmed less from any domestic argument, so much as from a reading of eugenics in other countries. In the USA and Europe, negative eugenics was gaining ground. Russian scientists had never had any time for this style of thinking, and now it was tainting the entire field.

The proponents of Soviet eugenics were perfectly aware of their growing image problem. Very early on, the word 'eugenics' was either qualified or simply dropped from discussions about human genetic health. In late 1925 Yuri Filipchenko – Koltsov's opposite number in Leningrad – added the word 'genetics' to the name of his Bureau of Eugenics. Three years later he removed the word 'eugenics'. By then the eminent geneticist's unease with the whole business had stopped him publishing any work on human heredity, and in May 1929, just a few months after the genetics congress was held in his city, he failed to renew his bureau's membership of the International Federation of Eugenic Organisations. Only his untimely death from meningitis the following spring prevented him from publicly announcing his decision to resign from the board of Koltsov's *Russian Eugenics Journal*.

Given the Russians' intellectual hostility to eugenic ideas – indeed, their long-nursed and war-hardened suspicion of any argument that involved checks on population growth – it is possible that Russian eugenics might have bumbled along as a mere footnote to the main work of genetics, or avoided public opprobrium by sheltering beneath some innocuous banner like social hygiene or health education. This is precisely what happened in Britain, another nation made queasy by the eugenic idea.

What brought eugenics to the forefront of public discussion in Russia, and made discretion impossible, was a discovery made half a world away in Texas, by a committed socialist and eugenicist who already, by his gift of laboratory *Drosophila*, had proved himself a friend of Soviet science.

Hermann Joseph Muller was born in 1890 in New York City and grew up in Harlem. At Columbia University he had joined Thomas Hunt Morgan's fabled 'fly room', studying the genetics of *Drosophila*. Though it made his name, and set him on the road to winning a Nobel Prize in 1946, work in the fly room involved more grind than glamour. The fly room was an actual *room*, and a modest one at that: just five by seven metres. Calvin Bridges and Alfred Sturtevant were already in there, and Muller was often shunted out for reasons of space. Their boss, Thomas Hunt Morgan, believed in teamwork, but his juniors were less convinced. Muller's sensitivity over who got to what idea first made the men's work difficult, and soured their later relations. Muller's sense that he was the poor cousin of the team was exacerbated by the fact that he was actually poor: while he was studying he had to teach English to immigrants and work as a runner on Wall Street to earn money for himself and his mother. (His father, who worked with fine metals, had died when he was ten.)

Eventually Muller settled at the University of Texas at Austin, where he met his first wife, Jessie Jacobs, a talented mathematician. Times being what they were, Jessie lost her position when she got pregnant. While her husband was conducting pioneering work on the effects of radium radiation on the genetics of fruit flies, Jessie juggled work as his lab assistant with care of their son and their house.

Sturtevant had once accused Muller of having a 'priority complex', and Muller's relations with his new colleagues at Austin lend credence to the claim. In the end, in an effort to avoid prickly encounters with his colleagues, Muller took to working at night.

Muller's politics did not ingratiate him with his fellows. He was sympathetic to the Communist Party, although he never joined. He advised the National Student League, a communist student organisation according to the FBI, and on the quiet he edited *The Spark*, a newspaper that campaigned for civil rights for African-Americans, equal opportunities for women, social security and other 'socialist' goals.

What sustained him, even as his marriage fell apart and his colleagues shunned him (and he them), was a major discovery. In a few months, using X-rays, Muller found a way to generate mutant genes. The practical benefits of the technique were legion. His induced *Drosophila* mutations were 150 times more plentiful than mutations arising spontaneously in the lab. But this was only the beginning. Muller noticed that his mutant genes were stable. Exposure to radiation didn't merely muddle them up. A particular exposure generated a particular effect.

The 'transmuting action of the X-rays is thus spatially narrowly circumscribed, being confined to one gene even when there are two identical genes close together'. Muller's conclusion is triumphant: 'the genes really do lie in the chromosome in linear arrangement, in the physical order in which we have theoretically mapped them . . .'[7]

Muller had proved that genes were physical things. They were made of something. Muller rushed to print, writing in the journal *Science* in 1927 to establish his priority. Now his reputation depended entirely on how, and how soon, he presented his evidence. Typically, he was still writing his paper and preparing slides on the train while heading to the Fifth International Congress of Genetics in Berlin.

His presentation that summer was, predictably, a mess. But it didn't matter. His results spoke for themselves. As he wrote later: 'Perhaps the most hopeful feature of the present data is that they show that mutation is indeed capable of being influenced "artificially" – that it does not stand as an unreachable god playing its pranks upon us from some impregnable citadel in the germ plasm.'[8]

Marxist sceptics of genetics – men like Agol and Levit at the Communist Academy – had thought of genes as playing cards, and genetic inheritance as a game played with a limited deck. They had thought genes were unchangeable, and for that reason timeless, and for that reason God-given – so, naturally, they didn't believe in them.

Muller's findings blew that prejudice wide open. Genes did change. Living things did adapt to their environment, and they did pass these changes to their offspring, and they did so through the mechanisms of their genes.

As soon as a copy of *Science* arrived in Moscow, Serebrovsky rushed to print – in *Pravda*, no less – with news of Muller's discovery. Under the title 'Four Pages That Shook the Scientific World' (a deliberate echo of the American Marxist classic *Ten*

Days That Shook the World by John Reed), Serebrovsky blew Lamarckian ideas out of the water. Genetics wasn't 'idealist'. It was a solid, down-to-earth material science. And in the manipulation of genes lay the promise of improving the health and well-being of the people.

Solomon Levit, who dreamt of a revolution in medicine, read Muller's results, talked to Serebrovsky, and realised he had been wrong to champion Lamarck's outdated ideas. More than that, he realised that, in the name of Lamarck, a doctrinaire element was seeping into his field, derailing new research and stifling new ideas. At the Communist Academy's conference in April 1929 he back-pedalled furiously:

History has played a trick on the Communist Academy. It happened that we in the Communist Academy set up a Lamarckian laboratory and that this laboratory represented our biological line to the outside world. And many got the false impression that in fact the Communist Academy was preaching Lamarckism, that Lamarckism was their slogan, that in the name of Lamarckism they would fight genetics. This is a myth. We have to do away with this myth, and the sooner the better, both for the academy and for science . . .[9]

Serebrovsky's second major convert at the Communist Academy was the political firebrand Izrail Iosifovich Agol. The son of a carpenter, Agol had followed his sister into revolutionary activities and fought with the Reds during the Civil War, before joining Boris Zavadovsky at his laboratory at Sverdlov Communist University. He had spent the whole of his young career researching the inheritance of acquired characteristics in axolotls and hens. Freshly elected to the Communist Academy, Agol added the zeal of a convert to his bullishly Bolshevik approach to scientific work. His strident call at the April conference for Lamarckism's official condemnation and its expulsion from the academy was politely ignored.

Nonetheless, the time had clearly come to rejig the research agendas of the Timiryazev Biological Institute to reflect the changing times. Agol was told to lay off some Lamarckian researchers and in their place to organise a genetics lab, to be headed by Serebrovsky.

Agol worked with Serebrovsky to replicate Muller's results and Levit did the same, assisting Serebrovsky at the Moscow Medical–Biological Institute (a research institution founded in 1924 under the Commissariat of Health).

Meanwhile the science section of the Commissariat of Education nominated Agol for a Rockefeller fellowship to study in the United States. As Solomon Levit was already in receipt of a fellowship, this meant the two men would spend 1931 together in Austin, Texas, where Hermann Muller, the beleaguered American communist sympathiser, was waiting for them.

Between late 1928 and his departure for Texas in 1931, Levit threw himself into the creation of a radical, truly Soviet eugenics. With Serebrovsky, he established an 'office of human heredity and constitution' at the Moscow Medical–Biological Institute. In its first volume of research papers, published in late 1929, Levit listed the office's interests. They would study human population genetics, and heritable diseases and conditions, using case histories, genealogies and twin studies. They would compile data on human chromosomes. They would lay the foundations for the mapping of the human genome. Levit's brief note is dizzying: written in 1929, it is nothing short of a roadmap of the next eighty years in global genetics research.

In his own paper, 'Genetics and Pathology (in Relation to the Current Crisis in Medicine)',[10] Levit explained just what genetics could achieve for human health and well-being. It wasn't just a matter of studying 'constitutional diseases' like cancer and Huntington's that had a well-established genetic component. If you looked at the statistics through Mendelian glasses, you saw that susceptibility to common infections like TB also ran in families. Genetic weaknesses and predispositions were everywhere. Short of mowing you down with a bus, there was little the environment could do to you that didn't involve a genetic component. 'We put an end', Levit announced, 'to simplified ideas about the almighty role of the environment, for which an organism is a kind of amorphous mass that is able to change in any direction.'

With Lenin's death, and the steady (if fiercely contested) rise of Stalin as his successor, a new approach to politics was emerging. The state would take charge of the country's development, accelerating its modernisation in massive coordinated campaigns, built around a series of five-year plans. In 1928, the Party called for proposals for the first plan. Levit's first volume of research papers was assembled with that official invitation in mind. Its lead article, by Alexander Serebrovsky, contained a remarkable practical proposal. 'Human Genetics and Eugenics in a Socialist Society'[11] advocated the mass voluntary artificial insemination of Russian women, using techniques already employed in the breeding of cattle and horses.

[Given] the tremendous sperm-making capacity of men [and] the current state of artificial insemination technology ... one talented and valuable producer could have up to 1,000 children ... In these conditions, human selection would make gigantic leaps forward. And various women and whole communes would then be proud of ... the production of new forms of human beings.

This, Serebrovsky asserted, would make it 'possible to fulfil the Five-Year Plan in two-and-a-half years.'

Was Serebrovsky writing science fiction? Of course. Thought experiments and speculative articles have always had their place in learned journals. And government purseholders for years had been encouraging just this kind of hyper-optimistic, future-facing writing. Between 1917 and 1922 – a time of martial law, famine and the devastation of industry – fabulous social projects were suggested: the total collectivisation of property, personal belongings and even spouses. In 1922 Ivan Stepanov's book *The Electrification of the RSFSR* planned 'for the transformation of nature and society', mixing popular science, Communist politics and optimistic science fiction.

Some of these works were purely fantastical. (Vladimir Obruchev's serialised novels of the 1920s began with *Plutonia*, describing the voyage through a hole in the Arctic ice to an underground world of rivers, lakes, volcanoes and strange vegetation, inhabited by monstrous animals and primitive people.) But they all at least tipped their hat to the real day-to-day business of doing science. There were novels about the Arctic whose heroes measured magnetic fields in impossible cold and made chloride titrations of water from under ice floes. There were novels about bridges and canals, chemists and doctors.[12]

A massive volume published in 1928, *Life and Technology in the Future*, edited by the philosopher Ernst Kolman, offered projections about the future by an assortment of distinguished scientists featuring commune-cities, personal flight, skyscrapers linked by bridges and air stations, and (in a piece by the psychologist Aron Zalkind) new organs, new minds and new senses.

If Kolman could publish *Life and Technology in the Future*, why was Serebrovsky's vision met with such opprobrium? He was officially reprimanded for his 'insult to Soviet womanhood', and the satirist Demyan Bedny published a poem in *Izvestiia* taking Serebrovsky's proposal to absurd lengths, envisioning Moscow clogged by 10,000 copies of the same bureaucrat.[13]

Serebrovsky did not take Bedny's satire quite seriously and even attempted to publish a versified response. It fell flat. Official disfavour and the public outcry eventually forced Serebrovsky to dissociate himself from Levit's office.

Serebrovsky's main idea had been to liberate sex from procreation. His eugenic future was a throwback to the heady days of the 1920s, when the family was just one more part of the capitalist past to be overthrown. In his view the 'bourgeois' family, in which the husband accepted only those children who were biologically his, was an entirely artificial and unnecessary state of affairs, stemming from the notion that a wife was a man's property. This view, if not commonly held, was certainly a familiar trope in speculative fiction and opinion pieces of the 1920s.[14]

But times had changed. The country had new problems to wrestle with: a falling birthrate, escalating abortions, paternal delinquency and a generation of homeless orphans – literally millions of them – now sustaining themselves by begging, theft and prostitution. Serebrovsky's idea was technically thrilling; politically, it was badly out of step.

Even this might not have mattered had 'Human Genetics and Eugenics in a Socialist Society' been pure fantasy. (No one was taking Koltsov to task for his Martian parable.) But this was the other problem with Serebrovsky's proposal: it seemed doable.

Since 1923, Serebrovsky had been running the department of poultry breeding at the Moscow Zootechnical Institute, so he was well acquainted with the latest work on artificial insemination, and particularly the achievements of his world-renowned colleague Ilya Ivanovich Ivanov.

Before the revolution Ivanov, who had learned his surgical techniques from the meticulous Ivan Pavlov, had won international acclaim for his research on artificial insemination, which up until then had been chiefly an experimental curiosity. He had developed a system of sponges, rubber catheters and syringes, inseminating ten times more mares than a stallion could when left to its own devices. This was no minor achievement at a time when horses were agriculture's main power source.

For Ivanov, artificial insemination was also an experimental technique. His lifelong ambition was to investigate the mechanisms of evolution, establishing the genetic relationships between different animals. He created a zeedonk (the hybrid of a zebra and a donkey), crossed cattle with wisent, antelope and yak, crossed mouse with rat, mouse with guinea pig, guinea pig with rabbit, and created a fine-haired breed of arkharo-merino sheep which one can still find grazing the hillsides of Kirghizia and Kazakhstan.

The revolution disordered Ivanov's work and robbed him of his patrons. In desperate need of new backers, in 1922 Ivanov asked his research associate Mikhail Nesturkh to start preparing abstracts for him on the subject of primate biology, and he began a correspondence with the American biologist Raymond Pearl concerning the possibility of cross-breeding a human being with a chimpanzee.

Interest in the project was intense. Sergei Novikov, the Education Commissariat's Berlin representative, called Ivanov's area of study an 'exclusively important problem for Materialism'. Lev Fridrichson of the Agriculture Commissariat thought 'the topic proposed by Professor Ivanov ... should become a decisive blow to religious teachings, and may be aptly used in our propaganda and in our struggle for the liberation of working people from the power of the Church'.[15] Ivanov presented his proposal to the Academy of Sciences on 30 September 1925. He played to a packed house: Sergei Oldenburg was there; so too were Fersman and Pavlov. Armed with $10,000 in roubles, and with his project overseen by the Academy of Sciences, Ivanov left Russia to drum up international support for his mission.

He garnered plenty of headlines. A provocative piece in the *New York Times*, 'Soviet Backs Plan to Test Evolution', prompted threatening letters from the Ku Klux Klan. And with the backing of the Pasteur Institute in Paris, Ivanov did visit Guinea twice to attempt the cross, in February and November 1926.

Inseminating chimpanzees proved an injurious and difficult endeavour (Ivanov's son, who had accompanied him on his trip, was hospitalised during one gruesome and rapine experiment), so Ivanov returned to his post at the Moscow Zootechnical Institute, intending to use sperm harvested from a laboratory primate to impregnate a Soviet woman. He had several takers.[16] But his breakneck and increasingly iconoclastic career was abruptly checked, once the Academy of Sciences discovered that he had already attempted to impregnate African women – without their knowledge.

On 19 April 1929 Gorbunov's Department of Scientific Institutions invited several scientists from the government-friendly Communist Academy for a one-day discussion on Ivanov's work. Alexander Serebrovsky and Solomon Levit were among the attendees. Though the meeting was fully aware of Ivanov's ethical transgressions, it reserved its criticisms for the mutinous and bourgeois Academy of Sciences, noting that it had sat on Ivanov's report for eighteen months since the last African expedition. If the Academy of Sciences was incapable of acting, then the project would have to be handed over to the Communist Academy in order 'to arrange a comprehensive review of the proposal by Prof. Ivanov ... and carry out the necessary experiments'. The Society of Materialist Biologists set up a commission 'on the interspecific hybridisation of primates' to oversee the work. Solomon Levit was chairman. Serebrovsky was a member.

As the tide of political opinion turned against eugenics, the two men could not have found themselves in a more exposed position.

As the 1930s advanced, extreme negative eugenics swept the Western world. Every Nordic nation in Europe adopted stringent eugenics legislation, but by far the most far-reaching statute was Germany's Law for the Prevention of Genetically Diseased Progeny, passed in 1933. Modelled in part on California's compulsory sterilisation law, it mandated the sterilisation of persons suffering from congenital feeble-mindedness, schizophrenia, manic depression, severe physical deformity, hereditary epilepsy, Huntington's disease, hereditary blindness or deafness, or severe alcoholism. 'We must see to it that these inferior people do not procreate,' the eminent biologist Erwin Baur declared. 'No one approves of the new sterilisation laws more than I do, but I must repeat over and over that they constitute only a beginning.' Between 1933 and 1945 around 300,000 German citizens were sterilised.

In the year before he left for Texas, Solomon Levit did what he could to refashion Soviet eugenics to distinguish it from the negative eugenics taking hold in the West. He was well placed for the task: in March 1930 he was appointed director of the Moscow Medical–Biological Institute, now housed in a new, luxurious, constructivist building complex on Kaluzhskaya, across from the mansion that, from the late 1930s, would house the Presidium of the Academy of Sciences.

Levit took advantage of the move to expand his 'office of human heredity and constitution' into a new Genetics Division – the bland new name was deliberate – and published a second volume of papers, prefaced with an editorial that drew a sharp distinction between eugenics and 'human genetics', which Levit defined as 'the science of human heredity'. He wrote: 'I would say that before about 1926 we lived through an infantile period in the history of our society. The reason I call it infantile is that an enormous number of theoretical errors were made, some factual, some methodological.'[17] The piece ended with a long, valedictory letter by Alexander Serebrovsky apologising for parts of his 1929 article.

All this proved not enough. In January 1932, while Levit was visiting the Timofeev-Ressovskys in Berlin on his way back from Texas, he was replaced as director of the Medical–Biological Institute by its acting director Boris Kogan, who suspended all genetics research there. Levit did not find out about this until he re-entered the country on 22 February. It was indeed a rude homecoming: he found his colleagues Agol and Serebrovsky under increasing ideological attack, and the three of them lumped together in hostile newspaper articles as 'Menshevising idealists' – a weasel formulation that more or less labelled them enemies of the state.

Levit fought back with aplomb. He wrote an article in which he fiercely attacked racism, fascism and 'social Darwinism' – the notion that human progress depends upon cut-throat competition – and this seems to have brought him back into political favour. Meanwhile he jockeyed for position in the new system of research institutions that had taken shape during his sojourn in Texas. We don't know exactly what strings Levit pulled, or precisely how he brought together the pieces of his enterprise, but by 15 August 1932, when the Medical–Biological Institute was re-opened, Levit was again its director.

Once more at the helm, Levit acted quickly to establish his institute's socialist credentials, holding a conference to settle on the right language and approach with which to discuss human genetics and heredity. Papers by Levit himself, Nikolai Koltsov, Hermann Muller and others called for the establishment of a new discipline, to be called 'medical genetics', as a way of improving human health and combating fascist pseudoscience.

Levit pitched the rhetoric of the conference very carefully indeed:

Our great Union has an honourable task to transform the little, wilted greenhouse plant that was grown by the capitalist world and that we call medical genetics into a strong tree, a great science, free from ugly gnarls created by racists and bourgeois eugenicists, and serving the cause of socialism, the cause of Soviet health care.[18]

In July 1935 the Medical–Biological Institute's impressive fourth volume of research was sent to press. Published in 1936, the volume was arguably the best single collection of original research on human heredity that had appeared anywhere. Twenty-five original research papers included studies of hereditary factors in asthma, allergies, pernicious anaemia, diabetes, stomach ulcers and breast cancer. A remarkable series of papers on pairs of identical twins analysed their electrocardiograms, height and weight, and fingerprints.

But while Levit was boldly revamping and renaming eugenics within the purlieus of a state-run institution, Nikolai Koltsov was being much more circumspect, deciding at last to off-load his entire eugenics operation. The problem for Koltsov was not so much ideological as practical: independent institutions like his Institute of Experimental Biology were under increasing pressure to conform to the goals of the First Five-Year Plan, and their independence was threatened. From early 1930, in order to ensure their conformity to the goals of socialist construction, every scientific society was expected to present its charter and membership roll for review and approval to the NKVD – the Commissariat of Internal Affairs, an arm of the security services.

Koltsov saw little point in exposing the members of his Russian Eugenics Society to the attentions of the NKVD, and simply did not submit the required papers. The society ceased to exist and its journal was discontinued.

A few months later, preparing his institute's plan for the following year, Koltsov took a leaf out of Levit's book and renamed his department of eugenics a 'department of human genetics', 'studying the various phenomena of human heredity and variability, defined not only by heredity, but also by the influences of external environment'. Koltsov also saw to it that Levit's Medical–Biological Institute inherited all his institute's eugenics talent. Levit was delighted.

He should have read his Marx more closely.

Marxism was more than a political philosophy; it was a cultural operation that was supposed to bind all scientific disciplines into one. Marx had dreamt of creating a scientific politics. But in the rush to fulfil his dream, the whole project was getting stood on its head. Politics wasn't becoming any more scientific – but it was learning to dress itself in scientific motley.

Levit's friends were aghast when they heard that Levit was taking on more eugenic work. They told him he was committing political suicide. Levit ignored them. He had no idea how incendiary his work would become.

Part Two
POWER

(1929–1941)

Too much is destroyed to the point of madness, to the point that chronology is wiped out, but even more is begun with open naiveté and faith.

ALEXEI GASTEV, *How to Work*, 1923

Previous page: An aluminium model of the Magnitogorsk steel works, the world's largest, on display at the USSR pavilion of the 1939 World's Fair in New York.

8 : 'Storming the fortress of science'

A fortress stands before us. This fortress is called science, with its numerous fields of knowledge. We must seize this fortress at any cost. Young people must seize this fortress, if they want to be builders of a new life, if they want truly to replace the old guard . . . A mass attack of the revolutionary youth on science is what we need now, comrades.[1]
  Joseph Stalin, May 1928

Lenin suffered his first stroke on 26 May 1922. He worked through it, hacking out memos from his hospital bed in Gorki, a small town near Moscow. A lot of his time was spent compiling lists of intellectuals he wanted to see deported. 'Arrest several hundred and without declaration of motives,' he commanded on 17 July: 'Be gone, gentlemen!'[2]

On his return to duty, however, Lenin found that events had overtaken him. Joseph Stalin – a man so efficient in day-to-day business and so unforthcoming in person, he had earned himself the nickname 'Comrade Filing Cabinet' – had ousted him.

'Wipe this Trotsky-Zinovievist band of murderers off the face of the earth – that's the workers' judgement.' News of show trials meets with approval on the factory floor, 1936.

Joseph (or Iosif) Vissarionovich Dzhugashvili, later known as Stalin (from '*stal*', the Russian word for steel), was born in the Georgian town of Gori in the Caucasus on 18 December 1879. His mother was a devout washerwoman, his father a violent drunk. The story goes that he was expelled from Tiflis Theological Seminary for reading Karl Marx under the table. Between April 1902 and March 1913, he was arrested, imprisoned and exiled, not just once, but seven times, before being exiled to Siberia for four full years.

An editor of the Bolshevik paper *Pravda* and Trotsky's chief rival, Stalin worked hard for the Party, proved his capacity to lead during the Civil War, and was rewarded with two ministerial posts in the new Bolshevik government, as commissar for nationalities and as state inspector of workers and peasants. But it was his position as secretary general of the Party's Central Committee, from 1922 until his death, that really mattered.

Stalin was a member of the Politburo, the Party's supreme policy-making body. Work on numerous other interlocking and overlapping committees filled his remaining hours. His rivals Leon Trotsky and Grigory Zinoviev considered themselves above such drudgery. By the early 1920s Stalin not only knew the contents of every drawer in the Smolny; he also knew how to employ all that paperwork to his advantage. Since the revolution, he had been appointing his supporters as provincial officials. Some 10,000 of them now worked to administer the state on behalf of a triumvirate made up of Stalin as Party general secretary, Grigory Zinoviev (chairman of the international communist organisation Comintern) and Lev Kamenev (chairman of the Politburo).

No wonder Lenin suffered a second stroke. In December 1922 Stalin persuaded Lenin's doctors that the leader should be kept, for his own sake, in isolation, and there he stayed, alone and politically impotent, until his death in January 1924, while the state he had built underwent the greatest political crisis of its short history.

In 1922 there had been a terrible famine. The tsarist response to such a calamity would have been to conduct trade as usual, exporting grain regardless of the needs of the people. (Cold-blooded, certainly – but the tsarist state lacked the political instruments to do much else.)

The Bolsheviks tried to do things differently, but it nearly killed them. Without foreign exchange, they had no money to buy equipment to help rebuild an economy already flattened by war. Iron and steel production fell to a level Peter the Great would have recognised two centuries earlier. Strikes broke out all over the country, and Leon Trotsky and his party of old Bolsheviks argued for a return to the brutal command-and-control government that had sustained the regime during the Civil War.

Opposing them were Zinoviev, Kamenev and Stalin, who at least paid lip-service to the utility of Lenin's New Economic Policy. It was Marx himself, after all, who had taught that without a period of capitalism, however rudimentary, there could be no industrialisation, no proletariat; and so, ultimately, no true revolution.

The triumvirate responded to the economic troubles and strikes of the early 1920s by bringing bourgeois specialists more firmly into their confidence. And the loudest and most effective advocate of this policy was, incredibly, Felix Dzerzhinsky, the former head of the secret police; Bloody Felix, whose Cheka had become notorious for torture and mass executions.

Dzerzhinsky was now head of VSNKh, the Supreme Council of the National Economy, in charge of all big industry, and over-seeing the uneasy alliance of bourgeois factory managers and so-called 'red directors', who acted as political commissars in the workplace. These red directors were phenomenally ill-educated. To support them, cover for their deficiencies, and generally oil the wheels of industry, Dzerzhinsky brought in former secret policemen, whose ruthless efficiency made the work of the harried bourgeois managers a great deal easier. Dzerzhinsky, a fiercely practical man, turned out to be a champion of the NEP and a staunch defender of scientists and engineers, politics be damned.

Dzerzhinsky had a strong ally in Alexei Rykov, the rightist chairman of the Council of Peoples' Commissars – a high office that made him effectively head of state. In 1924 Rykov had gone so far as to offer specialists a 'bill of rights', insisting that 'the specialist, the engineer, the man of science and technology must have full independence and freedom to express his opinion on matters of science and technology'.[3]

To all appearances Joseph Stalin, too, was a staunch supporter of the NEP. But Stalin nursed an extraordinary plan of his own. Rather than merely return to the ramshackle command-and-control apparatus of War Communism, as Trotsky advocated, he wanted to leapfrog capitalism entirely, industrialise the country by fiat, and – by controlling the economy with iron precision – rattle through the stages by which true communism might be achieved. His was arguably the purest expression of Bolshevik impatience that ever was. Even at the outset, it was clear that the cost in lives would be horrendous.

That plan took shape as numerous strikes, plus Lenin's illness, pitted Stalin, Kamenev and Zinoviev against Trotsky, who was still holding out for a world revolution. The defeat of revolutionary movements in Germany and Hungary ought to have cured Trotsky of his quixotic notion, and at the Thirteenth Party Conference in January 1924, Stalin and his colleagues unsurprisingly won national support for their more practical approach: defend the nation, and secure national gains. Establish socialism in one country. The world can wait.

Dzerzhinsky's death in 1926 robbed the old specialists of their most powerful patron, and a very different kind of economic order arose in place of the old. On 3 November 1929 Stalin produced an article declaring 'a great break on all the fronts of Socialist construction'. This 'great break' eliminated unemployment at the stroke of the pen and drew peasants into the factories on an unprecedented scale.

The peasants who flocked to the cities to work in the factories were accident-prone, ignorant, hostile and clumsy, and not at all the kind of workers who 'had endured the famine during the civil war ... beaten the class enemy, and raised our economy from the verge of ruin'. At a Moscow textile factory one worker, angry at being transferred to lower-paying work, hurled a bolt into a printing machine, causing 20,000 roubles in damage. Workers in the Donbass burned themselves with acid in order to shirk, while new arrivals migrated from one mine to another in search of better housing and lighter work. According to one worker from a factory in the Ukraine, they were interested less in working than in gambling and drinking; 'fist-fighting' was their favourite pastime. 'Bumpkins', 'sandal-wearers', 'niggers': by 1929 they were lowering the pay rates for everybody else and arousing the hostility of old hands who treated them with contempt and abuse.[4]

The unions and Party had great difficulty teaching new workers the basics of industrial work. All manner of schemes were introduced to rein in, inform and discipline the unruly peasant workforce, from comrades' courts to socialist competition to training sessions conducted by Alexei Gastev's Central Institute of Labour. The most effective long-term solution was education. Military measures were brought in to train technicians and scientists in the shortest time and between 1928 and 1932, the student body trebled. In some places, the theoretical parts of physics and chemistry were abandoned altogether, while new methods of teaching and grading in 'brigades' saw students learning through 'continuous productive practice'.

In the Central Committee, Dzerzhinsky's old ally Alexei Rykov could make no sense of what was happening. The government was on the verge of launching a series of heroic big engineering projects. Surely what mattered, at this crucial time, was to improve and expand technical education. Why flood over-stretched institutions with ill-prepared students?

But Stalin's plan was to swamp the old technical intelligentsia, once and for all, with sheer numbers of proletarian engineers (dissolving them 'in a sea of new forces', as a State Planning Commission report of 1930 put it) and – being a great believer in the Bolshevik adage 'learn by doing' – he was quietly happy to see the distinction between work and education virtually eradicated. He even proposed turning the whole of higher technical education over to the Supreme Council of the National Economy. When it came to the vote, Stalin failed to get his way, but he was a master of bureaucratic attrition – Comrade Filing Cabinet, indeed – and by 1930 the Education Commissariat had ceded control of all higher technical education to the Supreme Council and other economic bodies. 'And nothing has happened,' Stalin quipped, 'we are still alive.'[5]

It was now important for Stalin that he ruin Alexei Rykov. Rykov had tied his flag to the fate of bourgeois specialists and engineers, and had consequently become the figurehead of right-wing opposition to Stalin and the First Five-Year Plan. And in eliminating Rykov, Stalin achieved a great deal besides.

Late in 1927 Vyacheslav Menzhinsky, head of the secret police, received a disturbing letter from his man in the northern Caucasus, Yefim Yevdokimov. Apparently a group of wreckers was operating in the mountain town of Shakhty, conspiring with their former, émigré owners to wreck the district's coal mines. Yevdokimov had evidence, too, in the form of letters written in code. Menzhinsky smelled a rat. He wrote back to Yevdokimov giving him two weeks to decode the letters. If he didn't, Menzhinsky threatened to have him arrested for false accusation.

Yevdokimov went straight to Stalin – they were friends and went on holidays together – and Stalin ordered the wreckers' arrest. Rykov and Menzhinsky both protested that Stalin had exceeded his authority. Stalin, in reply, produced a telegram from Yevdokimov that hinted darkly at high-level Moscow links to the conspiracy.

On 7 March 1928 five German nationals were implicated in the case and arrested. The timing was exquisite. Russia's national economy depended on successful negotiations with Germany, and in charge of these negotiations was none other than Alexei Rykov. Rykov was trapped: the more he tried to do his job, smoothing Germany's ruffled diplomatic feathers, the more he risked accusations of covering up the Shakhty case.

Rykov trod carefully: in a speech reported in *Pravda* on 11 March 1928, he berated everyone for not uncovering the conspiracy sooner and tried in vain to create an independent commission to establish the facts of the Shakhty case.[6] Of course he had no illusions about what was happening:

Without a doubt the existence of this conspiracy arouses extreme concern on all sides. But it would be unusually harmful and dangerous if the discovery of this conspiracy were to bring in its wake a development of specialist baiting.[7]

Of course, this was precisely the outcome Stalin wished for. He aimed to achieve a truly socialist state overnight by main force, so it was key to his strategy that class struggle would intensify, justifying a wider use of coercion and terror.

The Shakhty trial began on 18 May 1928 in Moscow, in the main hall of the House of Trade Unions, formerly the ornate dining room of the tsarist Nobles' Club, a three-storey marble building with Corinthian columns. The immense hall retained large crystal chandeliers and baby blue walls topped with a frieze of dancing girls. The judge, Party legal scholar Andrei Vyshinsky, turned up in hunting costume for the benefit of the cameras. And there were plenty of them: nearly a hundred reporters, Soviet and foreign, attended the trial, and a hundred thousand spectators, schoolchildren, Pioneers, visiting delegations of peasants … this when there were barely 1,100 mining engineers in the whole Soviet Union.[8]

Yevdokimov's letters were never presented. Evidence against the Shakhty 'conspirators' consisted entirely of affidavits and confessions. Driven to hysterical self-incrimination, some defendants implicated people who were already dead. The prosecutor, Nikolai Krylenko, demanded that nearly half of the fifty-three defendants be executed. The twelve-year-old son of one of the accused demanded death for his father in a letter, published in *Pravda* and read out in court:

I denounce my father [Andrei Kolodoob] as a whole-hearted traitor and an enemy of the working class. I demand for him the severest penalty. I reject him and the name he bears. Hereafter I shall no longer call myself Kolodoob but Shaktin.[9]

In the end, only five of the accused were executed. Several were acquitted. The sentences were beside the point. The Shakhty trial was but the first salvo in a carefully orchestrated three-year-long class war that saw the prosecution of thousands of engineers – the most qualified, the most experienced, the most likely to oppose Stalin's First Five-Year Plan on technical grounds.

The Shakhty case and subsequent trials succeeded in uniting unruly factory workers under the twin flags of paranoia and xenophobia. By raising the spectre of foreign-backed saboteurs, the campaign gave the workers a pressing reason to regiment themselves.[10] In a speech to the Party on 12 April 1928, Stalin explained:

Formerly, international capital thought of overthrowing the Soviet power by means of direct military intervention. The attempt failed. Now it is trying, and will try in the future, to weaken our economic power by means of invisible economic intervention, not always obvious but fairly serious, organising sabotage, planning all kinds of 'crises' in one branch of industry or another, and thus facilitating the possibility of military intervention. It is all part and parcel of the class struggle of international capital against the Soviet power and there can be no talk of any accidental happenings.[11]

Stalin's carefully orchestrated trials of the late 1920s and early 1930s put an entire class on show. Alexei Rykov was simply the headline act. Opposition to Stalin crumbled under the onslaught. Rykov was accused of organising a rightist faction within the Bolshevik Party, and a loyal Stalinist, Vyacheslav Molotov, replaced him as the head of the Soviet government. Gorbunov was dismissed at the same time. So Russia's scientists lost their two most powerful patrons, and Stalin became as a consequence a sort of *über*-patron around whom, like it or not, they were all obliged to gather. By the end of the 1930s they were calling him 'the Great Scientist', and 'Science's Coryphaeus': the leader of the band.

Since the reign of Peter the Great, Russia, lacking any civic life worth the name, had been making do with networks of clients and patrons. Stalin, harnessing these networks and coming to dominate them, could only find, in what few institutions there were, a threat to his authority. They were the surviving trace of a liberal, democratic path not taken.

There were two significant redoubts, shielding bourgeois liberals from the regime. One of them, TsEKUBU, the Central Commission to Improve the Living Conditions of Scholars, had been created with Lenin's blessing and on the Bolsheviks' dollar. The commission, originally established to distribute the special 'academic ration' during the Civil War, had prospered along with its clientele during the easier years of the NEP. These days it was providing supplements to salary, rest and recreational facilities, a Scholars' Club in Moscow (a venue for concerts by the finest Russian artists), a couple of country estates, two health clubs, and a functioning church.[12]

It was a relatively easy business to replace TsEKUBU, a state-sponsored organisation, with another state-sponsored body, more geared to the times (VARNITSO, the All-Union Association of Scientific and Technical Workers for Active Participation in Socialist Construction in the USSR). More troublesome was Peter the Great's own Academy of Sciences, an unabashedly liberal, Kadet-dominated organisation which, from the outset of the Soviet project, had been dealing as an equal with the government – and sometimes more as though the Bolsheviks governed only through the Academy's patronage. At the Second Congress of Scientific Workers, held in February 1927, the Academy's leading scholars were not only confident but demanding, attacking the Commissariat of Education for failing to provide research and higher education with sufficient money. Sergei Oldenburg in particular, Lenin's familiar, treated the commissar Anatoly Lunacharsky with impatience bordering on contempt. For ten years, he had witnessed the commissariat's 'misfortunes' and tolerated its failure to sort out adequate financing. Lunacharsky had better make it clear to the leadership that the situation, on the eve of the First Five-Year Plan, was now critical.

This was more than patrician *noblesse oblige* having a gallop: Oldenburg and the Academy were conscientiously fulfilling the deal struck with Lenin; they were working sincerely in the national interest. If the Education Commissariat couldn't fund them properly, it was up to the government to find them an organisation that could. They assumed, in other words, that their relationship with the government was – or at least ought to be – an easy, collegiate one.

Lunacharsky had news for them:

The intelligentsia are waiting for an invitation from Soviet power for the most valuable elements of the aristocracy of the mind to enter the highest organs of government . . . But they must not be surprised if the Revolution, which has to defend itself against its enemies meticulously and ruthlessly, has also produced organs that look on such things from a completely different point of view.[13]

In April 1927 a group of leftist professors with connections to the highest echelons of the Bolshevik Party met to draft the outlines of what would soon become VARNITSO. They sent letters to the science section of the Education Commissariat, insisting that 'the Academy of Sciences was a right-leaning organisation, concentrating too much power in its own hands'. Their letter campaign was a long one. A year later they were warning that the Academy opposed 'Sovietisation' and was developing contacts only with Western scientists hostile to socialism.[14]

The professors argued that new communist organisations should end the Academy's dominance over Soviet science. That was a task-and-a-half: so far, communist replacements for the old institutions of bourgeois learning had been weedy affairs. One of the chief orders of business at the Communist Academy had been to convert scientists to Marxism, but the effort had hardly got off the ground. Even the Union of Scientific Workers, the least controversial of organisations, abstaining almost completely from the promulgation of Marxist science, boasted a membership comprising barely one in twenty scientists. Societies of materialist scientists were as numerous as mushrooms after a spring rain, but about as long-lived – they never boasted more than a few hundred members in total.

VARNITSO's birthplace was the Timiryazev Biological Institute, a major producer of Marxist popular science. Its leading light was Boris Zavadovsky, the Marxist biologist who had attacked Nikolai Koltsov so vehemently for his eugenic views. Over the years Zavadovsky's political concerns had come to dominate his scientific interests, and he was highly critical of science writers who left Marxist ideology out of their work. Through VARNITSO, Zavadovsky's classic Bolshevik impatience developed into classic Bolshevik vindictiveness, as he set the organisation to unseat an older generation of tenured academics.

Joining him as founders of VARNITSO were his elder brother Mikhail (head of the Biology Department at Moscow State

University) and some of the most aggressive Bolshevisers in science, including Alexei Bakh, a biochemist who a year before had been elected a member of the Academy of Sciences. Bakh was chosen to lead the organisation because of his impressive CV of practical achievements, particularly in the bread, tobacco and tea industries.

VARNITSO was unashamedly a lobbying group, fighting tooth-and-nail for money and support. By the end of 1932, it boasted more than 11,000 members. It represented the intellectual as well as the political ambitions of a new generation – young people who found their careers blocked by academics left over from before the Revolution.

VARNITSO's strategy was brutal: to end the collegiate relationship between the state and its scientists. In letters to Stalin and other Politburo members, VARNITSO used the revolutionary rhetoric of class war and international intrigue, arguing that science and technology were instruments in a struggle between two world systems – capitalism and socialism. They attracted attention. More than that, they exerted influence: for the next two decades, Stalin's own pronouncements on science might have been lifted wholesale from some letter or other from VARNITSO.

On paper, VARNITSO's purposes were straightforward. Dedicated to the welfare of scientists and scholars, it steadily replaced TsEKUBU as a lobbying channel for the science community. But VARNITSO also had a political agenda: to make it impossible for intellectuals to claim political neutrality. In April 1929 VARNITSO even challenged the secret police, the OGPU, to a competition in 'exposing wreckage'.

In 1929, VARNITSO dispatched two experienced provocateurs to Nikolai Koltsov's Institute of Experimental Biology – a bastion of pure research – to expose and force out its bourgeois element, which of course included Koltsov himself. They manufactured a series of scandals that all but ruined two of Koltsov's researchers,

Nikolai Belyaev and Boris Astaurov, and may well have been the cause of the bizarre arrest and internal exile of his brilliant colleague Sergei Chetverikov.

Right up to his death in 1959 Chetverikov – the expert in insects who pioneered population genetics – claimed not to know why he was arrested. Mostly likely, he was the victim of a smear that accused him of writing a postcard to the Academy of Sciences, congratulating it on Kammerer's suicide. That Chetverikov carried no flag for Kammerer's work was well known, but the postcard, which was never produced and of course also served to embarrass an already beleaguered Academy, made no psychological sense.

Chetverikov was exiled to Ekaterinburg and his celebrated seminar group, the 'screeching society', which at the time comprised about half the world's population geneticists, were scattered to breeding stations in central Asia and God knows where.[15]

VARNITSO existed to bring scientific expertise to the service of Soviet construction, and, specifically, to the First Five-Year Plan. With their state funding hanging in the balance, scientific societies went out of their way to show VARNITSO that they were not insular fortresses of pure research.

But the leaders of VARNITSO weren't interested in friendly gestures or even actual reform. They launched a press campaign against 'bourgeois' scientific societies, which was meant to climax in March 1931 at a meeting in Moscow, to which most of the nation's scientific and voluntary societies were invited. Wisely, leaders of non-Party organisations simply declined this toxic invitation, leaving Boris Zavadovsky to rail impotently in print against those 'citadels of bourgeois intellectuals' and their 'total inability and frequently wilful refusal to redesign their work and bring science closer to practice'.[16]

By April 1931 attacks on Koltsov and his Institute of Experimental Biology had got to the point where a decision was made to disband his institute. Four divisions had already split from the parent body and become independent. Was it worth preserving the core?

Obviously Koltsov thought so. His life's work had been devoted to his institute: experimental biology was a new and fertile field of research, and a centre for pure theory was essential if progress was to be maintained. Luckily, Maxim Gorky felt the same, and delivered a letter from Koltsov – his client and his friend – into Stalin's hands. The next day, Semashko's replacement as health commissar, Mikhail Vladimirsky, 'restored the undivided authority of the director and', wrote Koltsov, 'removed a number of annoying trifles which then were making my existence totally impossible'.

Koltsov replaced the exiled Chetverikov with a man who could not possibly offend VARNITSO. Nikolai Petrovich Dubinin was an orphan whose working-class credentials had earned him an accelerated education. Intellectually, he was a force to be reckoned with. Serebrovsky could testify to that, having taken Dubinin along on expeditions to Central Asia to map the spread of genes in domesticated fowl. Dubinin didn't just look good from the outside, he *was* good – a first-rate mind.

Koltsov's circumspection grew. When Stalin gave a speech entitled 'The Cadres Must Decide Everything', Koltsov delivered his own version to his institute. And when, at a 1936 conference the plantsman Trofim Lysenko went head-to-head with Nikolai Vavilov, director of the Bureau of Applied Botany, Koltsov firmly – though without malice – twitted Vavilov on his lack of 'practical achievements'.

The first public attack on the Academy of Sciences came on 15 May 1927, in the form of a satirical article in the daily *Leningradskaya Pravda*. M. Gorin's satire 'The Academy's Ark' neatly skewered the anomalous class make-up of that institution.

Certain myths clung to the Academy – that it was dominated by Germans, arch-conservative and elitist. But the truth was hardly more comfortable: that it had been dominated by foreign scholars until the second half of the nineteenth century, and had now become a magnet for nobles and so-called 'former people' who, unemployable elsewhere, found 'protection and the possibility to exist' in the Academy.[17] Twenty-three academicians, all former high government officials, either had titles or belonged to the nobility. Several of them were related to each other. The Academy, Gorin warned, 'cannot be used as an ark for has-beens.'[18]

Three further articles – less ironical, much more aggressive – followed within a month. The writers of these pieces – two of whom may have had close connections with the secret police – cooked up the most dangerous calumnies, including talk of a 'Cosmic Academy of Sciences', a (so far as we know, fictitious) religious and philosophic circle of academicians opposed to the Soviet programme.

These public attacks were a (presumably deliberate) extra pressure brought to bear on the Academy at a crucial moment. The Academy was fighting to retain its independence: in particular, the right to say who was worthy to carry the title of academician. There was a conflict of ideologies here, as well as a conflict of interests. Party scholars argued that, when it came to considering candidates, their textbooks and popularisations deserved consideration as scholarly works. For them, the act of binding the sciences together, and making them legible to each other, was the highest scientific calling. The Academicians, proud specialists to a man (the first female academician, Lina Shtern, was elected in 1939), insisted that only quality of research mattered.

Sincere cases could be made on both sides, and principled dis-agreement eventually dissolved into horse-trading. Oldenburg's fellow academician Ivan Pavlov was outraged: 'This is lackeyism that you are proposing!' he exclaimed.

Pavlov was almost screaming that we need to show ourselves to the Bolsheviks, that there is nothing to fear from them, that no preliminary deals are needed, that everyone should and must act individually and so forth. Sergei [Oldenburg] told him vehemently that he, Iv. Pav., can, and is permitted to, say what he likes, they will not touch him, since he is in a privileged position, since he is, as all know and as the Bolsheviks themselves say, the ideological leader of their party. Pavlov boiled over again. It was terrible.[19]

This 'terrible', messy, ad hoc dealing was reasonably successful until it came time to consider the Party's three most controversial candidates for election to the Academy: the dialectical philosopher Abram Deborin; Nikolai Lukin, a lacklustre historian of the French Revolution; and the Marxist literary critic Vladimir Friche. On 12 January 1929, an election by secret ballot rejected them. Straight away Oldenburg and a delegation of academicians were ferried in a curtained limousine to an emergency summit meeting in the Kremlin. Faced with the threat of the Academy's wholesale dismemberment, the delegation capitulated, and the three Party scholars were voted in at a special election. In return, the Academy's budget rose over 40 per cent on the previous year.

One more largely cooked-up conspiracy did for the old dispensation worked out by Oldenburg and Lenin; indeed, it ended Oldenburg's career. On 19 November a government commission was informed that the Academy was secretly storing documents 'of great political value': tsarist police reviews of revolutionary movements, personnel files of tsarist secret agents, secret military documents from the world war, Kadet and Menshevik party archives and even some Bolshevik material.

Never mind that the Academy had been tasked, in the chaos of the Civil War, with storing this material: wild accusations followed. Alexei Rykov – yet to be toppled as head of state over the Shakhty affair – wired the Academy demanding that Oldenburg be dismissed as permanent secretary. Oldenburg, on the brink of retirement, resigned in disgrace, while newly elected academicians Deborin and Nikolai Bukharin, a leading politician and revolutionary theorist, wrote the Academy a new constitution, bringing it under the absolute control of the Party.

Few openly opposed these reforms, and few had the opportunity, as Pavlov did, to tear strips off the government for its vandalism. In December 1929, Pavlov turned a centenary celebration of Ivan Sechenov's birth into a startling political spectacle, striding up to a large portrait of the 'father of Russian physiology' and declaiming: 'Oh noble and stern apparition! How you would have suffered if in living human form you still remained among us! We live under the rule of the cruel principle that the state and authority is everything, that the person, the citizen is nothing.'[20]

The audience was stunned. Pavlov told them to stand and salute Sechenov. What to do? Everybody rose, looking around nervously. Many communists walked out.

Not every elderly academician agreed with Pavlov. Unlocking the doors to the Party faithful had at least saved the Academy's bricks and mortar, not to mention its budget, which rose and rose in following years. Besides, these newly elected academicians were not monsters. Indeed, they now mounted a spirited defence of their new home. Leading the effort to rebrand and champion the Academy was, of all people, David Riazanov, founder of both the Communist Academy and the Marx–Engels Institute. It turned out that during his tenure at the Communist Academy, Riazanov had been growing sick to death of the combative and doctrinaire behaviour of his own researchers. His disgust with these people – mostly graduates from the Institute of Red Professors – and their constant 'hunt for deviations' had finally caused him to abandon any role in the Communist Academy's governance; now, from the purlieus of the Academy of Sciences, he was calling the Communist Academy a 'pale copy' of that august body to which he now belonged.

Now that its gates were open, the Academy of Sciences grew. The number of communist staff members went from virtually none in 1928 to almost 350 by 1933, and the following year, when it moved from Leningrad to Moscow, the Academy was given splendid new buildings originally planned for the Communist Academy. Two years later the Academy of Sciences swallowed the Communist Academy – a cruel blow for old Marxist scholars. Elderly academicians, meanwhile, were haunted by a painful awareness that you are what you eat.

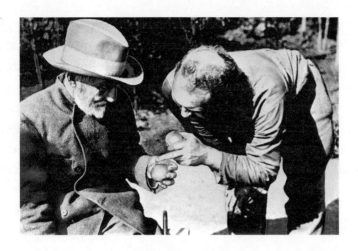

9 : Eccentrics

Natural science will . . . incorporate into itself the science of man, just as the science of man will incorporate into itself natural science: there will be one *science.*[1]

   Karl Marx, 1844

You can take the peasant out of the countryside, and Stalin did, literally, and on an industrial scale. It is harder – much harder – to take the countryside out of the peasant. Slave societies rarely breed fine feeling, and lives spent on the edge of starvation breed reciprocity faster than friendliness. (There is a word for debt in Russian – *dolg*, but no word for favour. The nearest equivalent, *odolzhenie*, still expects a return.)

   Traditionally, peasant relations in the countryside had been policed through reputation, and the ease, speed and savagery with which it could be destroyed. The threat of denunciation held together communities who could quite literally starve to death if too many of their members abandoned them. In tsarist times, people who left for side-earnings in the city and tried to break with their village were regularly accused of heresy. When

'Victory over the old fogies of horticulture': plant breeder Ivan Michurin and an assistant (Igor Gorshkov?) study one of his many varieties of hybrid fruit.

the Soviets took over, religious denunciations turned seamlessly into political ones. Being an 'enemy of the people' was not at all a Stalinist innovation. Stalin's genius was to harness and direct to his own ends the fierce belligerence of the peasant class.

In this he was supported by a new generation of Bolshevik official: earnest believers whose whole adult life had been spent under Soviet rule. These men knew their state was unfinished and unstable, but they had no experience of alternatives. For them, the bourgeois old guard represented, not a rejected alternative, but an obstacle to be removed. (The Party was by now, as it never had been under Lenin, a working-class party.) At a provincial conference on 27 May 1928 (and still some way off from real power), the Party chief of Nizhny Novgorod, Andrei Zhdanov, cast the gathering cultural revolution in apocalyptic terms, as the war of sons against fathers: 'The struggle for the cleanliness of our ideas, the struggle for the youth has to take an important place at the moment; it is necessary to develop within the Komsomol [the Party youth wing] a critical relationship toward the older generation [and] its shortcomings in its way of life and being.'[2]

Youth activists duly harried their elders and betters, and in particular attacked them for their religious beliefs. In a letter to the national teachers' newspaper, one reader complained:

My teacher in junior class, meeting me sixteen years after I left school, wept and told me that she is even afraid to live and work at the present time. She has no regrets for the Tsar – he drove her fiancé into the grave and so she is still unmarried at forty. But the icons that they threw out of the school – this was more than she could bear.[3]

This 'cultural revolution' silenced an entire educated class at the very moment the government, committed to breakneck industrialisation, thirsted after new practical ideas. So it was perhaps inevitable that it ended up backing ideas that turned out to be eccentric, to say the least. Every madcap professor shown the tsarist door rose clamouring for his day in the Bolshevik sun.

John Littlepage, an American engineer working in the Soviet gold-mining industry, wrote of 'the perpetual nuisance of so-called inventors, crack-brained persons who are convinced they have made some amazing mechanical discovery, a type that seems more numerous in Russia than elsewhere'.[4]

Linguistics discovered Professor Nikolai Marr's idiosyncratic 'Japhetic' theory of language. Music students embraced Professor Boleslav Yavorsky's theory of 'modal rhythm', which had been a standing joke for years within the Moscow Conservatory. These men weren't wrong or mad any more: they were radicals.

By the time the celebrated plant breeder Ivan Vladimirovich Michurin died of cancer in June 1935, he was not just considered an untutored genius, he was being hailed by the agriculture commissar, Yakov Yakovlev, as a heroic opponent of 'bourgeois science'.

Michurin rose to fame as a sort of Soviet Luther Burbank – a homespun genius who greened drought-browned valleys and filled the people's bellies by mixing together scientific thinking and folk remedies. Like the Massachusetts-born plantsman Burbank, Michurin grew up in genteel poverty. The produce of his home town – Kozlov in the province of Ryazan, 300 kilometres south of Moscow – included rather scraggy fruit, which ceased to sell once the new railroads brought in produce from Crimea in the far south. Rather than sell their orchard, however, Ivan Michurin's parents more or less destroyed themselves trying to hang on to it. A succession of misfortunes sent Ivan's father Vladimir steadily more crazy. Tuberculosis killed his wife; his half-mad mother terrorised the family, and of his seven children, only Ivan survived. At his wife's funeral Vladimir came out with a dance song instead of a dirge and was taken away for the first of several stays at the local asylum.

Ivan Michurin, though he aspired to an aristocratic lycée in St Petersburg, in the end enjoyed only one year of local elementary school education. He went to work in nearby Tambov as a railway clerk, then as a signal repairman. He married a mechanic's daughter, and there – but for his determination to realise his father's agricultural fantasies – Michurin's story ought really to have ended.

In 1888, in his mid-thirties and determined to make something of himself, Ivan Michurin bought fourteen hectares with borrowed money, quit his job, and set up a nursery. Rather than grafting southern varieties onto northern stocks, he decided to breed hybrids from seed – a highly dodgy tactic given that good fruit varieties are complicated and unstable hybrids to begin with, and unlikely to breed true. It is hardly surprising, then, that Michurin came to the conclusion that there was no regularity in hereditary phenomena, and no coherent science to be had from studying them.

In November 1905, taking advantage of the promises of new civil liberties, Michurin submitted a petition to turn his unprofitable nursery into a state experiment station. After an unexplained delay of two and a half years, the Ministry of Agriculture turned him down. He later earned two medals and a job offer from the ministry, but none of this quelled his resentment towards the know-all academics who had failed to see the value of his work.

In 1911 and 1913, a plant collector from the US Department of Agriculture, Frank Meyer, visited Michurin's nursery. Michurin priced himself out of an arrangement to sell regular fruit stocks from his nursery. He did not really need the trade. What he needed was the story: to hear him tell it, agents of the USDA had been visiting him since 1898, and had repeatedly begged him to come to the USA on an astronomical salary of $32,000 a year.

When the Soviets took control of the country, they agreed to turn Michurin's nursery into a breeding station. True, Michurin was both eccentric and cantankerous, but Nikolai Vavilov himself visited Michurin in 1920 and was impressed by the old man's notes. (Experimenters like Vavilov liked Michurin's stocks because they were so bizarre. They were not, however, particularly commercial. Even in 1931, when Michurin was being recast as a working-class hero, only one new variety of apple could be found worth certifying.)

Michurin's skills as a lobbyist were formidable. In September 1922 he got the President himself, Mikhail Kalinin, to visit his nursery in Kozlov. But the following autumn, in Michurin's seventieth year, the First All-Russian Agricultural Exhibition was held, and here, and at the very end of his career, things threatened to come unstuck. Michurin's view was that plants-manship was an art, incommunicable in abstractions or formulas: 'It is evident that nature, in its creation of new forms of living organisms, gives infinite diversity and never permits repetition.'[5] He was right, of course: a fruit nursery is not a laboratory, and the hybridisation of fruit trees is a tangled business, better handled as a craft, learned over the course of years, than as an undergraduate research topic.

But as science had taken hold of the European imagination in general – and the Bolshevik imagination in particular – so public funding had become dependent on delivering neat, simple, *scientific* explanations of one's work. And this, for Michurin's institute, was bound to end in failure. A young horticulturalist, Igor Gorshkov, sent by Michurin to wow the exhibition, was worsted in arguments again and again over the validity of some of the Kozlov nursery's hybrids. The one between a melon and a squash attracted especially negative attention. No, said the other experts, the stock onto which you grafted something would *not* alter the germ plasm of your fruiting plant, and all Michurin's talk of 'vegetative blending' and 'mentoring' made no sense.

When Michurin learned of Gorshkov's reception, he blew his top: if Gorshkov couldn't win extra support for the nursery, he might as well throw in his cards and accept the USDA's (fictional) job offer. Gorshkov pitched this threat to the editors of *Izvestiia*, who ran it under the headline 'Kozlov or Washington?' Where, the paper asked, would Michurin's work find support – the Soviet Union or the USA?

Michurin was so delighted with this coup, he presented Gorshkov with a watch, hand-engraved by himself with the words 'From I. Michurin to I. Gorshkov. For Victory over the Old Fogies of Horticulture. 14 October 1923.'[6]

Michurin remained an outsider. Few mentioned him in their papers. Trofim Lysenko took an interest, but the old man rebuffed him. He was far friendlier with Vavilov, and Vavilov was among those who elected him an honorary member of the Academy of Sciences a week before his death.

Most 'independent scholars' were like Michurin: opportunists seizing their last chance at glory. But some were closely allied to the Bolsheviks from the start. Olga Borisovna Lepeshinskaya had been a personal friend of Lenin and his wife. Between 1897 and 1900 they had all been banished to the same region of Siberia. As a consequence, she was terrifyingly well-connected and not remotely intimidated by power. On a personal level, she was charming. She fiercely opposed anti-semitism, and had dedicated her personal life to the orphan problem, bringing up at least half a dozen children as her own. As a scientist, however, she was a disaster. She once announced to the Academic Council of the Institute of Morphology that soda baths could rejuvenate the old and preserve the youth of the young. A physician, Yakov Rapoport, asked her sarcastically whether mineral water would work instead. Lepeshinskaya, oblivious to his tone, told him no. A couple of weeks later Moscow completely sold out of baking soda.

The same unstoppable cart that whisked Michurin and Lepeshinskaya to glory also carried along several articulate youths whose ambition far outweighed their talent. The simpler and more outrageous their schemes, the more they appealed to fond wishes, the more they were believed. (Science – real science, Marx's 'one science' – was supposed to be straightforward and practical.) Planning groups and building trusts approved plans for socialist cities of the future. A government commission thought about reforming the calendar (with 1917 as Year One). The Soviet patent office, the Committee on Inventions, reported one case of a 'scientist' entirely lacking in formal education who was given 200,000 roubles and an 'electrical biology' laboratory to show how bombarding seeds with ultra-high frequency radio waves would trigger bumper harvests. (The police eventually tracked him down to the cabaret clubs of Leningrad.) Considering the work of a physicist in Ashkabad who made rain with electrified smoke, the Marxist philosopher Isaak Prezent declared:

We are carrying out the grandest task, planned alteration of the climate ... A special grand institute for making and stopping rain is being organised ... The grandest, unheard of projects are now being worked out, in actual working plans with concrete economic calculation, for the irrigation of dry regions and an all-out assault on the desert. We are solving the problem of *heating Siberia*.[7]

Eccentric schemes tend to blow up in their backers' faces, and the state was aware of the risk. In 1947 Eric Ashby, a botanist attached to the Australian Legation, observed that,

there has been recently an increase in the efficiency with which the State separates genuine from bogus advances in agricultural research, and protects itself from being 'sold a pup' by enthusiastic and not too critical experts. There is, for example, an interesting and elaborate organisation for testing new crop varieties, known as the Government Commission for Seed Testing, under the chairmanship of Academician Tsitsin.[8]

But problems of this sort sprang from more than a handful of hucksters and 'independent scholars'. Old Bolshevik hands, Party leaders and key bureaucrats were themselves dedicated amateur philosophers of science. They ruled in the name of scientific government, and were honour-bound to pronounce on scientific issues. Their mistaken and wild schemes came stamped with the imprimatur of the state, and were much harder to challenge.

Stalin was himself a totally dedicated and self-declared 'Lamarckian', obsessed by the idea that it might be possible to alter the nature of plants. As the years went by this obsession grew, and became indeed his only hobby. At his dachas near Moscow and in the south, large greenhouses were erected so that he could enter them directly from the house, day or night. Pruning shrubs and plants was his only physical exercise. In 1946 he grew especially keen on lemons, not only encouraging their growth in coastal Georgia, where they fared quite well, but also in the Crimea, where winter frosts obliterated them. Stalin was not discouraged. He asserted that oaks and other deciduous trees, if planted as seeds, would adapt to the most hostile conditions, flourishing in the dry steppe, and in the salty, semi-arid wildernesses near the Caspian Sea.

Leaders, politicians and bureaucrats have their hobby horses, of course. The problems start only when these people assume for themselves an expertise they do not possess, when they impose their hobby horses on the state by fiat. The Bolshevik tragedy was that, in donning the mantle of scientific government, the Party's leaders felt entitled to do this. More: they felt *obliged* to do this. Stalin's Lamarckian beliefs and utopian fond wishes regarding the plasticity of living forms were rather ordinary for his day. Realising these ideas in policy would have extraordinary and often catastrophic effects on the lives of millions.

At the very end of 1929, at a conference of agricultural economists, Joseph Stalin clarified the Party's line on what science was, and how it should progress. Science was a human activity, not a mystical communion with nature, and the measure of theory ought to be how usefully it could be applied. In the field of economics, especially, theory lagged behind practice. Party workers in the countryside were pushing collectivisation far faster than the economists had thought possible.

From his narrow chivvying of a bunch of economic specialists, Stalin then extended his argument to cover all science and philosophy. Philosophy and all other branches of theory must be refashioned to be of immediate service to the revolution. The central principle of philosophy was now to be *partiinost*: partyness.

Under this dismal scheme, scientists at least got to be glorified engineers. Philosophers had no obvious second string to their bow: they were now, in all essentials, simply a branch of the media, there to explain and celebrate practical improvements wrought by the Party. They were yes-men at best; at worst, a species of thought police.

Stalin himself, in an effort to defend Lamarckism against its critics, instigated meetings of the Society of Materialist Biologists on 14 and 24 April 1931 to 'organise the study and unmasking' of the 'mechanistic school' of Nikolai Koltsov. Official hostility mounted against the Institute of Experimental Biology because it was run by a bourgeois intellectual was one thing. An official attack on its *science*, based on Stalin's strong *scientific* opinions, was quite another.[9]

There was a semblance of balance: the Society of Materialist Biologists included powerful proponents of genetics, men like Levit and Agol, and the president himself, the philosopher Isaak Prezent, had declared himself a 'Morganist'. As later events would reveal, however, Prezent coupled a belief in the functional role of philosophy with a quite gobsmacking lack of personal integrity.

Prezent was from Toropets, a town 400 kilometres west of Moscow. He was one of those enthusiastic working-class students given an accelerated education under the Bolsheviks. His progress within the Party was impressive – from Communist Youth League secretary in 1920 to political commissar in the Red Army in 1929. He began his higher education in 1925 in Leningrad, teaching at Party schools while attending lectures in social science at the university.

By the time of his graduation, Prezent's notorious personal style was well established. He compensated for his lack of physical stature by wearing shoes with built-up heels and a tall green hat. He was mind-bogglingly petty. A self-appointed classroom vigilante, he once tried to get a young student expelled because she couldn't properly articulate Lenin's important thoughts about cognition in *Materialism and Empiriocriticism*. He denounced a Leningrad teacher for writing a happy poem about the May Day holiday because it demeaned that 'holiday of struggle'.[10]

This political activity left Prezent little time for difficult work, for which, in any case, he had no aptitude. Mathematically illiterate and given to gestures over applied effort, Prezent gravitated, naturally enough, towards biology. Mathematics had yet to make serious inroads into university-level biological science, and Prezent expended much effort over the years in keeping his field free of sums. 'It is impermissible', he asserted, 'to allow mathematics to usurp the content of biology . . . We are interested in concrete knowledge, not in algebraic symbols.'[11]

Biology was a natural field of endeavour for Prezent because it impinged so directly upon the political world, sometimes in the guise of agricultural policy, sometimes under the banner of eugenics, sometimes as the discipline that promised to unlock the nation's vast untapped natural riches. The questions biology was asking were urgent, politically charged, and not yet weighed down too much by hard data and complex model-making. In biology, then, Isaak Prezent found his ideal base of operations.[12]

At first, Prezent's tendency to champion whatever argument seemed the most quintessentially Marxist played to the advantage of Mendelian genetics. In the hands of Levit and Agol 'Mendelism' (chromosomal genetics) was a radical new idea, a truly materialist explanation of how life develops and evolves. At Leningrad University, in his widely attended seminar on the dialectics of nature, new graduate Prezent attacked Lamarckian ideas, insisting that genetics provided as good a demonstration as you could want of dialectical materialism in action.[13]

Prezent's partisan declarations helped genetics get a toe-hold in Russian scientific circles. But his arguments won him no personal advantage. Established biologists didn't quite know what to do with him. He sought their patronage, but when they set him to work he proved useless. Even Nikolai Vavilov gave him a job, but they fell out very quickly.

Prezent had better luck creating his own fortune. As president of the Leningrad branch of the Society of Materialist Biologists, he acquired notoriety for 'unmasking' Boris Raikov, a leading educator in biology, as an 'agent of the world bourgeoisie' and a wrecker. (This was in 1930, around the time the first arrests at Shakhty were announced: Raikov and his colleagues were speedily arrested.) In 1931 Prezent turned his seminar into a bona fide Department of Dialectics of Nature and Science – and it was around then that his own position towards genetics began to change.

In the absence of Agol and Levit (who were in Texas) genetics ceased to be considered a radical Marxist idea; indeed, people were beginning to look askance at its foreign provenance and cosmopolitan, globe-trotting community. To begin with, Prezent simply dropped the chromosomal argument altogether in favour of some very dull political correctness, compiling a 500-plus page *Reader on Evolutionary Theory*, nearly half of which was taken up with excerpts from works by Marx, Engels and Lenin.[14]

Had Isaak Prezent simply been self-serving, there would be little enough to say about him. What made Prezent so energetic, and therefore powerful – and therefore dangerous – was his sincere conviction that, by dovetailing his own interests to the interests of the Party, he was actually furthering philosophy and science. In an interview with the historian David Joravsky he was able to pinpoint the exact moment these happy sentiments were drawn to a point: in October 1931, at an All-Union Conference on Drought Control, the commissar of agriculture, Yakov Yakovlev, was discussing Lysenko and his work on 'vernalisation'. Vernalisation was a technique for increasing crop yields by manipulating the temperature of germinating seeds. Lysenko's technique, Yakovlev announced, had a revolutionary significance that far exceeded mere agriculture: it was the very model of a new, Soviet way of doing science.

In a pamphlet titled *Class Struggle on the Front of Natural Science*, Prezent dashed down the substance of the revelation granted him that day:

Only productive practice is the criterion of truth and the essence of concrete cognition. And this same practice, which produces an object in accordance with a postulated law, makes possible the production of this object on the *mass* scale necessary for society. And only such socioeconomic practical mastery is the true meaning of cognition.

(Joravsky, who is nobody's fool, nailed this well as 'the boss knows best'.)[15]

Prezent, committed to the wholesale acclimatisation of exotic species and creating a world in which 'all living nature will live, thrive, and die at none other than the will of man and according to his designs',[16] rallied Soviet biologists to become 'engineers' and 'inventors' in a thoroughgoing transformation of nature.

Objects of nature have ceased to be objects of contemplative study . . . Soviet faunists must become inventors. They must develop concrete projects for the planned transformation of animal communities and for their geographical redistribution. We must master fauna and not only make it work for us, but we must reconstruct it as well so as to enhance its productivity.[17]

Underlying this grand purpose was the belief that the 'gigantic fodder base' of the Soviet Union was going to waste because there were not enough game animals to create a rich ecology. One 'faunist' who heeded Prezent's call, Boris Fortunatov, claimed that nature was riddled with empty niches begging to be filled with exotics or even with newly created life forms; another, Peter Manteifel, insisted that the country's network of nature reserves – a rare and enlightened hangover from late tsarist days and strongly supported by Lenin – ought to become 'staging areas' for the rearrangement of nature: 'We must regard *zapovedniki* [reserves] and hunting grounds as production units, and not as institutions cut off from life,' he wrote. Fortunatov's 'General Plan for the Reconstruction of Economically Important Fauna of European Russia and the Ukraine' arranged for exotic species to be acclimatised across the European part of the USSR.[18]

The repurposing and eventual dismantling of the Soviet Union's system of nature reserves ought to have raised much more protest than it did. 'Morganist' geneticists and the entrepreneurs who ran their institutions would have run very little risk in criticising a man like Isaak Prezent. Even his political allies had him down for a blow-hard. Alas, their silence was complicit; they sensed extraordinary opportunities in Prezent's acclimatisation plans. Acclimatising species was a difficult, slow, but eminently possible task – and one that would show genetics to advantage. The task of creating successful hybrids could only increase the standing of genetics – or so the geneticists thought. At the end of 1931 Vavilov was among the inspectors sent by the Commissariat of Agriculture to the world-famous breeding station and reserve at Askania Nova. 'Askania', he declared, at the end of his visit, 'must be an Institute for acclimatisation and hybridisation as its basic profile.' By mid-1932, pathbreaking research at Askania Nova had been shut down, and its reserve converted into the All-Union Institute for Agricultural Hybridisation and Acclimatisation of Animals. Heading the institute, its 150 scientific workers, its staff of 2,000, and a budget of just under 5.5 million roubles, were Isaak Prezent and the man who had by now become his brother-in-arms: Trofim Denisovich Lysenko.

From 1934 on, all kinds of exotics were introduced to the Russian steppe. Sika deer. European bison. The racoon dog from

Manchuria was introduced to Baikal, the central Urals and parts of the Volga Basin. Grey mullet and Black Sea invertebrates were dropped as food stock into the Caspian Sea. Meanwhile the nation's nature reserves set about exterminating their wolves.

The result was disaster, of course. With no wolves to pick off the weaklings, the health of deer, moose and other ungulates collapsed, even as their populations skyrocketed. Their diseases passed to domestic cattle, causing a huge death toll. And feral dogs quickly filled the niche the wolves had left behind.[19]

10 : The primacy of practice

*People jump about, play at dialectical materialism
(a substitute for philosophical thought) but they do not
work. It is a great shame that Russian scholars have not
learned how to behave properly . . . Scientific work right
now is close to a catastrophe.*[1]
Vladimir Vernadsky to Alexander Fersman, March 1933

It was just before the Bolsheviks seized power in 1917, and they
were passing their leader around between them like a bomb with
a lit fuse. On a couple of occasions Vladimir Lenin took shelter
in the apartment belonging to Margarita Fofanova, a dedicated
Bolshevik who lived in Leningrad, on the north side of the Neva
River delta. From her bedroom, behind chintz curtains and surrounded by flowered wallpaper, Lenin scribbled urgent notes to
shape a coup that was now barely a fortnight away.

Reading was his only relaxation, and luckily for him, Fofanova
had a decent library.

One day, after dinner, Lenin went up to the bookshelves and said:
'I've dug up a remarkable book in your collection. Simply amazing.

A collection of wheat from the Bureau of Applied Botany's collection –
still the world's biggest and richest seed bank when this photograph was
taken in 1973.

And a very convenient size, too, you can slip it in your pocket. When we take power we shall definitely reprint it . . . Look what the author says in the first chapter: "In the midst of this sad predicament science came to his help, that sensible science of our advancing day, which has for its ultimate end not merely discovery, but application; which is not so delighted with the formulating of a new law as it is overjoyed at the lifting of a burden.""[2]

The book, translated into Russian in 1909, was *The New Earth* by the American biographer William S. Harwood. It was subtitled *A Recital of the Triumphs of Modern Agriculture in America*. And triumphs there certainly had been. Millions of hectares under maize had seen yield increases of up to a third, thanks to a new method of hybridisation.

Some cereals tend to self-fertilise. Wheat is an obvious example. It has sex with itself and the next generation of plants are clones of their parents. Maize, on the contrary, tends to out-cross. At around the turn of the century, agricultural scientists in the United States noticed something odd: if maize plants were forced to in-breed, it appeared to do them no harm whatsoever. This was unexpected, not least because of the all-too-well-recorded damage that inbreeding does to, say, people.

Further researches – of an educated trial-and-error variety – threw up a way of breeding maize that radically increases yields. In-breed two lines of maize, then cross them. Then cross the result with *another* hybrid of inbred lines. The plant that results is quite spectacularly fruitful. The effect falls off in the subsequent generation, so farmers who go down this route will have to buy fresh seed each year. This is difficult for small farms, but easy for bigger, more industrial concerns, because they have better lines of credit. They can borrow money against the next harvest to obtain the seed they need.

So it was that Harwood's book sowed in the minds of Lenin and his fellows the notion that agriculture would be more efficient if it was collectivised. And they weren't wrong: along with massive yield increases, American agriculture saw 17 million people leave agriculture altogether between 1910 and 1920. Collectivisation wasn't a Soviet invention. It was an American one.[3]

Harwood's book was influential in another way, too, and one that was absolutely disastrous. In his desire to cast modern science in a democratic and practical light – science for everyone, that everyone might understand – Ronald Harwood picked, for particular praise, the efforts of his friend, the plantsman Luther Burbank.

Burbank was famous almost as much for his lack of formal education as for his varieties of vegetables, nuts and berries. A talented self-publicist, he made many outrageous claims for his products. One of the plates in his lavish twelve-volume *Methods and Discoveries* (1914–15) illustrates a plum that, he claimed, dried itself while still on the tree, becoming an instant prune.

Burbank was the original barefoot scientist, a man whose practical plantsmanship outperformed by miles the slow, plodding work of academics in universities. He was the new type of citizen–scientist Marx had predicted, and that Stalin was determined should lead his industrial revolution – and he was American.

And at the beginning the Bolsheviks, in awe of the pace of American industrial growth – the cars, the planes, the factories – were only too happy to follow American examples. Margarita Fofanova later recalled how:

In 1918, as soon as the Soviet Government moved to Moscow, Lenin rang me up and asked whether I had with me Harwood's *The New Earth* and whether I could send it to him at the Kremlin. Lenin sent the book to Professor K. A. Timiryazev, requesting him to look it through and to write a preface for it. The book was to be prepared for printing urgently. Timiryazev changed the title of the book to *Regenerated Land*, wrote a preface, and the book was published early in 1919.

The industrialisation and collectivisation of agriculture began long before Stalin made it over into an aggressive (indeed, death-dealing) means of internal control. Ford tractors imported in their thousands from America were so popular that, according to the Russian-American correspondent Maurice Hindus, writing in 1927, 'more people in Russia have heard of Henry Ford than of Stalin . . . I visited villages far from railroads, where I talked to illiterate peasants who did not know who Stalin was or Rykov or Bukharin, but who had heard of the man who makes the "iron horses"'.[4]

At first some peasants crossed themselves devoutly when they saw one and 'spat out three times as they would on the appearance of the devil', but before long the Fordson tractor had become so popular that peasants were naming their children after it.

The modernisation of agriculture in drought- and famine-prone Russia could not come fast enough. Nikolai Koltsov once told the American geneticist Leslie Dunn

how he traveled to Leningrad during the famine of 1920 with Lenin and some other members of the Central Executive. Lenin was to urge upon the responsible committee the diversion of some of the funds set aside for famine relief to the construction of a research institute for seed selection and plant breeding. 'The famine to prevent,' said Lenin, 'is the next one and the time to begin is now.' He carried his point and there was built with emergency funds the great Institute of Applied Botany which under the direction of Nikolai Vavilov became the center of the greatest plant breeding and seed selection service in the world.[5]

Nikolai Vavilov was the natural choice to lead the Bureau of Applied Botany. His utopianism was hardly less strident than Lenin's own. Even as Lenin was preparing to seize power in Petrograd, Vavilov was lecturing to his class at the Saratov Agricultural Institute, outlining the task before them: 'the planned and rational utilisation of the plant resources of the terrestrial globe'.

He was loud, but he was not alone. Geneticists at the beginning of the twentieth century liked to think their intellectual break-throughs would translate relatively quickly into practical applications, rather in the way stem cell researchers in the early 2000s were hinting at overnight cures that remain elusive today. Vavilov, who had studied Darwin's papers in Cambridge during his years overseas, was impressed by Darwin's tracing of 'the evolution of the gooseberry from a wild one weighing half a gramme to one weighing 53 grammes, or the amazing range of variation in the pumpkin where a cultivated variety was a thousand times larger than the wild one'.[6] Genetics, he declared, would go yet further, giving the next generation the ability 'to sculpt organic forms at will . . . Biological synthesis is becoming as much a reality as chemical.'[7]

Vavilov was a man of action as well as words. In 1921 he was invited to attend a conference by the US Department of Agriculture. Famine had consumed all the seed stocks of the Volga region, and the rouble was worthless, so Vavilov arrived in the United States carrying solid gold roubles and ingots of platinum, bought over 6,000 packets of seeds from twenty-six different seed companies in less than a month, and persuaded Hoover's American Relief Administration to ship them – all two tonnes of them – to the Baltic port of Riga along with its food aid.

He also set up a branch of his bureau in New York City, and for five years Dmitry Borodin, a Russian agronomist and long-time US resident, kept Vavilov supplied with seeds and scientific literature.

Vavilov even found time to visit the laboratory of Thomas Morgan at Columbia University; it was his invitation to visit and lecture that brought Hermann Muller and his valuable *Drosophila* stocks to Russia the following year.

Vavilov returned to Russia in February 1922 and, at only thirty-five, was elected a corresponding member of the Academy of Sciences. As director of the Bureau of Applied Botany, Vavilov held his research council meetings in the Kremlin. Newspapers followed his travels.

His workaholism was legendary. A constant stream of visitors and public duties awaited him at the bureau's offices, often long after closing time. If there were still people to see after 10 p.m., he would invite them to his house, a ten-minute walk away. In the drawing-room, at a long table, he laid on tea, pastries and cakes, and wine (though he himself was teetotal). Past midnight, having seen his visitors to the door, Vavilov sat down to work some more, rarely getting to bed before two or three o'clock in the morning. If you asked him about his impossible schedule, he invariably replied 'Life is short, my dear, one has to make haste!'[8]

'Spread the Achievements of Science Among the Masses!' So read the banner draped along the front of the hall.

The All-Union Congress of Genetics, Plant Breeding, Seed Production and the Raising of Pedigreed Livestock took place in Leningrad in January 1929: a gala event which launched the agricultural goals of Stalin's First Five-Year Plan. About 300 papers were given at the event; overseas visitors included Richard Goldschmidt and Erwin Bauer from Germany, Harry Federley from Finland, and many other illustrious names.

Nikolai Vavilov, the event's chairman and chief architect, greeted his government patrons warmly. Nikolai Gorbunov, then commissar of agriculture, was his boss. Sergei Kirov was one of Stalin's right-hand men. All three confidently predicted that science could achieve the goal set by the First Five-Year Plan: an increase in grain yields per hectare of 35 per cent.

The United States had achieved as much; why not the Soviet Union? Even Nikolai Koltsov, head of the Institute of Experimental Biology and never the Bolsheviks' easiest customer, wrote enthusiastic articles about the good times the collectivisation of agriculture would usher in. In terms shockingly similar to those used by Isaak Prezent, he described collectivisation as a bold national experiment in practical science, enriching both academe and agriculture. Koltsov's Party-minded colleague Alexander Serebrovsky, meanwhile, actually helped plan the collectivisation campaign.

And at the Sixteenth Party Conference in 1930, when the Party cast out its right-wing critics, Vavilov was there to help, endorsing the agrarian offensive: 'Enormous energy is needed to get our enormous country moving, the whole peasant mass of it, to rouse to wilful action all the creative forces of our country.'[9]

It is daunting, with hindsight, to find so many great figures of Soviet biology queuing up to endorse collectivisation – a movement that starved millions to death and was used quite deliberately as a weapon to obliterate an entire class of moderately well-off peasant, decimate the Ukraine, Russia's troublesome satellite, and subjugate the Soviet countryside.

Though the campaign's monstrous nature could not have been predicted, you would think its sheer impracticality would have set alarm bells ringing. But enthusiasm is catching. Vavilov and his colleagues, far from blowing the whistle on the idea, demanded more funding and easier foreign travel to speed the project along. (The money, at least, proved forthcoming.) Vavilov also oversaw the loose amalgamation of numerous institutes whose administration he had been juggling alongside work at his own bureau. The Lenin All-Union Academy of Agricultural Sciences (VASKhNIL) was to be, in Vavilov's own words, 'the academy of the general staff of the agricultural revolution' – nothing less, then, than the scientific and intellectual arm of the Commissariat of Agriculture.[10] Made a full member of this academy, and given administrative control of its 111 research institutes and 300 experimental stations, Vavilov was far too busy to ask awkward questions of a campaign that gave Russian agriculture all that reformers had dreamt of and campaigned for since the end of the eighteenth century.

Vavilov was first and foremost a bureaucrat. Yes, he was a talented scientist, of that there is no doubt, but his was a culture where dominance over a field comes with the obligation to administer it. Vavilov's was a bureau of *applied* botany, and quite different in its workings from other scientific institutes. Scholars never enjoyed much autonomy there, the way they did under, say, Koltsov, at the Institute of Experimental Biology. The Commissariat of Agriculture kept Vavilov's bureau under tight control. Its presidium was made up of administrators, not scholars, and its minutes are a catalogue of administrative matters, not scientific ones. It was perfectly usual for official meetings to involve ministry officials alone, with not a scientist in sight.

The tug of war between his scientific concerns and his administrative responsibilities was something Vavilov felt keenly, and he was not particularly stoical about it. In January 1930 he wrote to Georgy Karpechenko, who was studying alongside Theodosius Dobzhansky with Thomas Morgan's fruit fly team in California:

I am completely squeezed here. On top of all the work [at the Lenin Academy] with its tens of institutes, I suddenly turned out to be my own supervisor, as I was made a voting member of the Collegium of the USSR Commissariat for Land. We want to build Washington, nothing less. They told me, 'We will let you go in a year.' All in all, I have eighteen posts. My cranium will soon explode from these layers of rubbish on all sides.[11]

The huge expansion of the Lenin Academy of Agricultural Sciences at the end of the 1920s, from paper plans to a federation of real research centres, was predicated on the crazy belief that the peasants, once collectivised, would be able to harness the latest scientific techniques overnight. But the academy's promotion of Vavilov's grand project – to gather, organise and exploit a world-spanning collection of useful plants – could only widen the gulf between scientists and farmers.

Agrarian revolutions cannot be won with seed alone. On the average peasant farm, infested with weeds, rarely fertilised, and lacking a seed drill, improved seed was a complete waste of money, and the peasant farmer was quite right to spurn the government's improved seed in favour of the tried and tested, locally adapted mongrel seed he had been using year in, year out.

The answer was to turn ordinary farmers themselves into scientists, and to shape collectivisation around what actually worked in real fields. Collectivisation was the first great experiment in 'citizen science' the world had ever seen.

No one took citizen science more seriously, or had a more sincere belief in its potential to transform science, than Yakov Yakovlev, editor, since 1923, of *Bednota* (*Peasants' Gazette*), the state's mouthpiece to the vast majority of Russia's population and the only news organ most Soviet farmers ever saw. To drive modernisation forward, the *Gazette* organised an army of peasant scientists in 'hut labs', claiming over 23,000 participants by 1929.

A *Time* article from 1935 neatly captured Yakovlev's importance: 'Imagine sombrero-wearing William Randolph Hearst editing with Communist zeal the *Great American Farm Newspaper* and you begin to have some faint idea of Comrade Yakov Arkadevich Yakovlev.'[12] Yakovlev devoted his efforts as editor-in-chief to education, publishing articles on weed control, the application of manure, the introduction of clover, sprouting potatoes before planting, and other solid, practical advice. The *Peasants' Gazette* also promoted the sale of improved varieties in the hope that, as modernisation set in, these varieties would turn out to be worth the money.

The *Gazette* was anything but dry. It was a campaigning paper, and wanted its readers' active participation. Most of its pages were given over to self-promotional articles penned by the readership. Every crank plantsman in the country wrote in to the *Gazette* to share his 'sure-fire' agricultural technique. The contribution of these articles to the modernisation campaign was marginal; what they did contribute was a creeping hysteria, just waiting to be put in harness.

As a boy, Trofim Denisovich Lysenko was bright and industrious. His father even allowed him to spend an extra couple of years at the village school. By the age of thirteen he could read and write. Still, being the son of a peasant, Lysenko would not have amounted to anything very much had it not been for the revolution.

This talented youngster made the very most of the new opportunities given him. In 1917 he entered the Uman Horticultural School. He studied there for four years, while the Civil War raged about him. (Uman changed hands several times between the Whites, the Greens and the Reds.) His higher education came via correspondence courses from the Kiev Agricultural Institute, while he worked at a small plant-breeding station in the town of Belaya Tserkov in the Ukraine. In 1925 he graduated and left for Gandzja (now Kirovabad) in the Caucasus (now Azerbaijan). There, at a small experimental station, he was given the job of acclimatising beans as a green-manure crop. The results of his labours were promising. Practical, working-class, ambitious and working for the common good: was he not the very model of the new Soviet scientist?

Pravda thought so. Hungry for agricultural good-news stories – and arriving before Lysenko's experiments could be repeated and the results confirmed – the journalist Vitaly Fedorovich arrived in Gandzja to profile this *Wünderkind* of beans.

At twenty-nine, with no postgraduate training or higher degree and no formal claim to the title of scientist, Lysenko explained to Fedorovich how he solved practical problems by a few calculations 'on a little old piece of paper'. The encounter was not a particularly easy one:

If one is to judge a man by first impression, Lysenko gives one the feeling of a toothache; God give him health, he has a dejected mien. Stingy of words and insignificant of face is he; all one remembers is his sullen look creeping along the earth as if, at the very least, he were ready to do someone in.[13]

Still, Fedorovich was impressed. 'The Fields in Winter', published in *Pravda* on 7 August 1927, cast Lysenko as a 'barefoot scientist', who 'holds a plough with one hand, a flask with the other'. Rather than studying 'the hairy legs of flies', this sober young man 'went to the root of things.'

This article not only turned Lysenko's head; it derailed his scientific career. His experiments in acclimatising beans to new conditions were genuinely interesting and he should have persevered with them. He should have repeated them. The hype which engulfed him, however, took his mind off method. He was still at the stage of his education when science seems easy – a few calculations on a piece of paper. And all of a sudden, important organs of the state, bureaucrats and pundits were rushing to agree with his findings. Vavilov himself took an interest, dispatching a colleague to look into Lysenko's work. Lysenko was very excited, and even gave up his bed to his visitor and slept on the floor.

The report Vavilov got back was uncannily prescient: Lysenko was 'an experimenter who was fearless and undoubtedly talented, but he was also an uneducated and extremely egotistical person, deeming himself to be a new Messiah of biological science'.[14]

Lysenko's work on acclimatising plants to new conditions focused on the calendar. He studied how different varieties of the same crop responded to being planted at different times of the year.

Wheat was of particular interest. Normally, winter wheat gets established as the weather grows colder, then waits till early spring to produce its grain. There's the risk that a harsh winter can kill it. If you plant winter wheat in the spring, though, it probably won't ear (develop grain-producing spikes) at all. This is because it is waiting for a signal – a period of winter cold – that has already passed.

Lysenko studied the way a spell of artificially induced cold can fool a winter wheat variety to ear. This process is called vernalisation. By moistening and chilling winter wheat, the seed experiences a tiny amount of germination and growth – just enough to allow it to detect the chilly conditions in which it is stored. The seed will then do *all* its growing – earing and all – in a frantic burst once planted out in early spring.

The question is: why bother? What is the economic value of turning winter wheat into spring wheat? True, you avoid the risk of having your crop killed by a severe winter, and you get an early crop – but are these real advantages, or just paper ones? The Ohio State Board of Agriculture reported experiments in vernalisation as early as 1857, but could find no economic value in the practice: it was a scientific curiosity, nothing more.

Lysenko's 'Monograph No. 3 of the Azerbaijan Experiment Station: The influence of temperature on the length of the developmental period of plants' made no bid for originality. It was work in a well-established research tradition. But Lysenko did admit to an ambition: to evolve a precise, comprehensive theory to explain how plants adapted to the annual temperature cycle. For all the work that had already been done, no one really knew why plants responded to temperature in such a regular and predictable way.

Vavilov's institute was already devoting serious effort to the problem. Its resident vernalisation expert, Nikolai Alexandrovich Maksimov, had been working away since 1923, and courteously but firmly set Lysenko straight about the weaknesses in his paper. Lysenko's statistics were a mess, and further reading would have saved him the bother of reinventing a lot of ideas that were already in the literature.[15]

Maksimov considered the paper too weak to include at the All-Union Congress planned for January 1929, but Vavilov disagreed. Lysenko's work, for all its weaknesses, was original and promising, and delivering a paper would give him useful experience.

Come January, then, Lysenko presented his paper, and Maksimov, sitting through Lysenko's errors a second time, lost some of his courtesy. Lysenko seemed to think that plants came with little clocks inside them, when all the evidence suggested it was temperature, not time, that dictated how plants organised their development. As for his actual results, they did little more than confirm the results of the German researcher Gustav Gassner. 'The results obtained by Comrade Lysenko do not represent anything new in principle, [and] are not a scientific discovery in the precise sense of the word.'[16]

Maksimov's own presentation was much more convincing. One of the more stirring headlines to come out of the Congress – 'It is possible to transform winter into spring cereals: an achievement of Soviet science' – was referring to Maksimov, not Lysenko.

If Lysenko did not particularly shine at the congress, he had little to complain about. He had made good on *Pravda*'s hype by presenting his first real paper at a national congress, and the leaders in his field were taking a critical interest in his work. Only a man of Lysenko's gloomy stripe would interpret this modest success as a snub.

But this is what Lysenko did. He never touched statistics again, relying ever afterwards on crude theories 'proved' by arbitrary examples. And he avoided the science press; virtually everything he wrote after that was published in newspapers or in journals created for him by the government.

We do not know whether this extreme reaction of his had anything to do with Isaak Prezent. We do know that it was at this congress that this pair of mathematical illiterates met for the first time, and the most destructive partnership in the history of science was born.

*

The 1929 Congress could not have been more timely. Ukraine was suffering, for the second year in a row, a calamitous failure of its winter wheat harvest. About 7 million hectares of winter wheat had perished – 90 per cent of the entire crop.

The story goes that Lysenko's father – faced, like all his fellow farmers, with ruination – took a risk and tried to vernalise his own seed. (He did this in secret, afraid that his neighbours would sneer at him – a telling detail of peasant life, and one which lends the story some credence.) Lysenko's father soaked forty-eight kilos of Ukrainka, a winter wheat, in water and buried the moist seed in a snowbank to keep it cold until the spring. In the spring, he planted it alongside a field of spring wheat – and got a better yield from the winter wheat.

That, anyway, was the claim. A commission sent by the Ukrainian Commissariat of Agriculture investigated, and became markedly enthusiastic, ordering large-scale production tests of vernalisation. And after two terrible winters, and with Bolshevik leaders breathing down their necks demanding grain to feed their monstrous industrialisation drive, the commissariat went further. Even before their tests began, the commissars were telling the papers that a solution to the problem of winter crop destruction had been found.

One season's success on half a hectare made Lysenko a hero. Even Maksimov – as hungry as anyone for a quick fix to the famine problem – came on-side. On 1 September 1929 he attended a lecture Lysenko gave at the Bureau of Applied Botany in Leningrad, and was much more positive than he had been at the big congress in January. 'Both Gassner and I,' he wrote later, 'being plant physiologists, went no further than laboratory experiments. Cold germination appeared to us to be too complicated for direct application in field farming.' Lysenko had simplified the method to the point where ordinary peasant farmers could use it. It was 'certainly impossible not to acknowledge this as a great achievement'.[17]

The trouble with Lysenko, however – you might even say his tragedy – was that he could never see when he had won. Insisting that his theory of 'winterness' was quite different from Gassner's and Maksimov's theory of 'cold germination', he contrived to take offence at Maksimov's welcoming words. He did not want to be part of Maksimov's scientific community, or any other scientific community. He did not want his work considered as part of an honourable tradition. He wanted credit, all the time, for everything he touched.

In October 1929, Lysenko started work at the All-Union Institute of Plant Breeding in Odessa, the most important centre of agricultural research in the Ukraine. The following month Yakov Yakovlev, whose newspaper had made much of Lysenko's vernalisation, replaced Gorbunov as commissar for agriculture. No reader of the *Peasants' Gazette* would have been surprised at what followed. Following a glowing field report by Yakovlev himself, mass national trials of the vernalisation of wheat were ordered for 1932. Lysenko was given his own journal, the *Bulletin of Vernalisation*; the opportunity to construct his own courses; and 150,000 roubles. Vavilov's Bureau of Applied Botany was ordered to cooperate with the mass vernalisation scheme. Yakovlev himself told Vavilov to offer Lysenko 'every assistance' and 'personally look after him'.

Vavilov responded smartly. To his vice-director he wrote: 'Lysenko's work is remarkable and forces us to take a different view of many things. It is necessary that the World Collection is worked through with vernalisation.' His enthusiasm was genuine. Vavilov needed vernalisation to work. He had spent years abroad, collecting the world's largest collection of cultivated plants. He had risked life and limb in the wildest and most remote regions, beginning in Afghanistan in June 1924, where he was shot at, arrested, and even threatened with execution. In the years since he had travelled through the war-torn Middle East, through Spain (whose exhausted spies brokered a 'gentlemen's agreement' with him that let them stay in bed while he went climbing), and most recently to Japan, where he marvelled at 'numerous kinds of bamboo, edible in various forms; Chinese yams; and an enormous variety of radishes, turnips, roots, mustards, edible Japanese burdock, water chestnuts, lotus, arrowhead and edible bulbs of lilies . . . and peculiar vegetables such as "udo", rhubarb, perennial Chinese "tsyo-tsai" onions, "ou sen" or asparagus lettuce, peculiar white eggplants, colossal cucumbers, edible luffa, edible "miso" chrysanthemums, tuberous asparagus and so on.'[18] But having assembled a collection of such useful plants, Vavilov – not himself an expert plant breeder – now confronted the task of turning them to practical use.

Most cultivated plants come from mild latitudes. As much as nine-tenths of Vavilov's collection would have struggled to flower under Russian conditions. Many would not germinate at all, and crossing them with local varieties often proved impossible. Only through vernalisation could Vavilov hope to adapt his collection quickly to Russia's cold, dry, unpredictable climate.

This dream was a vain one. In a very few years, it became clear that vernalisation was not the panacea Lysenko and Yakovlev had promised. But this conclusion took time to reach. It took time to investigate, repeat and measure every positive result. It takes time, always, in even the most sceptical and measured study, to demonstrate the absence of an effect. Vavilov was entitled to hope for positive results. He even began to call himself an 'agronomist', distinguishing himself from laboratory scientists like Maksimov. In October 1931, at a meeting of biologists in the Communist Academy, he twitted Maksimov for underestimating Lysenko:

Lysenko not only developed Gassner, he went much further, he took the most different objects, he found that not only lowering of temperature, but also many other factors can be used to speed up development. His approach is very serious, and very new. We agronomists feel that a real revolution has started . . .[19]

11 : Kooperatorka

Lysenko is a careful and highly talented researcher. His experiments are irreproachable.[1]
  Nikolai Vavilov to the Lenin Academy, 17 June 1935

Perhaps it was only his perseverance, his extraordinary thirst for knowledge and his undeviating pursuit of the road he had chosen that distinguished [Lysenko] from the rest. And one other very characteristic feature: for him, knowledge was something that was immediately put into practice . . . The fact that, having arrived in Ganja in the autumn, he did not wait until the spring to commence work on his legumes, already revealed the 'Lysenko style'.[2]
  Vadim Andreevich Safonov, *Land in Bloom*, 1951

The talented Soviet geneticist Theodosius Dobzhansky had come a long way since the Civil War, when he had smuggled food parcels out of Kiev for Vladimir Vernadsky. In March 1927 he had applied for a Rockefeller Foundation fellowship to spend a year at Thomas Morgan's lab. The following autumn he had followed

Children in Donetsk dig potatoes out of the frozen ground for transportation elsewhere. A decree in August 1932 forbade peasants from eating their own crops.

Morgan to Caltech, extending his fellowship for another year. Dobzhansky adored America, He used to cross the country by car each summer.[3] In October 1930, over the course of a walk in California's Sequoia National Forest, Nikolai Vavilov did his best to tempt Dobzhansky back to the Soviet Union, but Dobzhansky had little desire to return home. There was no lab space for him in Leningrad, and Vavilov was quite open about him having to lecture and write textbooks if he hoped to make ends meet.

These privations were not the sticking point. Quite simply, Dobzhansky did not trust the Soviet government. He feared for the future of his country, and could not imagine his own future within its borders.

'We have to ignore, we have to leave out of consideration, the political matters with which we do not agree,' Vavilov conceded.[4] But even as he tried to argue Dobzhansky out of his position, the Soviet authorities were undercutting him. Vavilov's ambitions for a US-style extension service for his Bureau of Applied Botany were falling apart through lack of funds and trained staff. While he was away, the New York office had its funds cut and Borodin was dismissed, and it was only with the greatest difficulty that Vavilov got the operation restarted. Then, on 11 October, a telegram recalled him to a high-level political meeting.[5] Vavilov's confident, patrician response made an impression on Homer Leroy Shantz, president of the University of Arizona and a friend and fellow explorer, who set it down:

But I am employed by the Communists to work for the welfare of the people of the USSR, so I am still free to judge what is best . . . it is more important for the future of the people of the USSR that I visit the centres of origin of cultivated plants in Central America than I attend any state dinner that can be arranged.[6]

Vavilov was in a vulnerable position – a bourgeois absentee collecting plants of no immediate usefulness at a time of national famine. Even Shantz, having listened to him lecture, detected 'a wonderful capacity to confuse plan with accomplishment, and to conclude that what is decreed is already accomplished.'[7] It would take a determined enemy to brand as trivial Vavilov's breakneck schedule and devotion to duty. But Vavilov, like any major patron, attracted as many disgruntled enemies as he did friends and suitors – and one somewhat inept colleague, Alexander Kol, was to prove a particularly nasty opponent.

Kol was a seed curator, charged with coordinating the bureau's plant introduction programme. He was ten years older than Vavilov, and jealous of the younger man's reputation. Worse, he was not very good at his job. Vavilov had had to reprimand him repeatedly for losing or mislabelling the fruits of his expeditions. Finally, Kol had been demoted.

With a nice sense of timing, Kol waited till Vavilov was abroad before launching his attack. 'Applied Botany, or, Lenin's Renovation of the Earth' appeared on 29 January 1931 in the influential newspaper *Economic Life*. Not all its criticisms were baseless. Kol had worked briefly in the United States, where agricultural scientists tended to focus on just a few cultivated plants, especially wheat and maize. Vavilov's was a more comprehensive approach, and Kol took Vavilov to task for this 'separation from practice'. Why collect and maintain exotica that were useless to the breeder?

The overall tone of the article was something else again. It was nothing short of a political denunciation, branding Vavilov an enemy of the working class: 'Under cover of Lenin's name a thoroughly reactionary institution, having no relation to Lenin's thoughts or intents, but rather alien in class and inimical to them, has become established and is gaining a monopoly in our agricultural science.'

Vavilov did not take this accusation lying down. Replying through the same magazine, he made public the catalogue of Kol's ineptitude and spite. Remarkably, he made no effort to examine or criticise his own actions. This was risky: it was expected that bureaucrats regard any attack, however vexatious, as a valuable learning experience. In fact this was a point of honour among Bolsheviks, and was supposed to preserve the democratic spirit in a one-party state. Vavilov's commitment to the state was genuine. The trouble was, the more harried and overworked he became, the more he let his patrician side show.

On 3 August 1931, Vavilov's impossibly heavy work schedule collided with Yakovlev's hysterical agricultural planning. This was the day *Pravda* published a truly ridiculous Party and government resolution 'On Plant Breeding and Seed Production', in which the government posed the Lenin Academy and the Bureau of Applied Botany the following superhuman targets: for wheat and a few other crops, the Commissariat of Agriculture and the Lenin Academy would complete the transition from the usual peasant mongrel seed to certified varieties within two years. Wheat was to be improved so it could replace rye in the north and east. Potatoes were to be improved so diseases wouldn't ravage the crop when planted in regions with dry summers. And the capstone: to achieve all this, breeding times for improved varieties would be reduced from ten or twelve years, to four.

The majority of geneticists present regarded these totally unrealisable goals as ludicrous, and the following month, a conference was held at the Lenin Academy to discuss the decree. The plant breeder Georgy Meister spoke for many when he warned that the demands of the decree were impossible to fulfil. To produce a new variety in four or five years was not practically possible. 'We cannot wait for ten years,' was Yakovlev's terse reply.[8]

That being the case, several speakers suggested the state gear itself up to create a vast and expensive system of greenhouses in order to insulate experimental crops from the vagaries of the weather.

Yakovlev considered this as special pleading: collective farms could provide all the controlled conditions the scientific community could possibly want. All they had to do was descend from their ivory towers and muck in with the collectivists.

Again and again, Yakovlev approached Trofim Lysenko, encouraging him to lead the campaign for accelerated plant breeding. At the All-Union Conference on Drought Control in October 1931, he told Lysenko off for underestimating the changes his experiments would bring about in agriculture. A resolution was passed declaring that Lysenko's research was so important that his works must be rushed into print, that he should be freed from other duties and given 'maximally favourable research conditions'. Even Lysenko's inflamed ego balked at such a crazy task (even today, new varieties take around a dozen years to enter production) and, for some years afterwards, Yakovlev's poster boy remained carefully silent on the subject.

Nikolai Vavilov was careful to invite Trofim Lysenko to an upcoming congress in America:

In August there will be an International Congress of Geneticists and Breeders in Ithaca, USA, and the Commissioner has informed me that if you were willing to attend, the Agricultural Commissariat would apply all efforts to support your trip, so that you could make a presentation of your work there and prepare a display of your activities for the exhibition.[9]

But whether or not Lysenko actually wanted to go, the trip proved impossible to arrange: in December 1931, the Politburo's Departure Commission recommended that all foreign travels of Soviet scientists be cut back in order to save hard currency for industrialisation.

Instead Vavilov – the country's only high-level attendee – praised the absent Lysenko to the skies, saying his techniques would make it possible to grow alligator pears and bananas in New York and lemons in New England. News of Lysenko's 'discovery' made quite a splash in the American papers, who predicted the 'growing of subtropical wheat in cold climate' and the use of vernalisation as a 'weapon against drought'. Several US geneticists tried to repeat Lysenko's experiments.[10]

That Vavilov was able to attend the congress at all was thanks to some risky bureaucratic manoeuvring – inserting a side-trip to Ithaca as part of yet another collecting expedition to North and South America. From his cabin aboard the steamship *Europa*, about to sail from Riga, Vavilov wrote to his partner Elena: 'I want to bring an immense amount of seed from America this time because it's unlikely I'll go again.' Vavilov's itinerary was frantic. In less than eight months he visited Cuba, Trinidad, El Salvador, Costa Rica, Honduras, Panama, Columbia, Suriname, Brazil, Venezuela, Peru, Bolivia, Argentina, Uruguay and Chile. His schedule was so tight that he only ever slept while travelling. A fellow geneticist, Carlos Offerman, once watched him fall asleep during a plane ride through a lightning storm over the rainforests of the Brazil–Suriname border. The other passengers spent the journey screaming their heads off.[11]

At Cornell University in Ithaca, Vavilov's positive stories and indefatigable energy cheered along a conference convened under more than one bad star.

For one thing, America was already in the grip of the deepest and longest-lasting economic downturn in the history of the Western industrialised world. By the time the Great Depression reached its nadir, some 13–15 million Americans were left unemployed and nearly half of the country's banks had failed.

There was a gathering environmental crisis, too. During the drought of the 1930s, the soil of the US and Canadian prairies, loosened by mechanised agriculture, turned to dust. Prevailing winds blew 40 million hectares of topsoil into huge clouds that sometimes blackened the sky.

Innumerable crises much closer to home affected the congress. The conference president, Thomas Morgan, was still convalescing from an appalling car crash. Almost every delegate was poor, and there were no grants to pay for travel or accommodation. Those coming by car were told that 'there are several very attractive camping places within thirty minutes' drive ...'[12]

The trick, among these privations and annoyances, was to lose yourself in the work of the Congress. 1932 was a gala year for the new science of genetics. Timofeev-Ressovsky, visiting from Berlin, caught the mood nicely: 'We geneticists are in a very happy condition: our science is young, its "development curve" is rising rapidly and the future will bring us the most interesting facts and views concerning the gene problem.' Living organisms, charts, photographs, and hundreds of microscopes filled the rooms of the congress, while a 'living chromosome map' of mutant maize plants, planted out in positions corresponding to the genetic locations of their mutations, proved a hugely popular exhibit. As Vavilov put it, 'Genetics has, so to speak, "broken down" the species of maize into "building blocks" and presented an absolutely unusual aspect to the botanist.'

Unless you counted Timofeev-Ressovsky, visiting from the Vogts' Institute of Brain Research in Berlin, Vavilov was the only high-level Soviet geneticist at the congress. He did not come alone. With him was Vladimir Saenko, the head of the agricultural section of the Soviet trade agency in the United States. Saenko did not have much to say for himself, and attendees got the distinct impression that he was there first and foremost to keep an eye on Vavilov. Leslie Dunn, a Columbia geneticist and social campaigner, managed to invite Vavilov to his house in Riverdale for dinner. There, Dunn, the British geneticist John Haldane and three other interested guests tried to get out of their Russian colleague the reason for Saenko's presence, and why no one else from the Soviet Union had been allowed to attend. But as Dunn later remembered, Vavilov 'never took down his hair entirely'.[13]

While he was in Ithaca, Vavilov truly confided in only one man – Theodosius Dobzhansky, the geneticist he had failed to bring back to the Soviet Union two years before. Dobzhansky and

Vavilov were unable to speak privately until they turned up for lunch in a crowded cafeteria and found only two seats available at a table; Saenko had to find a place elsewhere. They conversed in Russian, which no one around them understood, but Vavilov was still careful about what he said.

He told Dobzhansky, in so many words, that the Soviet Union had changed a great deal since they had last spoken, and that his own situation was not what it had been. 'Dobzhansky, do what you want. If you want to return, do so. If you do not want to return, don't. Stay here.'[14]

Living on the very edge of starvation, working to a plan that starved no one and enriched no one, Russia's peasant class had learned over the years to count its pennies. Peasants borrowed with great caution, lent reluctantly and at high rates of interest, and everyone lived in fear of debt. They borrowed from each other sooner than from outsiders, and every village had its moneylender. A Ministry of Finance study from 1894 found that 'the power of moneylenders is founded precisely on the fact that except for them poor people frequently have nowhere else to turn'.[15] Without the village grandees who issued a debt for each little article, from salt, matches and kerosene, to luxury goods like clothing, tea, and sugar, it was hard to see how a peasant could live even a month outside harvest time.

Though in the worst cases a bad debt could land a poor peasant in debt slavery, relations between moneylenders and their clients were reasonably good. Before 1917 peasants hardly ever used the pejorative *kulak* ('fist') to describe their fellow villagers. There was, in any case, very little wealth for anyone to argue over. In good times, rich and poor peasants lived almost indistinguishably; the differences between them were only noticeable when times grew tough.

Just a year before the 1932–3 famine, a former blacksmith bemoaned the Bolsheviks' troublemaking:

There was a time when we were just neighbours in this village. We quarrelled, we fooled, we sometimes cheated one another. But we were neighbours. Now we are bedniaks, seredniaks, koolaks. I am a seredniak – a middle peasant. Boris here is a bedniak – a poor peasant. And Nisko is a koolak – a rich peasant. And we are supposed to have a class war – pull each other's hair or tickle each other on the toes, eh? One against the other, you understand? What the devil.[16]

The First Five-Year Plan that had expanded industry, had also swelled the urban population. How were all these new city-dwellers to be fed? The Plan had been put into action without much capital behind it, and to pay for it the state had been deliberately depressing the market value of grain. In January 1928 Stalin visited Siberia and found that despite a good harvest, peasants were withholding grain as prices fell.[17]

Stalin's solution was draconian. Peasants who could afford to do so had been refusing to sell their grain at the ruinously low prices the government was offering; well then, it was their well-to-do-ness that was at fault. These wealthy peasants, these money-lending *kulaks* who ran things in the villages, were enemies of the state, and Stalin declared war on them. Once all the little power bases of the countryside were got rid of, then the cities would no longer be held to ransom by the peasants' refusal to trade, and factory workers would no longer go hungry.

The famine, when it came, naturally and inevitably worsened relationships between poor and rich peasants in the villages, and this provided the cover the government needed. Armed detachments were sent to areas where the secret police anticipated the greatest resistance to their ruthless campaign to remove community leaders from their villages. Hundreds of thousands were rounded up in all parts of the country and dispatched in boxcars to remote and inhospitable regions. Few ever returned. Those who resisted were executed or sent to concentration camps. By eliminating the big men in each rural community, the Soviets robbed the countryside at a single stroke of its organs of resistance and protest. They effectively regimented the countryside, along what they imagined were industrial lines, but which to an outside eye resembled nothing so much as the notorious rural work camps of the lampooned Count Alexei Arakcheev. And it was in these camps – these supposed hotbeds of citizen science – that the assessment of vernalisation was to take place.

The grand experiment began poorly. The winter of 1930–1 was too mild, so there was not enough snow to chill the grain, which therefore tended to sprout or rot. The first issue of Lysenko's journal admitted as much. Still, practical trials were expanded in 1932, to 43,000 hectares of vernalised wheat.

At a conference organised in February 1932, the Bureau of Applied Botany declared Lysenko's work a great success. Though Vavilov missed Lysenko's lecture, he was still full of praise: 'We are in fact approaching the art of transforming the plant according to our will,' he said; Lysenko's results would provoke 'serious revisions of some fundamental theses in genetics'.

It is an intriguing claim. As this book is being written, the study of epigenetics – the mechanisms by which the environment regulates gene expression – is developing hand-over-fist. There indeed are, in a growing number of cases, examples of living things inheriting characteristics acquired by their parents or grandparents. No geneticist of Vavilov's generation believed the environment had *no* effect on gene expression. Perhaps Vavilov thought that Lysenko had uncovered mechanisms of this sort – regulatory systems that would enrich and complicate the picture of genetics developed by Morgan and his team at Columbia.[18] If so, Vavilov had some funny ideas about how such a visionary understanding would be attained. Lysenko had no desire to enrich genetics. Genetics belonged to other men. Lysenko wanted, above all else, to be an original.

Prezent was the next speaker. Maksimov's admission that his work was of little practical use spoke volumes about the life sciences in general, Prezent said. He mentioned two recent books, one by Yuri Filipchenko, the other by embryologist Mikhail Zavadovsky. Compared to such people Lysenko had published next to nothing, but there was no doubt who had produced the most 'works'. Applause followed.

Conditions for vernalisation improved over the coming years, but there turned out to be serious practical problems with the technique. These were itemised by Tatiana Krasnoselska-Maksimova, Maksimov's wife and collaborator. The overarching problem was the amount of labour involved in vernalising an entire crop. Seeds had to be soaked in a shed for several days. You didn't just dump them in a pile and water them. You had to spread them out, on the floor or in trays, and turn them over constantly. The work involved was herculean. Keeping the seeds in uniform condition – not too hot, not too cold, not too wet, not too dry – over long periods of time was often impossible. In many places, there were no refrigerators. In many places, there was no electricity.

Suppose the seeds survived the soaking process and did not germinate too soon (the cause of many a crop failure); this was but the beginning. The damp conditions in which they had been kept were ideal for the spread of fungi and diseases. And the ever-decreasing amount of healthy seed represented one final, back-breaking hurdle for the keen vernaliser. The seeds were damp. They were swollen. They were *heavy*: costs of sowing were doubled because the seeder had to go over the fields twice to ensure the usual amount of grain was planted.

In fact, so many things could go wrong with vernalisation, the practice itself evaded criticism. If your crop failed, what were you going to blame: the inadequacy of your equipment, the failure of your workers – or a state-sanctioned agricultural panacea with the personal blessing of Stalin?

Commissar of Agriculture Yakov Yakovlev sincerely believed that collectivist 'hut labs' could replace expensive institutional science projects and develop new varieties in a third of the usual time. But he was a journalist, not a scientist, and he had no conception at all of how hard science is to do. In particular, it does not seem to have occurred to him that scientific honesty is difficult to achieve. In 1932, questionnaires were sent to thousands of collective farms. The chairmen of these farms were not held responsible for the accuracy of their reports, and they dashed off these questionnaires the way they dashed off so many other pieces of paper. 'Anti-vernalisers' were equivalent to *kulaks*, so no one in their right mind would report a *failure* of vernalisation. Consequently, Yakovlev's office was deluged with sensational communications about the millions of kilos of grain brought forth by vernalisation.

The Agriculture Commissariat had created a self-fulfilling success story. The more good news it got, the more it invested in vernalisation, the more glowing reports it received. Before long, the expansion of vernalisation was itself used as an argument for its further expansion. If so many people and resources were already involved in applying Lysenko's technique, how could anyone possibly doubt its scientific validity?

Soviet agriculture was by now just one small, pernicious step away from believing that willing a thing makes it so. In 1931 Stalin remarked: 'If there is a passionate desire to do so, every goal can be reached, every obstacle overcome.'[19]

Drunk on apparent success, Lysenko's flagship institution, the All-Union Institute of Plant Breeding in Odessa, proposed a further experiment in proletarian science: a competition to see whose approach would develop vernalisation the furthest. A brigade would be formed, made up of workers drawn equally from the Bureau of Applied Botany and the Odessa Institute. In the end the bureau did not take part – but the Communist Academy did. Prezent led its brigade to Odessa during the summer of 1932, and the following January Lysenko described this visit as a great inspiration for his general theorising about vernalisation. It taught him the dialectical method of experimentation and reasoning.

According to Lysenko, 'Prezent got so closely intertwined with my work that . . . not a single new issue of the theory that we've been developing [was made] without detailed discussion and participation of comrade Prezent.'[20]

Vavilov had returned home to find the Soviet Union in the grip of its worst famine since the revolution. Between 2.5 and 5 million peasants had starved to death. He knew what was coming. Who wouldn't? Every state, of every stripe, throughout recorded history, blames its misfortunes on individuals. If there was anyone to blame for the famine, it had to be Vavilov, the absentee, the will-o'-the-wisp collector who had abandoned 20,000 researchers, 155 experimental farms and 350-odd other research sites to go hobnobbing with European and American elites. There was nothing fair about the charge. But who else could possibly be to blame? Lysenko? Stalin?

1934 ought to have been a year of celebration: the fortieth anniversary of the Bureau of Applied Botany, the tenth anniversary of Vavilov's leadership, and Vavilov's own twenty-fifth year as a scientist. But with congratulatory telegrams pouring in from around the world and the offices already decorated, Yakovlev (moving up the ranks to the Party's Central Committee) cancelled the Bureau's celebrations.

In May, word had come from Stalin himself that Vavilov's bureau had let down the nation, and Yakovlev had better find out the reason why. How could Lenin's vast agricultural research infrastructure have failed to save millions from starvation?

There was, in fact, a simple answer to this question. In 1930, the bureau had proposed the introduction of hybrid maize.

Vavilov had had very good reasons for regarding maize as a panacea. In the USA, maize yields had increased by between 20 and 30 per cent when hybrids had been introduced. Vavilov's Bureau had already experimented with these plants and the results had been excellent.

Maize produces more grain per hectare than wheat, and when the wheat crop does poorly, it is very tempting indeed to abandon wheat for maize. This temptation was multiplied once high-yield hybrid maize was available. But maize, though more resistant to drought than wheat, needs a perfect summer. It is actually *more* sensitive to weather conditions than wheat is. Having invested so heavily in maize, the Soviets had reaped a dismal harvest.

There were clear lessons to learn here, and between them, Yakovlev and Vavilov would have been capable of learning them, had the political atmosphere let them. As it was, Trofim Lysenko came up with the kind of answer that high-level bureaucrats could accept. Rather than tackle the complicated business of why Soviet farming was so difficult to modernise, Lysenko gave vent to a typical piece of peasant scepticism: *these new-fangled hybrid lines of maize from the USA were no good.*

Lysenko was particularly exercised at the thought that these hybrids had been produced by inbreeding. Like many a peasant farmer before him, he believed inbreeding *always* led to the deterioration of a line. Not content with bad-mouthing America's incestuously begotten maize, Lysenko also decided that wheat, a natural in-breeder, was long overdue a moral lesson, and before he was done, millions of peasants were sent off to strip the florets by hand, forcing the wheat crop to outcross.

(Like most everything Lysenko did, this particular folly made no difference, for good or ill, to Soviet agriculture. Wheat being an inbreeder, all Lysenko's peasant labourers ever managed to do was cross fields of clones with each other.)

It is a measure of Lysenko's popularity, or Vavilov's vulnerability, or maybe just the chaos and desperation of the time, that Vavilov let these nonsenses go unremarked. Lysenko's status rose. He was elected to the Ukrainian Academy of Sciences. In 1934, at a conference on planning genetic-selection work, Vavilov himself declared that 'Perhaps in no other division of plant physiology have there been such profound advances as in this field . . . In this respect, we consider the work of T. D. Lysenko to be outstanding.'[21] Vavilov even recommended that Lysenko be elected Corresponding Member of the Academy of Sciences: 'Although he has so far published comparatively few works, his latest work represents such a major contribution to world science that it permits us to propose him as a candidate . . .'[22]

The more flak Vavilov took for his grandiloquent plant collection, the more he needed vernalisation to realise the collection's potential – the more he needed Lysenko, and was inclined to take his boosterism on trust. He wrote to Lysenko in 1934:

It seems to me a definite necessity that you, Trofim Denisovich, would yourself spare at least a week, two or three times a year, to come to Leningrad and see what we are doing here, and to help the younger workers especially to perform faster and more effectively the vernalisation tasks which are under way here on rather a large scale. You should understand well the significance of such an involvement by you in this work both for us and for yourself.[23]

Meanwhile, and with a huckster's monomania, Lysenko was strong-arming the whole of plant physiology into 'vernalisation'. Soon enough, vernalisation became detached from its underlying logic and was applied more or less by fiat to any agricultural situation. When it turned out that yields of winter varieties actually fell after vernalisation, spring varieties received the treatment.

Meanwhile Lysenko, never satisfied, found something new to complain about. Why did people keep referring to vernalisation as a mere 'technique'? Why were his theoretical pronouncements being ignored? In early 1934, he published a paper, 'Physiology of Plant Development and Selection Work', that included critical comments on genetics. Vavilov, for the first time, expressed irritation. Lysenko's genetics was so elementary, it was hardly worth publishing. Worse, it was muddled. 'If he had only opened [Danish botanist Wilhelm] Johannsen's treatise, he would have found a brilliant exposition of the theory of genotype and pheno-type.' Lysenko was neglecting his practical work to dabble in theoretical questions that were clearly beyond him: 'From you, comrade Lysenko, we expect concrete work on these matters.'[24]

But Lysenko would not be discouraged. In 1935 Lysenko and his friend and collaborator Isaak Prezent published two pamphlets: *Theoretical Bases of Vernalisation* and *Selection and the Theory of the Development of Plants Through Stages*. Prezent brought some rhetorical rigour to Lysenko's theory-making. Now Lysenko was not merely confused about genetics. He was hostile to it. Heredity, he claimed, was the property of the whole organism, and was dependent upon the environment. There was no gene.

Bits of cell, nucleus or chromosome [are] not what geneticists under-stand by the term 'gene'. The hereditary basis does not lie in some special self-reproducing substance. The hereditary base is the cell, which develops and becomes an organism. In this cell different organelles have different significance, but there is not a single bit that is not subject to evolutionary development.[25]

The genetics community responded with dismay. In June 1935 the Lenin Academy held a meeting in Odessa to discuss Lysenko's work, in particular his campaign to force wheat plants to outcross in order to counteract their 'degradation'.

Georgy Meister, a successful plant breeder with several important grain varieties to his credit, led the attack on Lysenko and Prezent. The Soviet Union had just taken on the job of hosting the next international congress of genetics. How was it going to look now that Lysenko and Prezent's vulgar, inappropriate, 'marketplace' criticism of genetics was being read in newspapers 'sold on every street corner'? Their nonsense was even appearing in scientific journals. 'It is', Meister wrote shortly afterwards, 'completely incomprehensible that the official organ of the Soviet Ministry of Agriculture publishes an article that disorganises government institutions of selection and seed growing.'

Meister was a formidable opponent: an enthusiast of genetics who was also an adept philosopher and a sincere Marxist. (His habit of putting every scientific idea through the dialectical-materialist wringer used to drive his students mad.) Lysenko and Prezent responded in a style that was to become their signature: they ignored him, and redoubled their claims.

They claimed to have found new laws of inheritance that let them experiment with a much smaller number of plants than plant breeders considered necessary. This would greatly accelerate the speed with which new varieties could be developed. It also meant that their positive results would not have to be repeated and confirmed. They were now in a position to answer Yakovlev's call for new varieties of cereal. In 1934, with the vernalisation programme very obviously running out of steam, Lysenko announced that he would have a new variety of spring wheat ready for testing by 1935. And he had bred it without the use of expensive growth chambers, 'in five flowerpots in a corner of a crowded greenhouse'.

The announcement caused a sensation. The nation's favourite barefoot scientist had trumped the old fogies of agriculture and slashed the time it would take to develop better crops: from up to a dozen years, to less than three! Vavilov, called in for his annual review, faced a barrage of criticism from the Lenin Academy's governing council. None of them knew a thing about genetics but they could all read a headline. Faced with having to explain basic science to a hostile board, Vavilov for the first time found himself lost for words. Two months later the council released its report. Vavilov's approach to the famine crisis was 'utterly insufficient'. He had completely failed to learn from 'the mass experience of leading state farms, machine and tractor stations, and collective farms'.[26]

Lysenko, by contrast, was working miracles. Lysenko himself said so: 'With your support,' he wrote in his new journal *Vernalisation*, launched in 1935, 'our promise to breed in two and a half years, through hybridisation, a variety of spring wheat for the Odessa region which is earlier and more productive than the regional variety "Lutescens 062", has been fulfilled.'[27]

It is worth taking a moment to understand just what 'fulfilment' meant in Lysenko's book. He had taken two plants each of winter wheats Kooperatorka and Lutescens 329, and sowed them on 3 March 1935 into a single pot in a cool greenhouse until the end of April. The idea was to prevent vernalisation (the conditions were spring-like, not cold) and to keep the plants alive as long as possible without them heading.

The Lutescens plants lived until the late autumn and died without heading. In the middle of August 1935, one of the Kooperatorka plants died when pests ate its roots. The other one, however, did manage to head. Several paired seeds were collected from the surviving plant on 9 September 1935 and sown beside ordinary Kooperatorka plants in a hothouse. Unlike their neighbours, these unvernalised seeds grew happily in warm conditions, proving that they had been transformed into spring varieties.

Ludicrous as it was to base anything on the behaviour of just one plant, Lysenko concluded that the plant had passed its spring form to its descendants, in classic Lamarckian style.

This was too much for the staff of Vavilov's bureau, and they began to lobby Vavilov to stop encouraging the kid from Odessa. Mikhail Zavadovsky, in the August 1936 issue of the leading Soviet agricultural journal *Socialist Reconstruction of Agriculture*, shredded Prezent for not knowing the difference between individual development and the evolution of species: 'his activity is in fact obscurantist'. Lysenko, who was studying individual development, not genetics or evolution, had clearly been misled into 'careless generalisations'.[28]

But Lysenko was operating at a political level now, not a scientific one. At the Second All-Union Congress of Collective Farmers and Shock-Workers, Lysenko called for the mobilisation of the peasant masses to achieve miracles of vernalisation. His speech, 'Vernalisation Means Millions of Pounds of Additional Harvest', published in *Pravda* on 15 February 1935, is full of demagogical calls for class struggle and demands to cast off the fetters of the scientific method.

Comrades, kulak–wreckers occur not only in your collective farm life. You know them very well. But they are no less dangerous, no less sworn enemies also in science. No little blood was spilled in defence of vernalisation in the various debates with some so-called scientists, in the struggle for its establishment; not a few blows had to be borne in practice. Tell me, comrades, was there not a class struggle on the vernalisation front? In collective farms there were kulaks and their abettors who kept whispering (and they were not the only ones, every class enemy did) into the peasant's ears: 'Don't soak the seeds, it will ruin them.' This is the way it was, such were the whispers, such were the kulak and saboteur deceptions when, instead of helping collective farmers, they did their destructive business, both in the scientific world and out of it; a class enemy is always an enemy, whether he is a scientist or not.[29]

At the same time, Lysenko conceded his debt to Prezent:

I often read Darwin, Timiryazev, Michurin. In this I was helped by my collaborator Prezent. He showed me that the roots of the work I am doing lie in Darwin. And I, comrades, must confess here straightforwardly in the presence of Josef Vissarionovich [Stalin] that to my shame I have not studied Darwin properly.

He was apologising for his lack of ability as a speaker, saying he was only a 'vernaliser', not an orator or a writer, when there was an interruption. 'Bravo, Comrade Lysenko, bravo!'

The heckler was Stalin.

The audience broke into spontaneous applause, and the incident was publicised in all national newspapers.[30]

In October 1935, the Lenin Academy concluded that vernalisation had passed the experimental stage and that responsibility for the programme should now pass to the agricultural administration. In December 1935 the front page of *Pravda* showed Lysenko in the Kremlin, sharing the rostrum with Stalin at a public meeting.

12 : The great patron

*One feature of the history of old Russia was the continual
beatings she suffered for falling behind, for her backwardness . . . All beat her, because of her military backwardness,
cultural backwardness, political backwardness, industrial
backwardness, agricultural backwardness . . . Do you
remember the words of the pre-revolutionary poet: 'You are
poor and abundant, mighty and impotent, Mother Russia'?
. . . We are fifty or a hundred years behind the advanced
countries. We must make good this distance in ten years.
Either we do it, or they crush us.*[1]

Joseph Stalin, 'On the Tasks of Industrial
Administrators', 1931

In 1931 Maxim Gorky returned to Russia. Stalin had spent years
wooing back this revolutionary celebrity, though in the end it was
only the fading of his reputation in the West that persuaded him
to come home. The fashion for Nietzschean tramps was passing,
and Gorky belonged to the turn of the century. Returning to
Russia offered him the prospect of a fresh start.

Coal miner and national hero Alexei Stakhanov (centre) explains his
methods to fellow workers, 7 June 1935.

Gorky cannot have been very surprised by conditions in his homeland, since he had been a regular visitor since 1929. Still, the continuing decline in the social fabric was hardly encouraging. The early 1930s were dominated by shortages of food, clothing and housing. Rationing was in force from 1929 to 1935. At every point in the distribution chain, employees funnelled off goods for themselves or for sale on the black market. Suits, woollens and gramophones disappeared from regular stores and turned up in second-hand shops at exorbitant prices. When galoshes appeared in Kazan's main department store, speculators crowded out regular shoppers. When forty bicycles came in, the store manager funnelled them one at a time to friends. Shoes were impossible to get hold of through legitimate stores, not least because the livestock that supplied the leather had been slaughtered *en masse* by desperate peasants during the first years of collectivisation.

Despite all this, Gorky contrived to appear positive about his return and worked enthusiastically to celebrate the heroic practical achievements of the era. In a letter to Stalin he argued the need for propaganda about 'our achievements' and proposed introducing such a section in all the newspapers.[2]

He was particularly taken with the construction projects known as *stroiki*. These 'hero projects' included Belomorstroi, the White Sea Canal connecting the Baltic and White seas, which opened on 2 August 1933; Dneprostroi, the building of what was at the time the world's largest hydroelectric power plant; Magnitostroi, the construction of what was and still is the world's largest steel plant; and Kuznetstroi, the development of the coal fields and railroads needed to supply Magnitostroi.

Gorky's ecstatic glorification of big engineering stemmed from his hatred of the countryside. 'My sympathies', he wrote in a letter of 1926, 'have always been with the "insignificant handful" of the urban proletariat and intelligentsia': a sentiment that hardly begins to express the loathing he nursed for the land of his forefathers. As far as Gorky was concerned Russia's countryside was a harsh, miserable place where people starved half to death; a charnel-house where only brutes and their lackeys stood any chance of survival.

Maxim Gorky was rabid in his enthusiasm for the transformation of this ghastly Russian nature. In 1934, as he considered the utopian significance of the construction of the Baltic–White Sea Canal, he wrote:

Stalin holds a pencil. Before him lies a map of the region. Deserted shores. Remote villages. Virgin soil, covered with boulders. Primeval forests. Too much forest as a matter of fact; it covers the best soil. And swamps. The swamps are always crawling about, making life dull and slovenly. Tillage must be increased. The swamps must be drained . . . The Karelian Republic wants to enter the stage of classless society as a republic of factories and mills. And the Karelian Republic will enter classless society by changing its own nature.[3]

This was much more than mere propagandising. This was a long-awaited answer to Chernyshevsky's question *What Is to Be Done?* Gorky saw the *stroiki* as ushering in a heroic age that would fulfil centuries of utopian yearning. The Russian people had woken at last from their hypnotic slumber and were fast evolving toward total mastery of the planet. 'Our brave and mighty activities directing the physical energies of the people to the struggle with nature allow the people to feel their true purpose: to gain possession of the forces of nature and to tame its fury.'[4]

Significantly, this mastery would involve not only the subjugation of external nature, but also the absolute conquest of human nature, which in Gorky's estimation was 'nothing but an instinctive anarchism of a personality brought up through the ages of pressure placed on it by the class state'. You didn't have to get very far into *Belomor*, Gorky's celebratory account of the canal, to find this lesson spelled out for you. The motto on the flyleaf reads: 'Man, in changing nature, changes himself.'

*

235

How far human nature might be changed – by urban living, by regulated work patterns, and by machines themselves – was of course a question considered seriously from the very beginning of the revolution by men like Alexei Gastev and Nikolai Bernstein. But just as in agriculture – where 'the mass experience of leading state farms, machine and tractor stations, and collective farms' outweighed the scientific papers of Vavilov and Maksimov and Meister – so in industry, isolated achievements on the factory floor came to wield far more influence over government policy than Gastev and Bernstein's studies of ergonomic motion.

In 1927, a miner called Alexei Grigorevich Stakhanov went to work in the Tsentralnaya-Irmino mine in the Donbass region of the Ukraine. In 1933, he became a jackhammer operator. He was no superman, but he was energetic and intelligent. He saw a way of organising his work crew so as to increase the amount of coal they were able to dig in a single shift. On 31 August 1935, it was reported that Stakhanov had mined a record 102 tonnes of coal in 4 hours and 45 minutes – fourteen times his quota. Barely three weeks later, on 19 September, Stakhanov and his crew (who were rarely if ever mentioned in the press) more than doubled this record.

Stakhanov's achievement was made an example for others to follow. His story was so remarkable that on 16 December 1935 he appeared in the United States – on the cover of *Time* magazine.

Others rushed to follow Stakhanov's example, and newspapers and newsreels across the Soviet Union celebrated their efforts. In Gorki, a worker in a car factory forged nearly a thousand crank-shafts in a single shift. A shoemaker in Leningrad turned out 1,400 pair of shoes in a day. On a collective farm three female 'Stakhanovites' proved they could cut sugar beet faster than was thought humanly possible. Such workers were awarded higher pay, better food, access to luxury goods and improved accom-modation, and soon Stakhanovism was a mass movement. 'In factories and even in scientific institutes', wrote the American psychologist Richard Schultz in 1935, 'the workers' names may be posted on a bulletin board opposite a bird, deer, rabbit, tortoise, or snail relative to the speed with which they turn out their work. A great deal of prestige is attached to the "shock brigade" worker.'[5]

On the collective farms, Stakhanovism was greeted in the main with apathy. 'We have done enough fulfilling; our horses will all be dead by spring,' one exhausted farmhand complained.[6] In the factories, meanwhile, the push for extra production could as easily disrupt the running of a factory as improve it, with sections of the same shop unable to work together. Improved productivity itself served to accelerate the breakdown of aged, shoddily made and badly maintained equipment, so that along with the production problems came a shocking spike in the number of industrial accidents. Then there were the Stakhanovites themselves – young, quarrelsome egotists who, in their desire to get ahead, made life miserable for everyone else. Resentment towards Stakhanovite 'shockworkers' often boiled over into punch-ups and machine wrecking.

These problems might have spelled the end of the Stakhanovite movement, were it not for Stalin's extraordinarily adept way of orchestrating violence. If the Stakhanovites were causing trouble in the factories, were they not the perfect mechanism for getting rid of bourgeois factory managers? When trouble broke out between plant managers and would-be Stakhanovites in 1936, local press and party organisations were encouraged to take the side of the Stakhanovites. Several important plant and mine directors were fired; some were arrested for sabotage. And the optimism and get-go of Stakhanovism gave way, as the 1930s progressed, to a very different kind of campaign: one that highlighted the activities in industry of 'wreckers' and 'saboteurs'.[7]

Stakhanov was one of the more exceptional products of the revolution. But attempts to mass-produce him failed. In the end, Stalin's bureaucrats settled on a very different idea of how human nature might be industrialised. They learned how to use the labour camp.

Stalin's labour camps were more than simple borrowings from tsarist history. They evolved as a direct consequence of the hero projects of the Five-Year Plans, and they attained their grotesque scale and ubiquity thanks to that other great heroic project of the early Soviet era – the attempt to settle Siberia.

From the beginning, Soviet attempts to exploit the vast natural resources of that harsh and gigantic region had required more than volunteers.[8] Prison labour was considered essential from the very first. This conclusion was less harsh than it sounds. For generations the tsars had sent their criminals and political opponents to terms of exile in Siberia. The Soviet use of this system to establish new settlements and new industries in a hitherto neglected region was considered a rather far-sighted piece of penal reform. It was the sheer scale of the work required to settle Siberia which led the labour camp (or 'gulag') system to far outgrow anything conceived of by the tsars. At the time of the original 1929 order launching the Soviet labour-camp system, the USSR had a prison population of around 23,000. Less than five years later, half a million Soviet citizens were inmates.

There was an eerie idealism to the whole venture – one not that far removed from the ideas that powered the transportation of British convicts to Australia, as though moral correction would naturally follow from the rigours and rough satisfactions of the pioneering life. Henrikh Yagoda, deputy head of the OGPU, felt it 'necessary to convert the camps into colonising settlements'. Selected convicts would be sent to various regions to construct huts – 200–300 per settlement. Then they could send for their families (why anyone in their right mind would invite their family to the largely uninhabited wastes of Siberia was a point Yagoda did not address) and 'in their free time, when the forestry work is over ... they will breed pigs, mow grass, and catch fish. At first they will live on rations, and later at their own expense.'[9]

Needless to say, the future for these select convicts was not quite so bucolic in practice. Faced with the harsh production demands of the First Five-Year Plan, mine and factory directors clamoured for access to this new labour pool. In 1931 the Urals Metallurgical Trust sent a telegram to the local labour department complaining that out of the 2,700 recruits arriving that year, a thousand had already left. Far better, they said, to use prison labour for those jobs where the conditions were particularly harsh. Peasants were less likely to run away while they were being watched by armed guards. By the late 1930s forced labour was used in all major industries in the Urals, and a million people were prisoners in the gulag. It was their labour that created most of Siberia's vast industrial plants: the Norilsk copper–nickel factory, the works in Komsomolsk-on-the-Amur, Novosibirsk, and many other Siberian cities and towns.

Created in November 1931 and operating under the secret police, Dalstroi – the Main Administration for Construction in the Far North – prospected and mined for gold in north-eastern Siberia, particularly in the Kolyma basin. Up to the summer of 1937 Eduard Berzin, its first director, made the agency profitable while still keeping conditions, even for convict labourers, relatively humane. Then in June he was denounced by Stalin himself for 'the coddling of prisoners'. Berzin was shot, and Dalstroi's new leadership turned its mines and camps into the USSR's most brutal, making Kolyma and Magadan names to rival Auschwitz and Majdanek. The gulag and its pool of slave labour were by now essential tools in Soviet industrialisation. Prison camps were self-sufficient, even profitable. They were in no sense correctional facilities. They were state businesses.

Did Maxim Gorky know how his beloved Baltic–White Sea Canal was constructed? Did he know that from a total of around 126,000 workers, at least 12,000 died in its construction?

Did he know the cost Stalin's *stroiki* exacted in lives? That to build Dneprostroi, 10,000 villagers were forced out of their homes? That workers endured temperatures below –13 °C in the winter, and that tornado-strength winds whipped their tents away in the summer? That flour had to be delivered to the construction town at night under armed guard to prevent theft?

It is possible that Gorky's propaganda work was a sophisticated smokescreen, and he was dabbling with conspiracies to oust Stalin from power. There is plenty of anecdotal evidence to suggest that Gorky viewed Stalin as a less than ideal choice of leader. When the satirist and novelist Yevgeny Zamyatin was finally allowed to emigrate in 1931, he went round to say goodbye to Gorky, his old friend and patron, and Gorky told him: 'Leave, leave. We have yet to say who among us here will triumph, this' – and he made a gesture depicting Stalin's moustache – 'or our "Ivanoviches"', by whom he meant the leaders of the right opposition, Alexei Ivanovich Rykov and Nikolai Ivanovich Bukharin.[10]

By 1934 the leaders of that opposition were ready to act and were looking for a likely candidate to take over from Stalin as leader. They elected to approach the hard-drinking, foul-mouthed Sergei Mironovich Kirov, head of the Leningrad Party machine and one of Stalin's closest and most trusted lieutenants. Kirov, well read and garrulous, was hugely popular in government circles and had proved himself capable of arguing with Stalin – and besting him – in Politburo sessions.

Senior Politburo members asked Kirov to move from Leningrad to Moscow to focus full-time on his work as one of three Central Committee secretaries. Given Kirov's popularity, his move to Moscow would have greatly simplified Stalin's removal. But Kirov, a dedicated and loyal Stalinist, declined the offer and – fatally – reported the entire conversation to Stalin.

Late in the afternoon of 1 December 1934, Sergei Kirov was shot in the neck in a hallway of the Solnyi Institute. Efforts to resuscitate him proved useless.

The next day Joseph Stalin and a team descended on Leningrad. They returned to Moscow with Kirov's body late the next day, by which time the slain leader was already halfway to secular sainthood. The papers at the time of Kirov's death were filled with his praises, and with expressions of grief far exceeding those which followed the death of Lenin. For at least twelve days all Soviet newspapers devoted themselves to Kirov's life and death. By 5 December entire books about him, with reminiscences by former comrades, stories about his childhood and reproductions of his speeches, had gone to press.

Oddly, no one in the press thought to mention that a key witness to the murder, Kirov's bodyguard M. D. Borisov, was killed in an accident a day after the crime, while on the way to be questioned by Stalin. Although news of the accident spread quickly around Leningrad, the newspapers ignored this event completely. And while there is no direct evidence that Kirov was shot by the NKVD on Stalin's orders, it is certain that Kirov's death gave Stalin the opportunity to dismantle all the many networks of friendship and patronage that might otherwise have opposed him. It gave him *carte blanche* to disembowel Russia's second city, Peter the Great's 'window on the West' and home to Russia's most prominent intellectual and cultural figures. It let him obliterate the last vestiges of Russia's liberal tradition, and shift the nation's balance of power from Leningrad to Moscow, once and for all.

Whether or not Gorky really had been toying with conspiracy – and whether or not Stalin ever suspected him of doing so – in any event, Kirov's death marked the end of Gorky's welcome within Stalin's inner circle. In late 1935, a series of hostile articles in *Pravda* announced a marked change in the official attitude toward him. He found himself under unofficial house arrest in his country villa, about fifty kilometres from Moscow. From 6 to 17 June 1936, all the many newspapers that Gorky read daily were printed for him separately. These special editions omitted the bulletins about his health, with which the country and the world were being prepared for his death on 18 June, at the age of 68. Maxim Gorky died, 'the result of pulmonary congestion following grip', according to the *New York Times*: 'His heart weakness had already given rise to anxiety, which was greatly accentuated yesterday.'

The story of Gorky's death fooled the *New York Times*, as it fooled the world. And those who were not fooled needed only to wait two years before Gorky's doctor, Dmitry Pletnev, was accused of killing his client Maxim Gorky, on the instructions of a right-wing conspiracy. Pletnev was sentenced and died in a prison camp.

The fiendishness of the state's manoeuvring here is instructive. Whether you believed in the story of Gorky's heart weakness or not, whether you thought the man had died of natural causes or been murdered in his sleep, you were overwhelmingly likely to believe one or other official version of events.[11]

When Kirov declined the opportunity to replace Stalin as General Secretary of the Party, he sealed his own fate. He also left a position open in Moscow. The man who filled that position was Andrei Alexandrovich Zhdanov, the Proconsul (local Party chief) of the city of Nizhny Novgorod. An opportunistic survivor, cautious to the point of cowardice, Andrei Zhdanov was to set the tone of Russian cultural and scientific life for the next twenty years.

His background was patrician: his mother was the daughter of nobility, his father a student of theology. The family had gone to ruin with the death of his father, and Andrei had been a committed Marxist since his teens. His breeding showed through though in his musical tastes, his education and his manners. Lavrenty Beria, Party Secretary for Georgia and Stalin's most trusted subordinate, couldn't stand him: 'He can just manage to play the piano with two fingers and to distinguish between a man and a bull in a picture, yet he holds forth on abstract painting!'[12]

From behind his desk in Nizhny Novgorod, Andrei Zhdanov had herded the local peasantry into collective farms. The exercise, a comprehensive economic failure, had established Zhdanov's leadership and loyalty and marked him for high office.

For a week after Kirov's assassination, Zhdanov and Stalin held lengthy meetings in the Kremlin, thrashing out how Zhdanov was to divide his time between his work in Moscow as Central Committee secretary, and his new responsibilities as head of the Leningrad Party organisation.

No material evidence was ever presented to suggest anyone but the killer was involved in Kirov's murder, Nonetheless, on 16 December 1934, two important political opponents of Stalin – Grigory Zinoviev and Lev Kamenev – were arrested on charges of inspiring the act. By late December, fourteen people had been sentenced to death for organising the Kirov murder. By March 1935 the NKVD in Leningrad had arrested 843 people. The Great Purge had begun.

During the month of July 1937 *troikas* – a kind of emergency court – were appointed in all provinces and republics. These handed out death or labour-camp sentences to criminals and political opponents alike, using lists submitted by the NKVD. In August 1937 the arrest quota for Leningrad was established at 4,000 people for the death sentence, and 10,000 for terms in prison or labour camps. As elsewhere, the quotas were overfulfilled. The Leningrad Party leadership was purged, and 30,000–40,000 Leningraders were deported, effectively ending Leningrad's role as a political and cultural centre to rival Moscow.

Stalin's decision to eradicate and terrorise all opposition, real or potential, had begun in Leningrad but soon spread to the rest of the country. The notorious Order No. 00447 sent the NKVD after ex-*kulaks*, Socialist Revolutionaries, nationalists, criminals, Whites, sectarians and religious leaders. The campaign soon widened to include regional Communist leaders and their clients. Order No. 00447 supplied arrest quotas for all regions and republics of the Soviet Union and specified how many should receive the death penalty and how many should 'merely' be dispatched to concentration camps. It was usual for regional NKVD departments to overfulfil the plan, for which they retro-actively received Politburo permission. Of the quarter of a million people arrested under Order No. 00447, at least 72,000 were sentenced to death. The victims still represented only one tenth of the approximately 700,000 people shot in 1937 and 1938. All told, the Great Purge saw the arrest of some 8 million and the execution of about 1 million.

Even the Army General Staff was not immune. The army command included many a former tsarist officer, and it was common knowledge among Communists that the Soviet military had engaged in intensive military collaboration with the Germans since the Treaty of Rapallo was signed in 1922. The General Staff was liquidated in June 1937; 80 per cent of members of the Military Council were executed.

David Shoenberg, a physicist visiting from Cambridge, wrote later that the Great Purge was 'rather like a plague and you could never tell who would catch it next'.[13] But sciences centred in Leningrad suffered most. The small though highly prestigious fields of astrophysics and astronomy were devastated. Ten senior physicists, including leading specialists at the State Astronomical Observatory in Pulkovo, just outside Leningrad, were arrested for 'participation in a fascist, Trotskyist terrorist organisation which arose in 1932'.[14]

Physics too was changed out of all recognition, though here the catastrophe was rather longer in the making, the intellectual arguments mattered rather more, and the ultimate tragedy a matter more of fate than chance.

'This world is a strange madhouse,' Albert Einstein wrote to his friend Marcel Grossmann in 1920. 'Every coachman and every waiter is debating whether relativity theory is correct. Belief in this matter depends on political affiliation.'[15]

For more than a decade after the publication of Einstein's key papers, relativity was a political issue. Albert Einstein had done away with the notion that there had to be some invisible, ineffable medium – an aether – through which light travelled. That made his theory materialist: there was no longer any need to assume some hidden 'higher order' to the universe. On the other hand, relativity made observations hopelessly contingent upon local conditions, which themselves could not be ascertained with any certainty. In an Einsteinian universe, there are no stationary points, and no reliable clocks. All this makes materialist science very hard to do. Common sense is shown the door, and intuition fails. No wonder amateurs everywhere railed against Einstein's theories. Einstein had at a stroke made amateur work in the field immeasurably more difficult, even as the new, big laboratories and institutes were making amateurs themselves irrelevant.

What this boiled down to, politically, was a sense that Einstein was helping the sciences professionalise themselves at the expense of everybody else; that physicists were talking gobbledegook and getting paid ever greater sums of money for doing so; that scientists were becoming, in an ever more godless age, a replacement priesthood.

You didn't need to be a Marxist to deplore the passing of common sense in the field of physics. For the Bavarian occultist Max Seiling, the arrival of Einstein quite clearly spelled doom for the neighbourhood: 'Special research has become an end in itself. Since that time, it has been ridiculous, and along with it those who carry it out and take such delight in their rag collecting.' Nor was the Europe-wide Academy of Nations, established in 1921, a Marxist organisation. Nevertheless as an *anti-Einstein* organisation, it clung, Marx-like, to a dream of 'the unification and coordination of systems of knowledge, thus procuring the development of a synthesised body of knowledge as against the highly specialised condition now existing'.[16]

The face-off between 'academic science' and what amateur societies liked to call 'correct natural science' was not, then, an especially Russian phenomenon. But a few unhappy circumstances saw to it that the argument would rumble on much longer there and ultimately fuel the purges in physics.

Albert Einstein was, for one thing, quite vocal in crediting Ernst Mach with some of the philosophical ideas that inspired both the special and general theories – the same Mach whose theories, according to Lenin's deathless *Materialism and Empiriocriticism*, amounted to 'a jumble of idle and shallow words in which he himself does not believe'.

A further irritant was the way in which a prestigious group of theorists, following the work of Danish physicist Niels Bohr, had come to the conclusion that at a very small scale, measurements affect outcomes to the point where nothing useful can be said about the actual state of the universe. This hardly sat well with Lenin's lumpen 'reflection theory' of mind, and in 1936 a physicist, K. V. Nikolsky, wrote to a philosophy journal to say so, complaining of the Copenhagen interpretation's 'idealism' and 'Machism'.

Einstein's own spiritual inclinations added fuel to the fire in 1930, when the *New York Times Magazine* and the *Berliner Tageblatt* ran articles of his under the titles 'Science and Religion' and 'What I Believe'.

Resistance to the special and general theories of relativity was not exceptional, and everyone – but everyone – had an opinion. In Moscow, Arkady Klimentyevich Timiryazev was making hatred of Einstein into a political platform.

Arkady Timiryazev was a close associate of Maxim Gorky, and the son of the famous agronomist and biologist Kliment Timiryazev. At Moscow University he was referred to as a professor of physics. Behind his back he was dubbed 'the monument's son', because there was little enough, aside from his impressive parentage, to recommend him. He was not an outstanding scientist. His devotion to classical Newtonian physics was absolute, and absolutely uncreative. Timiryazev hated Einstein's theories of relativity with a passion that, so long as you did not get in the way of it, appeared positively farcical. At a public meeting in the mid-1920s he sarcastically conceded that he didn't want Einstein actually shot. Unhappily for his colleagues – natural allies and opponents alike – it was remarkably easy to find yourself the target of one of Timiryazev's scattergun denunciations. Timiryazev used politics as a blunt weapon as readily as any Bolshevik ideologue, and with considerably less care. Among the people he vilified were academicians including the godfather of Soviet physics Abram Ioffe and Nikolai Vavilov's brother Sergei – later the Academy's president. All in all he turned the physics department at his university into a kind of militant ideological camp.[17]

Only Stalin expressed affection for Arkady Timiryazev – possibly because Arkady was the only person we know of who ever bothered to search the leader's writings for references to modern physics.

Timiryazev did not have things all his own way, however. Einstein had many fierce champions in the Soviet Union, and none so brilliant as Boris Mikhailovich Hessen, a gifted mathematician born into a middle-class Jewish family in 1893 in the town of Elisavetgrad (now Kirovgrad), a city in central Ukraine. Hessen had studied physics at the University of Edinburgh between 1913 and 1914 and at Petrograd University during the First World War. After the revolution he became a soldier in the Red Army, and at the end of the Civil War he became a student of natural science at the Institute of Red Professors in Moscow. His own Marxism was sincere and sophisticated. Hessen saw off Timiryazev's reactionary objections to relativity theory in the pages of *Under the Banner of Marxism* with an argument that ran something like this: if you are a Marxist, and you claim that relativity theory is anti-Marxist, that's all very well – right up to the point where the theory is demonstrated to be correct. Then what are you supposed to do? Did Marxists really want to create a situation for themselves in which, if Einstein was right, Marx had to be wrong?

There was no need to get into such a tangle, Hessen said. The world was the world. Of course human explanations of how the world worked were coloured by one ideology or another, but what mattered to good scientists were the findings of science: the measurements, not the philosophical interpretations. Newton's physics supposed a 'divine first impulse' that set the solar system in motion. Were atheists and Marxists going to reject *Newton*, now?[18]

Hessen suffered for his cleverness. In 1928 his own head of department, Timiryazev's ally Alexander Maksimov (no relation to the vernalisation expert Nikolai Maksimov) called him a 'Machist' and a 'Right deviationist' – labels which even then packed a serious political and personal threat.

By October 1930, that threat was out in the open. At a conference on the state of Soviet philosophy, Hessen and his views on physics were heavily criticised. He was not even allowed to speak in his own defence as critics denounced him as a 'pure idealist', a 'metaphysicist of the worst sort' and as a deserter from the materialist cause. He even attracted the attention of Ernst Kolman, a Czech philosopher of science and arguably the most savage of Stalin's intellectual cheerleaders. Kolman had once been a man of action: twice between 1918 and 1923 he had been sent under cover to Germany to foment world revolution. Now riding a desk in Moscow, Kolman spent his days hunting down ideological heresy in his own field of mathematical and physical sciences. In an article published in 1931 Kolman issued a direct challenge to Hessen, calling on him to turn over a new leaf and correct his political mistakes: 'One must speak directly here, and say that there is no Bolshevism in Hessen's science, nor in that of his comrades.'

Hessen got his chance at redemption three months later when he was included in a delegation to London to attend the Second International Congress of the History of Science. The Soviet party, arriving in a special aircraft at the last minute, included both Party members and fellow-travellers. All were intellectual heavyweights. Abram Ioffe, founding father of Soviet physics, was there; so was Boris Zavadovsky. Vladimir Mitkewich attended, to speak about Faraday and that perennial favourite, the electrification of the USSR. Nikolai Vavilov, meanwhile, waxed positively millennial when describing his study of the history and prehistory of food crops:

In approaching this problem from the point of view of dialectic materialism, we shall be led to revise many of our old concepts and, which is fundamentally important, we shall gain the possibility of controlling the historical process, in the sense of directing the evolution of cultivated plants and domestic animals according to our will.[19]

Also on the delegation was Ernst Kolman, whom the Politburo had charged with supervising and reporting on the political behaviour of two suspect members of the delegation: Boris Hessen; and the leader of the party, Nikolai Ivanovich Bukharin.

Of the two, Bukharin was by far the more important, and the one in the more serious trouble. Lenin himself had dubbed him the Party's 'leading theoretician'. But the dead leader's support now marked Bukharin out as a possible opponent of Stalin. Making his position even more delicate, Bukharin had once been a pupil of Alexander Bogdanov.

Bukharin did not necessarily oppose Stalin's efforts to make science over in the service of the state – to make all science into applied science, to be applied moreover in the service of the Party. Explaining the Soviet scientific revolution in London, he described how 'the rupture between intellectual and physical labour' was being eliminated and scientific research was rising to a new level of efficiency. (His prime example was plant breeding.)[20]

But there were crucial and growing differences between Bukharin and Stalin. Stalin's 'Great Break' roused nothing but contempt in Bukharin. 'Ignorant nonsense', he called it: one of several remarks that, by December 1930, had got him expelled from the Politburo.

Bukharin's opening speech was enthusiastic, politically on-point and, for the international audience gathered in London, quite uncontroversial. The days of 'scholastic monasteries, the laboratories of alchemists and the quiet offices of individual university scholars' were gone, he announced. Science now had to be organised along the lines of big industry. Few ambitious scientists of that time would have disagreed with him, regardless of their politics.

The contribution of the other 'suspect' delegate, Boris Hessen, was of quite a different order. Written with an eye to saving his career (and possibly his neck), and delivered in the presence of Ernst Kolman, a man he knew was spying on him, Hessen's paper, 'The Socio-Economic Roots of Newton's *Principia*', ought to have been a piece of utter humbug. It turned out, however, to be the most influential paper presented at the conference. In Britain it brought together under a common purpose an entire generation of left-leaning scientific celebrities, from John Haldane to John Bernal. Internationally, it inspired an approach to the history of science which still holds sway today.

Hessen used Newton's work as an example of how good, valuable, proven science is at the same time a product of its time and class. Hessen demonstrated that Newton's accomplishments in physics were both astonishing glimpses into the nature of reality *and* historically determined documents, full of philosophical and religious conclusions which Newton could not help but draw from the economic order around him.

Hessen's paper contained a double message. As far as the audience in London was concerned, Hessen had expressed all that was valuable and timely and commonsensical about Marx's views on science. Science was not the product of great minds somehow isolated, priestlike, from the world and its practical demands. It was a cultural practice: it was what people *did*. For an audience steeped in the animistic and often quite reactionary writings of Sir James Jeans (in *The Mysterious Universe*, 1930) and Sir Arthur Eddington (*The Nature of the Physical World*, 1928), Hessen's talk was nothing short of a revelation.[21]

For his critics at home, meanwhile, Hessen had audaciously demonstrated why the political and philosophical tenor of scientific works didn't impinge on their scientific value. If you reject the findings of science because you don't like the philosophical or political conclusions the authors draw from them, then you not only have to reject Einstein: you also have to reject *Newton*. Ultimately, you have to reject everything that does not grow in your own precious back garden. Newton is celebrated for his laws of motion, not for his 'general religio-theological conception of the universe'. The only test of science is the reality it measures. Measurement is the valuable bit of science; the rest hardly matters.

Politically, Hessen could not have been more badly out of step. In June 1934, a special session at the Communist Academy was held to commemorate the twenty-fifth anniversary of the publication of Lenin's unlovely work of philosophy, *Materialism and Empiriocriticism*. During the months preceding the conference, such journals as *Frontiers of Science and Technology*, *Socialist Reconstruction and Science* and *Under the Banner of Marxism* ran a series of articles which 'exposed' Heisenberg, Schrödinger, Bohr and Born as lackeys of Western 'idealism', and subjected virtually every great name in Soviet physics, from Hessen and Ioffe to Iakov Frenkel and Igor Tamm, to the most violent abuse.

Physicists attempted to push back against this ignorant assault. Ioffe and Nikolai Vavilov's brother Sergei participated in the commemoration, arguing that contemporary physics, far from being 'idealist', was as clear a demonstration of dialectical laws in nature as you could hope for.[22] Ioffe borrowed from Hessen in his assertion that in field after field, seeming contradictions achieved a dialectical synthesis in the unity of opposites. Marxist philosophers found idealism under every stone, but only because they didn't have a clue about what the new physics actually meant. He was even more forthright in an article which followed: 'The Situation on the Philosophical Front of Soviet Physics'. Why, Ioffe asked, was Arkady Timiryazev siding with the European anti-Einstein lobby at the very moment it was acquiring a distasteful and frankly worrying anti-semitic edge? German Nobel laureates Philipp Lenard and Johannes Stark were going around calling relativity 'non-Aryan physics' – was Timiryazev really embracing that filth? Did he, too, consider Werner Heisenberg a *'Weisser Jude'*? And if he did, what did he have to show for it? The physicists he was so busy attacking – Fock, Frenkel, Tamm, Mandelstam and Landau – had produced theories in solid-state physics, come to a better theoretical understanding of metals, unpicked the photo-electric effect, explained the nature of magnetic polarity and contributed to the development of Heisenberg's quantum mechanics. All Timiryazev's party had come up with were 'fetishes of the aether', 'force tubes', 'electrical bagels' and like gibberish.

By then the debate had ceased to be about physics. What mattered now was affiliation. What saved you, or ruined you, was your network, the people you knew, the people who supported you and to whom you gave your support. To be identified with the European scientific community, with Leningrad, or with the wrong sort of Party scholar, was to be damned out of hand.

Of the eight-man delegation to London in 1931, all but Kolman would be purged. Six would perish at the whim of the state. Hessen himself was arrested in 1935 and died in prison in 1938.

Two years before Hessen's death, a special week-long session of the Academy of Sciences was convened to evaluate the Leningrad Physico-Technical Institute. The institute's founder, godfather of Soviet physics and practically a patron saint to many in the field, Abram Ioffe, described how in eighteen years his brainchild had grown into a system of fourteen institutes and three higher technical schools with almost 1,000 scientific workers, 100 of whom were first-rate physicists.

He was promptly ticked off for his 'aristocratism' and 'empire building'.[23]

The criticism that enveloped Ioffe from all sides stunned many. For Ioffe, the most troubling comments came from the notoriously tactless young physicist Lev Landau.[24] Landau pulled no punches, dismissing out of hand Ioffe's claim that Soviet physics was in any sense 'world class'. Granted, Soviet physics owed its very existence to Ioffe, for 'in tsarist Russia up to the moment of the revolution physics practically didn't exist'. But, said Landau, Ioffe's rosy assessments of the present were laughable. Landau counted no more than a hundred independent researchers in Soviet physics, and few of them were engaged in training the next generation. As for Ioffe's own research style, the best you could say about it was that it was careless. Ioffe's habits, prevailing in the profession, had led to a tremendous waste of resources. His projections for future energy technologies were farfetched; the work of his physicists was messy; and he had a tendency to make claims about Soviet priorities in discovery where Western physicists had earlier achieved similar results. Worst of all was Ioffe's over-blown, haughty and self-aggrandising leadership.

Landau's individual criticisms were cogent, and even drew applause, but the spirit in which they were made left a bad taste in the mouth. It became apparent that, by stitching up Ioffe, a younger generation of physicists was hoping to free up opportunities for itself. As the session progressed, more and more junior researchers took Landau's lead and unsheathed their knives. Only a few very senior figures were left to defend the man.

The price the field paid for bandying about cheap political accusations was high. Over 100 physicists were arrested in Leningrad between 1937 and 1938, as part of a general effort to extinguish Leningrad's intellectual and cultural life. Ioffe and his colleagues could not save his institute from decline, and in the end even Ioffe adopted protective colouration, expressing in a public meeting 'the anger and indignation of Soviet scholars at the ignoble work of Trotskyite bandits which demands from the proletarian court its destruction'.[25]

Landau ran no immediate risk from the Leningrad purges because in August 1932, having fallen out with Ioffe, he had left Leningrad for Kharkov to head the Department of Theoretical Physics at the Polytechnic Institute there. With his penchant for practical jokery more or less under control,[26] Landau built Kharkov into an important and innovative department, centred around low-temperature and nuclear physics. The country's first artificial accelerator of atomic particles was built there, while Landau's low-temperature laboratory was the first in the Soviet Union to work with liquid helium.

As the purges spread from Leningrad to the rest of the nation, however, inevitably Landau was drawn into trouble. If nothing else, Landau's grating personal style was more than enough to get him labelled as a troublemaker. In December 1936 he had a personal quarrel with the rector of Kharkov University, and told his friends he was likely to be fired. His friends rallied round, resigning from their part-time teaching jobs in protest, and Landau retained his post. Twenty months later, this episode – ugly, but fairly ordinary – got the rector arrested. More: he disappeared, never to be heard from again. Paranoia and fingerpointing ensued and Landau, aghast at the accusations and denunciations being hurled at him, fled to Moscow. A few weeks later friends in Kharkov received a message: Landau had decided to take a job in Moscow, at the new Institute for Physical Problems directed by Peter Kapitsa.

Landau had already worked with Kapitsa, albeit briefly, at the Cavendish Laboratory in Cambridge. (This was at the turn of the 1930s, when foreign travel was much easier.) Kapitsa, the son of a military engineer and general of the Russian Army, had come to the Cavendish during the Civil War and had never wanted to leave. To his mother he wrote: 'to go back to Petrograd – to struggle without electricity and gas, with the lack of water and equipment – is impossible. I have only now come into my own. I find success exhilarating and my work inspiring.'[27] Kapitsa's work, studying alpha particles, had required generating enormous magnetic fields, and the machinery he designed for his work had pushed the Cavendish Laboratory out of tabletop experimentation and into modern, large-scale machine research.[28]

Eventually the Soviet authorities, keenly aware of Kapitsa's practical, industrial approach to physics, decided his talent was an asset they could no longer afford to keep lending out. In 1934, on one of his usual summer trips home (visiting his mother in Leningrad and the low-temperature laboratory in Kharkov) Kapitsa learned that he would not be allowed to leave the country.[29]

After an unconscionable delay, the new Institute of Physical Problems was set up within the Academy of Sciences and Kapitsa was appointed its director – news he came across by accident while reading the newspaper: 'Without doing anything in reality, the Academy already published about the new institute. This is of course very touching and very characteristic of us; we always talk and do not.'[30]

After a year of humiliating privations – of holes in his shoes and worn-out clothes ('I am a useful man [so I was told]', he once wrote to his wife, 'I cannot go about without pants'), Kapitsa acquired an apartment in Moscow, a personal Buick, and a dacha in the Crimea for his family. Still, as he wrote to Niels Bohr:

In general the position of science and research people is somewhat peculiar here. It reminds me of a child with a pet animal which is tormented and tortured by him with the best intentions. But indeed the child grows up and learns how to look properly after his pets, and make of them useful domestic animals. I hope it will not be long to happen here.[31]

Within a year, Kapitsa offered the world perhaps his greatest scientific discovery – the superfluidity of liquid helium. The problem now was to understand it, and for this Kapitsa required the help of a theoretician.

Kapitsa hired Landau in March 1937. Six months later the purges swept Kharkov. Landau's colleagues Shubnikov and Rozenkevich were arrested, forced to confess to 'espionage' and 'sabotage,' and were executed after a short trial. Landau figured in some of their extorted confessions as co-conspirator, but his being in another city delayed his arrest for another half a year. He was arrested in Moscow on 28 April 1938.

Kapitsa immediately sent a personal letter to Stalin pleading for Landau's release. He got no reply. A year later, in April 1939, he wrote to Vyacheslav Molotov, the highly placed Old Bolshevik who had just been appointed Minister of Foreign Affairs, and claimed that he needed Landau's help to understand his recent discoveries. Molotov permitted Kapitsa to bail Landau out of jail so long as he gave a written promise that he would keep Landau from committing further 'counterrevolutionary' acts. (The deal paid handsome dividends in the end: Landau won the Nobel Prize in 1962 for his theoretical explanations of superfluidity and superconductivity; Kapitsa's practical accomplishments in the field won him a Nobel in 1978.)

Kapitsa's willingness to write personally to Stalin and his inner circle speaks to his courage, and also to his sense of his own importance. He once described himself wryly as 'a hot-house plant under the special care of the government'. Playing the holy fool, telling truth to power, required not just bravery but arrogance. His letters convey the strong impression that he wanted the whole science base turned over to him.

But his strategy was not nearly as suicidal as it might at first appear. The Party could not abide opponents, but it actively encouraged critics. Only through criticism and self-criticism could democracy survive the necessary period of dictatorship by the proletariat.[32] As long as you couched your argument as a grievance, rather than an opinion, you could get away with a great deal.

Even as the purges rolled across the country, Soviet citizens were being encouraged to write critical letters to public officials or to the newspapers. Kapitsa was hardly an ordinary citizen, but he acted as though he was as entitled as any factory hand to tick off officials who got in the way of his work. All in all Kapitsa wrote about forty-five letters to Stalin, dozens each to Politburo members Molotov and Malenkov, and several dozen to other political leaders. These letters were neither servile nor formal. He was not a bourgeois scientist claiming rights for himself he did not possess. He was a diligent worker struggling to get something done in the face of an inefficient bureaucracy:

But then, what sort of a Government are you, that you can't get a little two-storey house built on time and put its ten rooms in order after assembly? In that case you are simply wets! . . . Well, that's the picture of what's happening. Imagine you saw a violin at your neighbour's and you were able to take it from him. And what do you do with it instead of playing on it? For two years you use it to hammer nails into a stone wall.[33]

Only later, when Kapitsa began to wield real industrial power, did his complaints land him in trouble. That was the problem with having Stalin as a patron – an unavoidable problem, as Stalin rid the country of every patron but himself. You were safe only as long as you could demonstrate your powerlessness. And if Stalin raised you, it was inevitable that, sooner or later, he would cut you down.

By extinguishing the intellectual life of Leningrad, Stalin ensured that power in the Soviet Union was concentrated exclusively in Moscow. Everything was to be drawn to the centre. Obsessive centralisation was to be a defining characteristic of the rest of his rule. Centralisation took many forms, from Stalin's own neurotic insistence on his own expertise in every imaginable field, to the way unlike institutions were folded into each other, so that everybody had to compete for half as many jobs. Though the consequences of centralisation were often harmful, things rarely descended into chaos, and this was thanks to the party's system of *nomenklatura*. *Nomenklatura* was, literally, a list of posts that could not be occupied or vacated without permission from the appropriate Party committee. All Party committees, from the Central Committee to the remotest rural outpost, had a list of jobs for which they were responsible. Originally meant as a way of organising the Party itself, the *nomenklatura* system spread, as Stalin's influence spread, through all institutions of government. By the early 1930s the *nomenklatura* system had subsumed the scientific community.

The *nomenklatura* system was hierarchical. The higher your post, the more important the committee you answered to. The posts of president, vice-president, and scientific secretary of major institutions were the purview of the Politburo. Institute directors and the editors-in-chief of academic journals answered to the secretaries of the Party's Central Committee. Running a laboratory brought you onto the *nomenklatura* of a regional Party committee. Even librarians were hired or fired by local Party committees. Meanwhile a special government agency, the Supreme Certifying Commission, was created to approve every degree and title awarded throughout academia.[34]

Centralising intellectual work inevitably favoured those with a talent for bureaucracy. No matter the quality or originality of your thought, your career depended on how well you served your institution. Could you be trusted to approve research budgets? Could you procure enough paper for books? Did you know the right people to obtain permissions to travel abroad to conferences?

In such an environment, only bureaucrats flourished, and inevitably the best of them had learned at the feet of the Party. In January 1939, a number of Party functionaries were 'elected' to the Academy of Sciences, notably 'Stalin's Prosecutor' Andrei Vyshinsky and the philosophers Pavel Iudin and Mark Mitin, a Marxist nihilist who memorably declared that 'There is not and cannot be a philosophy that wants to be considered Marxist-Leninist philosophy while denying the necessity of ideational-political and theoretical leadership on the part of the Communist Party and its leading staff.'[35] Mitin's belief that science was nothing but an instrument for the development of technology endeared him to Stalin. In the 1930s, he was considered the leading Soviet philosopher.

No one was much surprised when, in 1939, Stalin himself threw his hat into the ring, establishing a prize in his own name for scientific research, the Stalin Prize. At the same moment the Academy of Sciences elected this great patron as an honorary member. 'The Coryphaeus of Science', they called him. The leader of the band.

13 : 'Fascist links'

*Like the young Ladies and Gentlemen of Boccaccio's
Decameron, we were brought together by withdrawing from
the terrors of a great plague to jointly consider some of the
riddles of life.*[1]

Max Delbrück to Nikolai Timofeev-Ressovsky,
1 October 1970

In January 1932 Hermann Muller, the American geneticist who
had brought *Drosophila* as experimental animals to the Soviet
Union, disappeared from his laboratory at Austin, Texas.

His wife Jessie called the university, asking anxiously if anyone
had seen her husband. Eventually Muller's students formed a
search posse and went out looking for him in the woods near the
outskirts of town. They found their professor walking about in a
confused state, dishevelled and spattered with mud. He had over-
dosed on barbiturates in an attempt at suicide.

The next day he turned up to class as if nothing was wrong.
There was a great deal wrong.[2] Muller had been working at
night to avoid colleagues he had irretrievably alienated, and as a

Nikolai Timofeev-Ressovsky leans between Cécile and Oskar Vogt
for a photo at Berlin-Buch. The man second left is unidentified;
Hermann Muller stands far left.

consequence his marriage had all but collapsed: he and Jessie had already discussed a separation.

Muller's work, far from suffering, bloomed: all his energies and hopes were now devoted to genetics and its practical, medical applications. Just days after trying to kill himself, he was at the American Museum of Natural History in New York, speaking to the Third International Congress of Eugenics.

'The Dominance of Economics over Eugenics'[3] was a landmark performance: a radical critique of the eugenic project, very much in the spirit of early critics like Peter Kropotkin, and armed with findings and insights from the latest research. American eugenics, Muller said, was based on false premises. Pauperism, vagrancy, feeblemindedness and criminality were almost certainly not innate. You couldn't begin to understand the genetic component of human social behaviour until you had established an egalitarian society. Only under socialism, where all children, regardless of gender or race, had equal opportunities, could you even consider a eugenics programme.

Not long after, the reckless professor from Texas was once again on the public stage, this time presenting 'Further Studies on the Nature of Gene Mutation'[4] to the Sixth International Congress of Genetics at Cornell. It was, many said, a work of some genius: a capstone of classical, 'Mendelian' genetics. Others, aware of his recent overdose, were troubled by Muller's confused and slapdash delivery. Thomas Hunt Morgan, himself convalescing from a bad car accident, knew a casualty when he saw one: 'Something is wrong with Muller,' he said.

What Muller desperately needed at this point, and what he got, was a change of scene. Awarded a Guggenheim Fellowship, in November Muller left for Berlin and Oskar Vogt's Institute for Brain Research.

By the time Muller arrived, the Timofeev-Ressovskys, Nikolai and Elena and their son Dmitry, had already spent four years in Berlin. They had lived in the centre at first, Vogt's institute at that time being little more than a shell company, with a lab in an old tenement building.

Since then, with backing from both the German state and the Rockefeller Foundation, a new six-storey building had risen in Buch, a leafy suburb about twenty-five kilometres from the centre of town, nine kilometres outside the city limits. It was not exactly a rural idyll. The new building stood among regimented conifer plantations, part of a 15,000-bed complex of hospitals and clinics. The site had been meant for a cemetery, until the ground had turned out to be unsuitable. But a short walk led you into farmland, giving a sense of relative seclusion that would prove very welcome as the political situation deteriorated.

Construction of the new building was completed by the end of February 1930, by which time it was already housing an imposing number of laboratories, study centres and animal departments using monkeys and other exotic animals. The Timofeev-Ressovskys and the Vogts shared a large house overlooking the main building where some ninety co-workers had rooms.

Nikolai and Elena's work on *Drosophila* was wide-ranging and attracted international interest. They had together studied wild populations of *Drosophila* in the grounds of Buch, and confirmed by experiment Sergei Chetverikov's revolutionary ideas about population biology. In the laboratory, Nikolai was exposing *Drosophila* to X-rays to determine the size of the gene.

Muller stayed at Buch between November 1932 and September 1933 to contribute to these X-ray studies.[5] The men set out to estimate the nature and extent of the gene by recording how many mutations of a particular gene were triggered in that tissue when exposed to X-rays.

Their results, acquired after the most painstaking study – hundreds of hours spent studying thousands of flies through the microscope – were suspiciously easy to interpret. Whether their X-ray dose was administered in a single shot, in several small shots, or continuously at a low level over a long time, a certain level of X-ray exposure invariably caused a certain number of mutations. And however weak the X-ray dose was, it always, eventually, triggered a mutation.

Put simply, genes were getting knocked sideways at an entirely predictable rate, almost as if they were coconuts in a fairground booth, and X-rays were the balls being thrown at them. From this, Timofeev-Ressovsky developed what he and his later collaborators nicknamed 'target theory': the idea that the gene really was a discrete, solid mass, which X-ray photons either hit or missed.

Measuring the gene effectively became a very complicated, small-scale game of Battleships. Timofeev-Ressovsky played this game throughout the 1930s, first with Muller, then with Karl Zimmer, a research radiologist, and Max Delbrück, a young physicist inspired by the new science of genetics. With these later researchers Timofeev-Ressovsky published a landmark paper[6] revealing that a *Drosophila* chromosome contained no fewer than 10,000 and no more than 100,000 genes – and that a gene was a sphere, one to ten microns across. This result, crude by modern standards, was of immense importance: it showed a generation of physicists what they might be able to achieve if they shifted their attention to biology. After the war, this new field of 'molecular biology' transformed the life sciences.

Muller's stay in Berlin brought him into contact with extraordinary people. The Timofeev-Ressovskys' circle included Bohr, Dirac, Schrödinger, Vernadsky, Darlington, Haldane . . . Nikolai was a regular in Copenhagen and other elite European seminars, and with the support of the Rockefeller Foundation, organised annual conferences on genetics, biophysics, and radiation biology right up until the outbreak of war in 1939.

Oskar Vogt, concerned that Timofeev-Ressovsky's celebrity would eventually lure him to the United States, recommended that he become the head of the institute's Department of Genetics. By signing a contract at the beginning of 1932, Timofeev-Ressovsky now enjoyed the tenure and salary of a university professor.

Everything Muller told Timofeev-Ressovsky about academic life in the USA persuaded the Russian that he had made the right decision by staying put in Germany. Mired in the Depression, American academics had seen their wages cut by a third. Professors were being dismissed from universities and laboratories by the dozen, and funding for scientific research was drying to a trickle. Muller's bleak accounts, and his fierce criticism of American eugenics policy, convinced Timofeev-Ressovsky not to move to the USA, even as the political situation in Germany deteriorated and war loomed.

On 30 January 1933, the National Socialists took power in Germany. Of the Brain Institute's one hundred workers, four were communists and seven were Nazis, and all hell broke loose, with Vogt firmly resisting calls to fire any Jewish, Social Democrat or communist workers. Suspicions, claims and counterclaims circulated wildly. One of the Nazis at the institute was a cleaner who, in a denunciation of his employer, quoted Oskar Vogt's view that 'National Socialism is an unexplored toxic bacillus. Hitler is an uneducated man, and the party consists of murderers and criminals.'[7]

In May the Nazi civil service law was extended to the Kaiser Wilhelm Society. All Jews were immediately dismissed except for institute directors, who were allowed to continue in their posts until the Nuremberg laws were passed in 1935.

In March 1933 Vavilov visited Berlin on his return from the genetics congress in Ithaca. Seeing the political situation for himself, he offered Muller the chance to head the laboratory he had inherited from Yuri Filipchenko, his opposite number in Leningrad, and which had been grandly rechristened the Academy's new Institute of Genetics.

Muller needed little further encouragement. Levit and Agol had already painted him a glowing picture of Russia's socialist future while working with him in Texas. In Buch, meanwhile, Nazi stormtroopers were repeatedly raiding the institute, forcing their way through the windows and breaking furniture as they hunted, they said, for an underground member of the Communist International. (Inevitably, they detained Muller, though he was soon released.)

In September 1933, having been elected to the USSR Academy of Sciences, Muller moved to Leningrad.

He did not travel light: 10,000 glass vials, 1,000 bottles, two microscopes, a suitcase packed with food-making equipment, a 1932 eight-cylinder Ford, two bicycles, trunks of clothing, books and personal effects arrived with him. Accompanying him were his long-suffering wife Jessie and his lab assistant Carlos Offerman.

The Soviet Union afforded Muller a hero's welcome. Muller in his turn published articles in popular magazines praising the collective farms and the government's generous funding of science. Late in 1934, when both the Academy of Sciences and its Institute of Genetics were moved to Moscow, Muller moved too.

It is possible that Muller had no idea just how unlikely his greeting was, and how fragile his welcome. He seems to have had a talent for bad timing. Having arrived in Berlin in time for the ascension of the National Socialists, he now came as an honoured guest of the Soviet state even as the country was gearing for another world war with an isolationist zeal that bordered on xenophobia. Beginning in 1934, the exchange of reprints with foreign scholars became officially possible only through the All-Union Society of Cultural Relations with Foreign Countries (VOKS), in all but name a branch of the secret police. Papers submitted for publication abroad were censored by the Main Directorate on Literature and Presses (Glavlit). The Main Directorate also controlled the distribution of foreign literature, including scientific journals and books, and periodically removed 'harmful' literature from libraries. The Academy of Sciences stopped electing foreign members, and in 1936 a broad press campaign was organised against 'servility to the West'. By 1939 Soviet science's international contacts were almost completely severed.

Muller seems to have taken no note at all of the worsening political situation. He had come to the Soviet Union hoping to create a socialist eugenic society, and saw no reason to divert from his grand purpose. In his book *Out of the Night*, published in Britain and the United States in 1935, he resurrected, without attribution, Alexander Serebrovsky's discredited proposal from 1929 recommending a large-scale programme of human artificial insemination, using the sperm of eugenically selected male donors. 'How many women,' he enthused, 'in an enlightened community devoid of superstitious taboos and of sex slavery, would be eager and proud to bear and rear a child of Lenin or of Darwin!'[8] Neither the Great Purge nor the press attacks on fascists deterred him. In the spring of 1936 he sent a copy of *Out of the Night* to Stalin.

Dear Comrade Stalin! As a scientist with confidence in the ultimate Bolshevik triumph throughout all possible spheres of human endeavour, I come to you with a matter of vital importance arising out of my own science – biology, and in particular, genetics . . . The matter is . . . the conscious control of human biological evolution. This is a development that bourgeois society has been quite unable to look squarely in the face.[9]

Muller asserted in his letter that

it will be possible within only a few generations to bestow the gift even of so-called 'genius' upon practically every individual in the population – in fact, to raise all the masses to the level at which now stand our most gifted individuals, those who are helping most to blaze new trails to life.

Most people could have 'the innate qualities of such men as Lenin, Newton, Leonardo, Pasteur, Beethoven, Omar Khayyam,

Pushkin, Sun Yat Sen, Marx, or even to possess their varied faculties combined.'

'After twenty years', he wrote,

there should already be very noteworthy results accruing to the benefit of the nation. And if at that time capitalism still exists beyond our borders, this vital wealth in our youthful cadres . . . could not fail to be of very considerable advantage for our side.

This claim put in the shade even Lysenko's bid to grow new varieties of maize in two-and-a-half years. Muller's promise of quick results reflects the same kind of opportunism. But his flattery of Stalin and use of Marxist rhetoric all came to nought: Stalin hated the book[10] for the very same reasons Bolshevik commentators and headline writers had hated Serebrovsky's 1929 paper. Unless you had a firm grasp of the science, and a deep knowledge of the political background, *Out of the Night* appeared to be peddling the very racial theories spouted by the fascists. With the rise of the Nazis in Germany, Russian leaders and thinkers were frantically back-pedalling on their own Nietzschean excesses. Talk of creating new forms of human being – a commonplace of Bolshevik rhetoric just a decade before – was now considered fascist. Muller had no idea just how much damage he had done to himself, to his friends, and to the field of Soviet genetics.

The first intimation that something had gone wrong came on 14 November 1936, when the Politburo decided 'to cancel the convocation of the Seventh International Genetics Congress in the USSR in 1937'. Muller was involved in organising the congress, which was meant to be an important political coup for the Soviet Union. He was bitterly disappointed.

The congress's ambitious programme had included numerous excursions to universities, research institutes, laboratories and museums, not only in Moscow, but also in the provinces. A centre-piece of the show was supposed to be the brand new home for the

Academy's Institute of Genetics. But its construction, begun in Moscow in April 1936, was taking much longer than expected. Turning the cancellation into a mere postponement, an official Soviet announcement explained the confusion: the congress was simply not going to be ready in time. This explanation was probably true.[11]

Nonetheless, in the autumn of 1936, Muller privately let his friends know that he 'was not overenthusiastic about recent developments in the USSR' and if he 'could get a decent job somewhere else after the congress', he would be 'glad to accept'.

The whole eugenics issue was becoming especially sensitive, open to misinterpretation and false assertion. The Spanish Civil War was bringing the Soviet Union and Germany to the brink of open conflict. The USSR was the only state openly supporting the Republicans, while Germany backed Franco. Spying for Germany was a typical charge during the show trials of that time, and the Soviet press had launched an extensive campaign of anti-fascist propaganda.

On 13 November 1936 the Science Section of the Moscow Party called a meeting of biologists and physicians at the House of Scientists to unmask 'the fraud of Fascist and para-Fascist scientists' and the 'racist falsification of biology'. Medical genetics was denounced as 'fascist' – invective that found its way into print in *Under the Banner of Marxism* through its editor Ernst Kolman. An article under the uncompromising headline 'The blackguard nonsense of Fascism and our Medical–Biological science' attacked Levit and his co-workers for holding 'Fascist views' on human genetics. On 4 December Levit was expelled from the Party 'for smuggling hostile theories into the publications of the Institute, and for Menshevist idealism', and found himself being followed about the streets – none too subtly – by agents of the NKVD.

On 14 December, the *New York Times* published a dispatch from its Moscow correspondent covering the cancellation of the Seventh International Congress of Genetics.

An interesting story of a schism among Soviet geneticists, some of the most prominent among whom are accused by Communist party authorities of holding German Fascist views on genetics and even being shielders of 'Trotskyists,' lies behind the cancellation. The fact that so many of the Soviet Union's most distinguished geneticists are under fire is believed to be motive for the government action.

The dispatch also announced that Nikolai Vavilov, the president of the forthcoming congress, and Izrail Agol, the former Rockefeller fellow, had been arrested. On 17 December, the *New York Times* added fuel to the fire, publishing a sarcastic editorial on the cancellation of the genetics congress under the title 'Science and Dictators'.

TASS, the Soviet news agency, routinely prepared internal government summaries of important Western news, and on 19 December, its 'Bulletin of Foreign Information' landed on Stalin's desk. Two days later *Izvestiia* published an unsigned editorial article, heavy edited by Stalin, under the title 'A Response to the Slanderers from "Science Service" and "New York Times"'.

'The New York Times simply lied about Vavilov's arrest,' the editorial declared. True, Izrail Agol had been arrested, but his arrest had nothing to do with genetics or the congress; he had been arrested for 'his direct connection to Trotskyist murderers'. The next day a lengthy 'Telegram to the New York Times' appeared, in which Vavilov angrily denied the story of his own detention.

These controversies came to a head just a few days later at the fourth annual session of the Lenin Academy, that vast conglomerate of all things agricultural. The organisers of the session, which ran from 19 to 26 December, had never planned it to be some kind of extra-judicial venue to try genetics. The session's main purpose was simply to sum up the results from the Lenin Academy's 'socialist competition' in plant breeding. But the relatively small number of people involved in genetics meant that a political controversy in one corner very quickly coloured people's responses to the entire field.

The unmasking of apparently 'fascist' scientists with strong professional and personal links to the Lenin Academy, even as the Soviet people were recovering from the worst famine in living memory, was a press story worth pursuing. Daily reports in *Pravda* built up tremendous public interest in the session, which had to be relocated half way through when the audience of 700 expanded to more than 3,000.

Alexander Ivanovich Muralov, the new head of the Lenin Academy, tried to maintain an even-handed debate. Both Lysenko and his allies and the geneticists had reason to trust him. Muralov had replaced Vavilov as president of the Lenin Academy eighteen months before, delivering a speech entitled 'Do Not Lag Behind Life', which praised Lysenko's Odessa Institute as a model of close contact between science and farms. On the other hand, relations between Vavilov and Muralov were good. Losing the presidency to Muralov had been a demotion, obviously, but not an unwelcome one for Vavilov, given the bureaucratic pressures on him. (He was now one of three vice-presidents, which gave him time for science; the presidency had been a full-time bureaucratic post.) Indeed Vavilov, as president of the forthcoming genetics congress, had put Muralov at the head of the organising committee because Muralov's political connections made him the better spokesman.

Muralov's reputation was one of warmth and fairness – but he was in serious trouble, and very distracted. Trotsky had been his friend and champion: 'Muralov is a magnificent giant, as fearless as he is kind,' Trotsky had written. 'In the most difficult situations he radiated calm, warmth, and confidence.'[12] That sort of affiliation was by now quite enough to get a person killed. Muralov's older brother Nikolai was already in prison as a Trotskyite and an 'enemy of the people' and was about to appear at the second Moscow show-trial in January 1937. (He was shot;

Alexander Muralov was arrested that autumn, and executed in March 1938.)

There was another problem with Muralov – one which came to shape the Lenin Academy's session in a way that sent a chill through Vavilov and the field of genetics: Muralov knew nothing whatsoever about genetics. Indeed, he made Lysenko look knowledgeable. And he believed (or at any rate fervently hoped) that the 'criterion of practice' would solve all controversies.

This belief in the 'criterion of practice' was knocked sideways from the first when Lenin Academy members Peter Konstantinov and Peter Lisitsyn, together with the Bulgarian geneticist and plant breeder Doncho Kostov, repeated charges they had already made in print about Lysenko's scientific work in Odessa. This account was so negative as to suggest that Lysenko's vernalisation, to which the state had already devoted huge resources, was nothing more than a fraud. They cited cases where vernalised grain had been given preferential treatment in trials; and others where negative results had been discarded. They considered Lysenko incapable of holding a 'rational discussion'.

This should have been a knock-out blow, and the focus of the whole session, but Lysenko blindsided Muralov by simply brushing these accusations aside. Such quibbling belonged to old and outdated bourgeois science, he said. Why were they even talking about this? The successes of vernalisation could 'be read in the press every day, centrally, regionally and locally'. As far as Lysenko was concerned, the test of practice was to be measured in point sizes and column inches. Anything else was obfuscation.

Muralov, president of a vast government organisation that had staked its reputation on vernalisation, let the matter go.

The second half of the Lenin Academy's session was devoted to genetics. Lysenko's recent pronouncements – embracing Ivan Michurin as the founder of a 'correct' homespun biology, denying the reality of the gene, and so forth – were hugely embarrassing for Vavilov, who still wished to be considered Lysenko's supporter and patron. In his opening speech, Vavilov attempted to smother the controversy, but his equivocations convinced no one. His insincerity actually drove one of his students to Lysenko's camp, and in declaring his defection, the student lampooned his old teacher for failing to declare himself. Vavilov was 'a vegetative chimera, whose individual parts are incompatible. He is both . . . a Michurinist and an anti-Michurinist; he is both a Lysenkoite and an anti-Lysenkoite.'[13]

That speech – and that desertion – wrought a change in Vavilov. Clearly, the centre would not hold. Agrobiology was splitting, irrevocably, into two camps. He was going to have to take sides.

In one corner stood Lysenko, Prezent and a group of Lysenko's associates from the Odessa Institute; on the other, a group comprising Koltsov, Muller and a small but very high-functioning caucus of biologists who had dedicated themselves to establishing a Marxist position in their field long before the effort had become both compulsory and vacuous. Serebrovsky, Mikhail Zavadovsky, Dubinin and Anton Zhebrak – like Lysenko, a peasant's son – were all sincere Marxists, and their opposition to Lysenko, who seemed determined to pit Marxism against good scientific practice, was loud and fierce.

The urgency these men brought to the debate over genetics was new to Vavilov. Regardless of what they believed, Vavilov and Muralov and the plant breeders of the Lenin Academy had a vast bureaucracy and a huge range of activities not related to genetics to fall back on, should genetics become politically dangerous. Serebrovksy and his fellows had no such easy escape route. They were animal geneticists. They were committed wholeheartedly to the reality of the gene. Their jobs, careers and reputations depended on it. They had everything to lose.

Serebrovsky spoke for them all: 'Under the supposedly revolutionary slogans "For a truly Soviet genetics", "Against bourgeois genetics", "For an undistorted Darwin", and so forth,' he declared, 'we have a fierce attack on the greatest achievements of the twentieth century, we have an attempt to throw us backward a half-century.'[14]

If the passions of the animal geneticists startled Vavilov, their effect on Lysenko and Prezent can only be imagined. The pair resorted to ever more extreme theoretical positions, the more the men from the Communist Academy laid into them.

This, however, in a debate chaired by Muralov, who refused to consider these matters in detail (and who was, in fact, wholly incompetent to do so), merely gave Lysenko and Prezent the opportunity to grandstand. The fact was, Lysenko and Prezent *needed* opposition. They defined themselves by it. By the time he delivered his speech 'Two Trends in Genetics', Lysenko's method was clear: reject any tenet held by the geneticists and embrace its opposite.[15] Lysenko indignantly rejected accusations of denying the existence of genes and in the same breath he stretched the scope of the terms 'genes' and 'genotype' to the point where they became meaningless. One by one he appropriated genetic terms and rendered them absurd. One fed-up geneticist remarked it was like trying to talk to a builder of perpetual motion machines.

In several speeches, people offered Lysenko the opportunity to back down from his increasingly crazy position. Clearly, Isaak Prezent was responsible for this mess! Lysenko would have none of it. Praising Prezent, he went so far, finally, as to deny the existence of genes:

We deny little pieces, corpuscles of heredity. But if a man denies little pieces of temperature, denies the existence of a specific substance of temperature, does that mean that he is denying the existence of temperature as one of the properties of the condition of matter? We deny corpuscles, molecules of some special 'substance of heredity', and at the same time we not only recognise but, in our view, incomparably better than you geneticists, we understand hereditary nature, the hereditary basis of plant forms.[16]

The Lysenkoists were on firmer ground when it came to politics. It did not matter to them whether you were a bourgeois like Koltsov or a Marxist like Serebrovsky. Exploiting ties between the early development of Soviet genetics and eugenics, and tying that to Nazi concepts of a 'higher race', the Lysenkoists accused their opponents of nurturing 'fascist links'.

The session had from the outset and on purpose omitted human genetics from its agenda. But there was no avoiding it now, and it was in fact the geneticists who brought up the subject, when Nikolai Koltsov read aloud a Russian translation of Hermann Muller's speech.

That Muller characterised Lysenko's ideas as 'quackery', 'astrology' and 'alchemy' surprised no one. As he had himself written to Stalin, when delivering to him that ill-starred book *Out of the Night*,

There is one means and only one whereby a worthwhile beginning may be made in the direction of providing more favourable genes. This is not by directly changing the genes, but by bringing about a relatively high rate of multiplication of the most valuable genes that can be found anywhere. For it is not possible to artificially change the genes themselves in any particular, specified directions. The idea that this can be done is an idle fantasy, probably not realisable for thousands of years at least.

When Muller accused the Lysenkoists of unacceptable views on human genetics, however, the temperature in the hall rose sharply. In a letter to Julian Huxley a little later, Muller explained how he had 'called attention to the fascist race and class implications of Lamarckism, since if true it would imply the genetic inferiority, at present, of peoples and classes that had lived under conditions giving less opportunity for mental and physical development'. The audience 'applauded wildly, but there was a terrific storm higher up and I was forced to make a public apology, while the statement was omitted from the published address'.[17]

So much for the main body of the speech: Koltsov then gave the floor to Muller for his concluding remarks. Muller dealt with Lysenko's accusations of 'fascist links' with aplomb. If genetics was fascist, how were they to explain the message he had just received from the English geneticist John Haldane, who was dropping his laboratory work in order to go to Madrid to defend it against Franco's forces? Indeed Muller too had made plans, already advanced, to offer his scientific and medical support to the International Brigades prosecuting the Spanish Civil War!

The next morning Muller met Vavilov in his hotel. Vavilov was extremely anxious. Muller had effectively accused the Soviet government of fascist sympathies, and a public retraction of some kind was urgently required: 'Neither Vavilov, nor others, said any more to me about the matter,' Muller later recalled, 'but there were evidences that the chasm between the opposed groups had permanently widened.'[18]

Muller issued a rather mealy-mouthed retraction and left the Soviet Union temporarily for Berlin. *En route*, he wrote to his friend Julian Huxley, noting unhappily that the Soviet Union 'is hardly the place, at present and probably for some years to come, where one can hope to develop genetics effectively, let alone the application of genetics to man which I had hoped might gradually be introduced'.[19]

Arriving in Berlin, Muller headed directly to the Timofeev-Ressovskys' home at the Brain Institute. At this point he probably did not know that Izrail Agol, the Soviet geneticist who had studied under him at Texas, had been shot. But the news he did bear was bad enough: Nikolai Timofeev-Ressovsky's colleague Vlad Slepkov, who had been recalled to Russia from Buch, had been shot. Nikolai's brother Vladimir, who had worked in Leningrad with Sergei Kirov, had been shot. Another of Timofeev-Ressovsky's brothers, Dmitry, was imprisoned, and so were many of his wife's relatives. Muller passed on to Timofeev-Ressovsky Koltsov's and Vavilov's explicit wishes that he stay put for the sake of his own and his family's safety.[20]

Muller returned to Moscow briefly in September to pack up his things, then left for good.[21] 'As time passed, the sad conclusion was forced on me that in the USSR genetics was passing under too much of a cloud for my return there to be of any help.'

For Solomon Levit there were no exits to take, no gestures to make. Stalin, as patron, permitted no rivals, no alternatives, and the Bolshevik idealists of the Communist Academy were particular targets of his Great Purge.

On 5 July 1937 Levit was removed as director of the Medical–Biological Institute. On 17 September his institute was closed down 'for the purpose of organising a truly scientific study of medical genetics'. Levit's internal passport was suspended. He had nothing to do but remain in Moscow and await arrest. At home, he hid the newspapers in case his wife and daughter came across printed attacks on him. Every day he left home, pretending that he was going to work; he headed instead to the Lenin Library. The NKVD shadowed him closely. One day he managed to get away and made a payphone call to a close friend, asking him to look after his wife and daughter. The friend snubbed him.

Levit's arrest was scheduled for the night of 10 January 1938. To let his daughter sleep Levit led his police escort on a walk through night-time Moscow. In the morning NKVD agents found him at his apartment, talking with his daughter who was still tucked up in bed.

They took him to the Lubyanka prison and accused him of being an American spy. He held firm for four months, then signed a false confession in return for a phone call with his daughter.

On 17 May Levit was sentenced to death for terrorism and espionage. He was shot twelve days later.

14 : Office politics

They did not want to make room for a simple peasant like me.[1]
   Trofim Lysenko

Lenin Academy vice-president Georgy Meister was a crop breeder who actually did understand something about genetics. When he summed up December's Lenin Academy session, he was punctilious in his even-handed account. But he could see well enough the danger implicit in Lysenko's contrarian behaviour. 'We cannot permit a groundless condemnation of theories and methods whose practical value has been established through broad scientific experience,' he warned. The press worried him too:

We know that some not so well-considered statements by Lysenko on pure lines, inbreeding and genetics have in certain places been interpreted to mean that work with inbreeding is almost a counter-revolutionary activity and some of our newspapers and publishers refuse to print articles about inbreeding and even about genetics.[2]

Trofim Lysenko measures the growth of wheat in the field of a collective farm near Odessa in the Ukraine.

Away from the headlines and the popular press, the geneticists had won a considerable victory. Experimental work on 'issues of heredity' was expanded. Additional funding was found for genetics research. Even the cancelled International Genetics Congress was revived, when the Politburo agreed to reschedule it for Moscow in 1938.

But the popular press could not be ignored. It mattered. It was fast becoming the final arbiter in the controversies its own headlines whipped up. The fate of Leningrad's intellectuals, the disappearance of Levit and the death of Agol were proof enough of that. A cold wind of popular opinion was blowing through science, and it made an oppressive impact on the pioneering biologist Nikolai Koltsov. He wrote to the editors of *Pravda*, criticising them for their 'biased and often completely illiterate publications about the meetings of the session'. Meister's own summing-up had been so badly distorted in the editing that 'this "truth" undermines faith in *Pravda*'.[3]

He asked Muralov to support the publication in *Izvestiia* and *Pravda* of 'extensive articles, written by genuine geneticists in defence of their science', but Muralov, with his brother facing the firing squad and his own arrest looming, simply reiterated his (by now obligatory) faith in citizen science. The problems of genetics had to be discussed in the wide circle of scientists and production workers, not only in a closed circle of narrow specialists. Koltsov was being elitist.

Meanwhile Lysenko and Prezent and their patron Yakovlev used the popular press to heap up the pressure on the geneticists. For them, any study of human heredity was *ipso facto* fascist. Prezent wrote in *Pravda*: 'It doesn't matter who has taught whom, the fascists Koltsov or Koltsov the fascists. The fascists, following Koltsov's programme, are physically destroying thousands and thousands.'

*

Towards the end of March 1937, the Lenin Academy summoned a special session to discuss how its scientists and bureaucrats were going to work together in the light of the new Soviet constitution. Passed on 5 December 1936 this constitution had greatly strengthened the system of *nomenklatura* on which everyone's job depended.

The March *activ* was essentially an informal local meeting for Moscow employees. That it opened its doors to the public was not significant: Russian science had always had a public face. Public lectures and public defences of dissertations were commonplace. What was significant on this occasion – and shocking – was the speed with which this informal talking-shop generated into a shouting match between two very clearly demarcated camps, with Lysenkoists on one side and Koltsov and the geneticists on the other.

Even this might not have mattered had the Lenin Academy's President Muralov – by now fighting for his freedom and his life – not used his position to demonstrate his political credentials. He went on the attack immediately, using his opening remarks to single out Koltsov, whose letter to the Lenin Academy following the Fourth Session had shown, he said, that 'we have not organised our cadres to fight the bourgeois worldview'. He then proceeded to define this 'worldview' by quoting from Koltsov's *Russian Eugenics Journal* articles of 1929. To hear Muralov tell it, Koltsov's playful Martian fantasies were now evidence of his innate fascism.

The threat to Koltsov was clear, but what comes across equally clearly from the transcript is Muralov's own insecurity. The chief victims of the purges were party bureaucrats, and the higher your position, the more risk you ran. Koltsov and the geneticists were not actually enemies of the state so much as whipping boys for officials whose own lives hung in the balance.

For Koltsov himself, there was an obvious and reasonably easy way out: he could disarm. Under Bolshevik rule, public disputes and discussions of this sort had acquired a theatrical form that had very little to do with the subject at hand and a great deal to do with showing who was boss and who owed loyalty to whom. Self-criticism and repentance were all part and parcel of the show. They had very little bearing indeed on what people actually did and believed, and virtually no bearing at all on their future action. All you had to do to survive was lie.

It was a lesson that would save entire disciplines after the war, and keep brilliant minds employed and out of prison for years. Koltsov, however, was the child of a very different generation. He did not know how to bend. Muralov, he said, had wildly and deliberately misinterpreted his 1929 work. For their time and context, his claims were legitimate and scientifically well founded; why should he take back a word of what he had said?

Koltsov's Galileo act, posing proudly in defence of science, infuriated the room. The battle now became who held the whip-hand over the work of the Lenin Academy. Muralov had been seeing research into rust-resistant wheat through research institutions for months: 'Should I just keep away from this?' he asked Nikolai Vavilov, the Lenin Academy's former president and, next to Koltsov, the senior representative of the genetics camp.

'Absolutely,' replied Vavilov from his seat.

'Is it not the task of the presidium to organise research so that new varieties can be produced as quickly as possible?' Muralov snapped back. 'You call this bureaucratic interference, but we call it organisation of research.'

'You should take advice from the best specialists,' Vavilov replied, reeling off some names.

'But you did not mention Lysenko and Tsitsin,' Muralov observed.

Vavilov had nothing to say to that. No one needed him to say anything: battle had already been joined.

*

In the autumn Muralov was arrested, and Georgy Meister appointed as acting president of the Lenin Academy. Meister's record as a plant breeder was second to none: varieties of spring wheat produced under his direction were growing on 7 million hectares of Soviet farmland. But even practical successes were not sufficient to save him from the purges. On 11 August he too was arrested as an enemy of the people, lost his mind in prison, and died there the following year.

Nikolai Vavilov himself took the reins of the Lenin Academy for a few months. Then, on 28 February 1938, the inevitable happened. Trofim Lysenko was appointed president, and his close follower Nikolai Tsitsin became a vice-president.

The Great Purge had turned the world upside-down. Lysenko had ascended to the top position in Soviet agriculture, and not because the purges were particularly directed against genetics, but because, in the end, the geneticists had lost more key patrons to the firing squads than Lysenko had. (Lysenko's champion Yakov Yakovlev was killed in July 1938; but the geneticists lost Grigory Kaminsky, the commissar of health; Karl Bauman the Central Committee's science adviser; and several others.)

It did not really matter whether you were a friend of Lysenko or not; your removal made room for him and his star kept on rising. With the Lenin Academy under his control, geneticists now had only one institution to call their own – the Bureau for Applied Botany, with Nikolai Vavilov himself the director.

Lysenko did not have the power directly to remove this prominent opponent from the bureau. Worse, he had to put up with Nikolai Vavilov and Mikhail Zavadovsky as vice-presidents of the Lenin Academy. Official policy still held that all controversies could be solved through practice, and keeping Vavilov and Zavadovsky in office was supposed to encourage 'socialist competition'. In reality, the entire bureaucratic behemoth that was the Lenin Academy came to a lumbering halt.

So Lysenko resorted to underhand means to oust Vavilov

from the bureau. Those means were already in place: Grigory Nikolaevich Shlykov had been a researcher in the subtropical department of the bureau since 1931, and had risen to the position of vice-director, from which lofty position he had published an article titled 'Formal Genetics and Consistent Darwinism', calling the bureau's work 'a grandiose fiasco'.

Shlykov's reports to the NKVD were of another order entirely. Yakovlev having fallen in the Great Purge, Shlykov claimed that Vavilov was part of Yakovlev's group, Yakovlev's 'outwardly negative attitude' towards him being 'a cover for their actual relations as accomplices . . . The baseness and cunning of these people . . . has no bounds.'

To supplement Shlykov's efforts, Lysenko appointed a young specialist (and NKVD major) Stepan N. Shundenko as the bureau's deputy director for science. Shundenko quickly befriended Shlykov and together they set about disordering the bureau's operation. We know at least some of the institute's staff guessed the men's real loyalties because they made up a jingle about a 'weedy little devil' (Shlykov) and a 'pint-size Napoleon' (Shundenko), both of them fashioned from shit.[4]

John Hawkes, visiting from Cambridge in September 1938, got a glimpse of the state of Soviet genetics. He visited Vavilov and found him exhausted, the subject of attacks both professional and personal. It was no secret that these attacks originated from Lysenko. Hawkes wrote:

I can realise now, I think, why the government thinks so highly of Lysenko. He represents not the man so much as the idea. Of common peasant stock he has risen to the highest place of honour in Soviet intellectual life – that of Academician. Whether he is worth it or not is another matter. They are overwhelmed by the fact that under the new regime a man can reach the highest place from the lowest by virtue of his own efforts alone. That is a grand thing but I have a feeling it may be something of a wish fulfilment for the people who voted him there. Vavilov can lay hold to no such claims as this, for he had some of his education in England and America before the revolution. He does not claim, as Lysenko does, that Communism has done everything for him. He would have been great in any case.[5]

It is unlikely that Vavilov opened up to Hawkes to any great extent. He had learned to be very circumspect in conversing with foreigners. In letters abroad, he communicated by code. The American botanist Jack Harlan remembers wanting to study with Vavilov, who was a great friend of his father and an occasional house guest. Jack even studied Russian during his school days in preparation for the trip. But when at last Harry Harlan wrote to Vavilov about Jack studying in Leningrad, the reply came immediately: 'My Dear Dr. Harlan, what you said about Chinese barley is very interesting . . .'

'My Dear' was a code between the two men that something was wrong. Besides, Harry Harlan had said nothing about Chinese barley. So Jack would not be going to Russia after all.[6]

In May 1938 the government decided to reorganise the Academy of Sciences completely, to increase the number of its divisions, enlarge its membership, and 'strengthen' it with 'young scientific forces'. The establishment of a new Philosophy Division, and the election to the Academy of Stalin's favourite philosopher Mark Mitin, was an obvious draw for the ambitious philosopher Isaak Prezent, while Lysenko and Tsitsin, president and vice-president respectively of the Lenin Academy, stood an excellent chance of being 'elected' under the Academy's new constitution.

Among their natural opponents, as the January 1939 elections neared, was Nikolai Koltsov, head of the Institute of Experimental Biology. A bizarre chain of events had put Koltsov's name on the rolls. His institute had only just been assigned to the Academy from the Commissariat of Health – a bit of bureaucratic tinkering that to Koltsov's mind threatened the survival of his 'independent, entire, and unfractured institution'. The move also meant that the institute was the only bastion of genetics within the Academy.

Koltsov, mindful of his anomalous status within the Academy, did not put his own name forward for election. Unfortunately, he omitted to explain his reasoning to Maria, his wife, who felt she owed him because back in 1915, he had turned down an offer for ordinary membership of the Academy because it would have entailed a move to St Petersburg and Maria hadn't wanted to leave Moscow. In his absence, then, Maria put his name forward.

Koltsov's popularity as a candidate made a discreet withdrawal from the contest impossible. Forty-five colleagues rushed to second his nomination. He sent the president of the Academy an official request to withdraw from the election, but by then the Lysenkoists had already launched a furious attack on him, Prezent exhuming, yet again, the articles Koltsov had contributed to the *Russian Eugenics Journal* in the 1920s. 'In the institute directed by Koltsov and in the journal edited by him,' Prezent wrote, 'his collaborators dragged in all sorts of pseudoscientific trash, and at times open fascist homilies, in the guise of genetics.'

Worse was to come. A few days before the election, on 1 January 1939, *Pravda* carried an article under the headline 'Pseudoscientists Have No Place within the Academy'. Signed by Lysenko's backers, including academicians Alexei Bakh and Boris Keller, it described Koltsov as 'a counterrevolutionary' and a 'fascist'.

The *Pravda* article oppressed Koltsov greatly. He wrote personally to Stalin, pointing out that he himself had closed down both the Russian Eugenics Society and its journal when the first indications appeared in Germany about the connection between fascism, racial ideology and eugenics.

That letter – cogent, dignified, and ungainsayable – kept Koltsov out of further trouble. But it did not keep Lysenko and Tsitsin from being elected to the Academy, and it did not save the institute to which Koltsov had devoted his career. After their election to the Academy of Sciences, the Lysenkoists began a direct assault. On 4 March 1939 the Presidium of the Academy of Sciences set up a commission headed by Bakh and conducted by Isaak Prezent to investigate the 'pseudoscientific deviations' of Koltsov's institute.

The committee held four meetings, the most important of which was on 15 April 1939, when Koltsov was present to answer questions. These questions were phrased in such a way that Koltsov could not win – or that, anyway, had been the intention. But Koltsov was in fighting mood. He knew a witch-hunt when he saw one and, naming no names, he let everyone know he blamed Isaak Prezent for the current farrago. Far from disarming, Koltsov actually got two of the judges, Alexei Bakh and Khachatur Koshtoyants, to admit that they had not in fact read the articles that they had been criticising so savagely in *Pravda*.

The Academy's new president Vladimir Komarov, a plant geographer and a quiet convert to genetics, did not have the stomach to cook up criticisms of his own, and instead quoted Prezent's accusation 'that you did not pay enough attention to the influence of the environment on the hereditary process'.

Koltsov shrugged: 'He says that by feeding one can turn a cockroach into a horse.' (It is said that Stalin laughed out loud when he read the transcript.) But there was nothing more that he could do. He was dismissed from his directorship, and left working in his personal laboratory while his institute was taken apart around him. On 3 June 1939 he wrote to his old friend, Secretary Academician Leon Orbeli:

It looks as though I will not be ejected from my apartment and my small laboratory (both are completely separate from the other rooms of the institute) . . . Do I still have a salary? Will my wife, M. P. Koltova (Doctor of Biological Sciences), and my personal technician, E. P. Kumakova, be permitted to carry on working with me? How am I going to be paid and funded? (I need very little money for experiments

– I can cover those sums from my salary.) I would prefer to get this money from the Academy directly, but if this is not possible, you could affiliate my small operation (three persons) to any institute within the Academy . . .[7]

Koltsov died of a heart attack on 2 December 1940. The next day his wife settled her affairs and drank poison. Her deathbed letter was read out at the funeral. She recalled that as Koltsov lay dying he had a moment of complete lucidity and said: 'How I wish that everybody would wake up. That everybody would wake up.'[8]

15 : 'We shall go to the pyre'

*Clearly sparks were flying. We entered the room and saw
that Vavilov was holding Lysenko by his suit collar, cursing
him for ruining Soviet science.*[1]
A worker at the Bureau of Applied Botany

The Great Purge wore itself out. By February 1938 Andrei
Vyshinsky, perhaps with an eye to his decorous second career
as an academician, used his reputation as prosecutor in the
Moscow show trials to condemn any future use of torture. The
man notorious for shouting down defendants was weary of
interrogations and tortures, and the 'direct fabrication of cases'.
In March the Procuracy Council, under his direction, deplored
the 'beating of honest Soviet people'. Vyshinsky's hypocrisy was
jaw-dropping, but hypocrisy is not the worst vice: many memoirs
report that by the autumn, torture had become a rarity.

Controversies conducted during the Great Purge had not gone
away; eventually it seemed possible to revisit them in a more
measured way. A group of Leningrad biologists wrote to Andrei
Zhdanov, Stalin's right-hand man in Leningrad and the Party

The last image to survive of Nikolai Vavilov is this prison photograph,
probably taken at Saratov in 1941.

secretary responsible for science, about the lack of open public debate on genetics. 'Conditions for work in the field of genetics', they complained, 'are absolutely abnormal at the present time.'

The request was well-judged. Zhdanov's role in the Great Purge, and his famous mistrust of Western foreigners, made him an unlikely ally of genetics. But he was a practical, plain-speaking man who had seen through the forced collectivisation of agriculture in his district and he frankly despised Lysenko, whose assurances about bumper harvests and high-yield species had come to nothing. Zhdanov saw to it that a new conference, 'On the Controversy in Genetics and Breeding' was held in October 1939.

Reopening the debate around genetics made sense, given the desperate state of Soviet agriculture. What the geneticists had not caught on to, however, was the decline in their own importance. Scientists, whether bourgeois hangovers or militant Marxists from the Communist Academy, were used to being taken seriously. Was the Soviet Union not supposed to be the first truly scientifically run state?

Since the First Five-Year Plan, however, the balance of power had changed. 'Practical results', gathered through questionnaires, discussed in public debates and disseminated through the popular press, now trumped the academic mill of specialist publication and peer review. And philosophy, of the militant kind practiced by Mark Mitin and Isaak Prezent, ensured that debates remained comprehensible to non-specialists, and ended with clear declarations and calls to action.

In line with the new orthodoxy established at the Academy of Sciences, the terms of the genetics debate were set by philosophers, not scientists. The meeting was held under the auspices of the party's theoretical journal, *Under the Banner of Marxism*, and chaired by its editor-in-chief, the nihilist philosopher Mark Mitin, for whom science was simply a factory for the production of useful technology.

The meeting was held between 7 and 14 October 1939, at the Marx–Engels–Lenin Institute, a handsome Constructivist block in Moscow's Znamenka quarter, a stone's throw from the Kremlin. *Under the Banner of Marxism* fielded four 'judges' to oversee the proceedings: Mark Mitin and Pavel Iudin, both recently appointed to the Academy of Sciences; Ernst Kolman, the spy-turned-philosopher of science who had launched the first press attacks on Solomon Levit; and Victor Kolbanovsky, head of the psychology department of the Institute of Philosophy.

Under the aegis of such men, it was impossible to push the debate into specialist territory, or even ask hard factual questions. All the geneticists could do was argue for the 'practicality' of their own research, while their opponents the Lysenkoites barracked them for studying a useless fly, *Drosophila*, while they, the Lysenkoites, busied themselves with tomatoes, potatoes and other useful plants and animals.

Vavilov turned up at the conference grim-featured and close to collapse from exhaustion. He held out little hope for his side. Some months before, at a meeting of the Bureau of Applied Botany, he had spoken in the most hopeless terms about the controversy in which Lysenko had embroiled them: 'We shall go to the pyre,' he predicted, 'we shall burn, but we shall not retreat from our convictions.'

His pessimism soon proved justified. The debate, such as it was, was soon swamped in acrimony. Vavilov's saboteur-employee Shlykov was so abusive towards his boss that the shocked chairman shut him up and deleted his charges from the transcript.

Lysenko once again responded to every criticism with a belligerence that staggered sense. The *Drosophila* geneticist Julius Kerkis quoted Lysenko's words back at him, hoping for a retraction, or at very least an explanation. What were they supposed to make of a statement like 'In order to obtain a certain result, you must want to obtain precisely that result; if you want to obtain a certain result, you will obtain it . . . I need only such people as will obtain the results I need'?

'I spoke correctly!' Lysenko shouted.

Kerkis was at a loss: 'It just doesn't fit in my head,' he complained.[2]

It didn't fit in Mark Mitin's head, either, and he could see the whole debate sliding out of his control if he let Lysenko reduce it to a broad-brush condemnation of genetics. If anyone was going to condemn anything, it would be his panel of professional philosophers. Behind Lysenko's wildness he saw the hand of Isaak Prezent, a man despised by the philosophers of his own side – who had to put up with him on a daily basis – quite as much as he was disliked by the geneticists. Mitin laid into Prezent, accusing him of 'conceit that passes all bounds' in trying to crowbar his 'scholasticism' and 'bombast' into the important practical work of Comrade Lysenko. (This was neatly done: *both* sides broke into loud applause.)

But prising Lysenko and Prezent apart proved impossible. Lysenko defended his colleague, and spent the remaining days of the conference hurling imprecations at 'the falsehood of the fundamentals of Mendelism' – a performance that left Lysenko looking as exhausted and defeated by the conference's end as Vavilov had at its beginning. Eleanor Manevich, a genetics student, bumped into Lysenko in the cloakroom after the end of the conference: 'We were standing there. He did not look well and was very hoarse. I felt pity for him. He had been criticised by all the geneticists. I said to him: "Trofim Denisovich you should take care of your health." His overcoat was not very warm.'[3]

No side can be said to have won the 1939 debate, if indeed 'debate' is the right term for such a slanging match. In the absence of any real substance, all Mitin and his colleagues could do was judge the rhetorical performance of each side, and naturally this swayed them towards Lysenko's side, as the more 'practical-minded' of the contenders. Lysenko's theories were 'progressive' and 'innovative', while the geneticists were 'a self-enclosed group that not only does not want to listen to the voice of practice, but reacts to this criticism in a very negative way'. Pavel Iudin, the philosopher who closed the conference, called upon the geneticists to reject the 'rubbish and slag that have accumulated in your science'.

The conference had achieved nothing; worse, it gave its official imprimatur to the very clichés and cant phrases that had dogged genetics since the beginning of the 1930s. Vavilov wrote to Mitin afterwards: 'The conclusions you drew at the conference on questions of genetics left us with a bitter aftertaste.'[4]

Lysenko was by now about as powerful as he would ever be: president of the Lenin Academy of Agricultural Sciences; member of the Academy of Sciences and its ruling presidium; and a deputy of the Supreme Soviet of the Soviet Union, the Soviet Union's highest legislative body.

His petty attacks on his former patron Nikolai Vavilov were boundless. He rejected a whole series of exhibits Vavilov had drawn up for a national agricultural exhibition in Moscow. Vavilov's temper finally broke: he stormed into Lysenko's Moscow office. 'Clearly sparks were flying,' a passing worker reported:

We entered the room and saw that Vavilov was holding Lysenko by his suit collar, cursing him for ruining Soviet science. Lysenko was terrified, he screamed that he, Lysenko, was an untouchable as a deputy of the Supreme Soviet, he would complain to the government and that Vavilov would be held responsible for an attempt to beat him up.

With the men's animosity spiralling out of control, Vavilov appealed to the highest authority. According to his bureau colleague Yefrem Yakushevsky, on 20 November 1939 Vavilov presented himself outside Stalin's office at the Kremlin.[5]

At around 10 p.m. he was let into the reception room. At around midnight, Stalin found time for him. 'So,' began the great patron, the Coryphaeus of science, 'you are the Vavilov who fiddles with flowers, leaves, grafts and other botanical nonsense instead of helping agriculture, as is done by Academician Lysenko.' He did not even invite Vavilov to sit down. Talking to Stalin was like talking to a brick wall, and after about an hour, Vavilov left empty-handed.

Yakushevsky met Vavilov eight days later. His boss's hopes were killed, Yakushevsky says. It was clear to Vavilov, following his interview with Stalin, that Lysenko had been given *carte blanche* to do whatever what he wanted.

As if to demonstrate the fact, at the end of 1939 Lysenko sent Vavilov on a scientific expedition to the Caucasus, and used his absence to replace his bureau's entire scientific council. Vavilov, outraged, appealed to the new commissar of agriculture, Ivan Benediktov, but Benediktov did nothing. By 1940, Vavilov was fully expecting to be arrested. Though his office in the Bureau of Applied Botany was only a couple of minutes away from his flat, he would ring his wife whenever he arrived or left, so she would have warning if he was whisked away.

Colleagues and friends overseas found it hard to gauge the true scale of the confusion and controversy consuming Soviet intellectual life. Indeed, very little information of any sort was emerging from a state hell-bent on isolating itself from every foreign influence.

The Soviet authorities had suggested that the postponed Seventh Genetics Congress might finally be held in 1938. But the news since then, such as it was, had been uniformly bad. Vavilov might or not have been arrested. Levit was missing. Agol had been most definitively shot. Muller meanwhile was licking his wounds in Edinburgh and telling the most ghastly anecdotes. Otto Mohr, the Norwegian geneticist whose job it was to oversee the annual international genetics congresses, had to make a decision: could he really stake a global meeting on the promises of a handful of Soviet bureaucrats, more than half of whom had since been replaced, and none of whom seemed capable of replying to letters?

On 20 May 1937 Mohr received a letter from Cyril Darlington, informing him that 'Koltsov and Serebrovsky have been arrested ... [Julian] Huxley suggests to me that we ought to organise a protest signed by all leading geneticists in this country.' American geneticists, meanwhile, were circulating information (which turned out to be false) that 'seventeen professors of several of Moscow's universities and institutions are imprisoned'. On 28 May, Mohr received a cable from Vavilov: 'Information you received about Koltsov and Serebrovsky completely wrong letter follows,' but by then he had had enough. The British Genetical Society had already offered to host the congress should circumstances require it. So be it: the Seventh International Genetics Congress would be held in Edinburgh.[6]

This was a coup for the general secretary of the British organising committee, Francis Crew. Crew had been filling his draughty genetics department in Edinburgh, one refugee at a time, by offering places to young geneticists desperate to flee mainland Europe. Peo Koller, Charlotte Auerbach and Guido Pontecorvo had joined him; so too of course had Hermann Muller (though, unlike the others, he never did adapt to the cold, and never learned the knack of working with gloves on). Now Crew sent Vavilov an official letter inviting him and his Soviet colleagues to Edinburgh. Cambridge, Oxford and London might have their Fishers and their Haldanes, but hosting the Seventh Congress would put Edinburgh's motley international community on the map.

Crew also told Vavilov that the organising committee had unanimously elected him president of the congress. This was the first time a delegate not of the host country had been so honoured, and Crew hoped this would mollify the Soviet authorities and make it more likely that their geneticists could attend.

On the eve of the congress, however, Crew received a letter from Vavilov: 'Soviet geneticists consider it impossible to take part in the congress held in Scotland instead of its originally planned location – the Soviet Union.'

The letter came as a complete surprise. A new programme was printed, bearing the legend: 'After this programme had been printed and only ten days before the actual opening of the congress, no fewer than fifty names and titles had to be removed and the whole programme hurriedly recast.'

The Congress convened on 23 August 1939. That same day in Moscow, Germany and the Soviet Union signed a non-aggression treaty that became known as the Molotov–Ribbentrop pact. The war that had been looming over Europe for over three years was quickly becoming a reality.

The next evening, British citizens in Germany were advised to leave the country. German subjects in Britain received similar advice, and the German delegation prepared to leave for home. Many other European geneticists started packing up, too, anticipating a difficult journey. Some geneticists from England also left Edinburgh to be with their families. The congress was cut short by a day and on the Tuesday evening the farewell party continued well into the night with toasts to absent friends and those in danger and distress.

The next day Polish forces mobilised against the massing German invaders, and the Second World War began.

'The announcement of the pact between Stalin and Hitler struck us all like a thunderbolt,' wrote the memoirist Gennady Andreev-Khomiakov. Everyone knew that representatives of Britain and France had visited Moscow to discuss a union against Hitler, 'officially characterised as our primary and most evil enemy'. But out of four million Muscovites, perhaps only a thousand knew that Germany's foreign minister, Joachim von Ribbentrop, was holed up in the Kremlin, signing a pact of friendship. 'This turn of events stunned everyone. We were all so bewildered for the first few hours that no one could collect his thoughts.'[7]

Stalin's ideas about the Second World War were shaped by the terrors and losses of the First. He had little interest in Adolf Hitler, and certainly no affection. He did understand that Hitler would touch off a second worldwide conflagration. He believed that, just as the First World War had led to a successful socialist revolution in Russia, so a second and more terrible war would lead to major socialist gains across the continent. This was Russia's manifest destiny, expressed by no less an authority than Lenin.

Stalin's pact with Germany was opportunistic. The French and British had refused to consider Polish territorial concessions to the USSR, but the Soviet Union's non-aggression pact with Germany contained a secret codicil allowing the Soviets to do as they pleased with most of the Baltic region, and gave them additional rights to parts of Poland and Romania. On 17 September, then, the Red Army duly began its 'liberation' of western Ukraine and western Belorussia from the 'Polish yoke'.

Ordinary Russians observed the seizure of these lands with cynical indifference. 'We will extend them a hand,' ran a popular joke, 'and they will be lying at our feet.' Meanwhile those who could get permits to travel – mainly writers, journalists and film-makers – followed the Red Army into Poland to help enlighten their 'liberated brothers'. They returned with furniture, musical instruments, leather coats and shoes, and not just for their own use, either: the film-maker Alexander Dovzhenko brought several train-car-loads of booty back from Poland under the guise of film equipment.

The Finns, unlike the Poles, did not have to fight on two fronts, and when the Soviet Union declared war on them on 30 November they put up a savage resistance. Disfigured corpses of Red Army troops littered Finnish forests, their ears or noses cut off, their eyes gouged out. The Finns set boobytraps, connecting landmines to pens, cameras, bicycles – whatever they thought a Russian soldier might want to take and sell. The weather wrought an even more terrible toll than the Finnish mobilisation: thousands of Red Army troops froze. More died from the cold than from combat. Train convoys arriving in Moscow from the front disgorged frostbitten amputees by the thousand.

By March 1940 Finland was in Russian hands. It was a pyrrhic victory, for the German high command now knew how ripe the Soviet Union was for invasion: how inadequate its transport network, how unprepared its men.

Soviet authorities, meanwhile, wanted a deal more from their new territories than a few car-loads of second-hand furniture. The western Ukraine, in particular, offered fantastic farming potential, and in May 1940 Nikolai Vavilov was appointed to lead a short expedition to the Carpathians to assess the agricultural potential of the area.

Vavilov travelled by train and met his staff and members of the Ukrainian Academy of Science in Kiev on 26 July. After a short and busy stay in the capital, visiting institutions and exhibitions, arranging a conference on the history of agriculture, and delivering a speech at a Rally of Pioneers, Vavilov and his staff headed for the foothills of the Carpathians, squashed into three small, black, Soviet-made cars. One of Vavilov's staff, Fatikh Bakhteyev, wrote later:

I remember his excitement when he viewed the vast fields under newly-bred wheat cultivars reaching the old border of the Ukraine, far beyond the horizon. His interest became even more lively after this frontier was crossed. Here we were facing wide fields resembling a patchwork quilt: each tract was under a different crop. As a plant breeder, Vavilov was enjoying it and, in spite of his haste, he very often stopped the car to gather samples of rye, wheat, barley and oats ...[8]

On 6 August Bakhteyev was bumped from a trip up to the highlands of Putila to make room for a local guest. (His side-mission – visiting a local brewery to ask about the barley cultivars they used – provided some consolation.) The journey, meanwhile, proved far harder than expected. One of the cars couldn't handle the gradient and had to turn back. In it rode another of Vavilov's colleagues, Vadim Lekhnovich. As they descended the mountain they met another car, whose passengers waved them to a stop. They were looking, they said, for Academician Vavilov; Moscow needed to speak to him urgently on the phone. Lekhnovich told the men not to try the road, as it was virtually impassable. The strangers ignored him and drove on.

'V. S. Lekhnovich and I were not at all astonished by the fact that N. I. was wanted in Moscow,' Bakhteyev wrote later. 'The scientist would quite frequently be consulted on important state affairs . . . I was inclined to consider this a good omen.'

When in the evening they returned to their hostel, however, Bakhteyev and Lekhnovich were told by the doorman that Vavilov had been driven off 'to Moscow' in a great rush – indeed, he had left all his luggage behind. 'Thus we parted from N. I. Vavilov for ever.'

Vavilov's arrest had nothing to do with Trofim Lysenko. Three or fours years before, certainly, Vavilov's run-ins with the Michurinist camp might have led to his detention, even his execution. But times had moved on, and life had become a lot less hysterical. Lysenko, for all his institutional clout, was no longer a favourite, now that Andrei Zhdanov was part of Stalin's inner circle. His recent performances had won him few friends among the philosophers, and the failures of the vernalisation programme were becoming for hard for the state to ignore.

What actually triggered Vavilov's arrest was most likely the determined way he had kept up his international network. Despite the fact that the Molotov–Ribbentrop pact had made Britain an enemy of the Soviet Union, Vavilov had continued his work with British geneticists, and in the spring of 1940 had even approved a plan by Cyril Darlington to arrange for the English translation and publication of the latest volume on genetics issued by the Bureau of Applied Botany. *The Theoretical Basis of Plant Breeding* summed up the main results of Soviet plant-breeding research (including, ironically, a broad and very positive chapter on Lysenko). Letters went back and forth between Vavilov and Darlington all that summer. Vavilov's readiness to provide 'British imperialists' with an account of the latest Soviet genetics research was all the NKVD needed to trigger a warrant.

A week after his brother's arrest, Sergei Vavilov[9] wrote: 'My diary is full of grief: mother's death, sister's death, and now the horror hanging over my brother, I cannot think of anything else. It is so terrifying, so pitiful, and it makes everything senseless.'[10]

Nikolai's first interrogation began at 1.30 p.m. on 12 August 1940.[11] It lasted for five hours, and NKVD Senior Lieutenant Alexander Khvat was only just getting started. On 14 August Khvat began interrogating Vavilov during the night, in sessions lasting between ten and thirteen hours.

After ten days of this, on 24 August, Vavilov signed his first confession: 'I admit that I was guilty of being from 1930 a participant in the anti-Soviet organisation of right-wingers that existed in the People's Commissariat of Agriculture. I am not guilty of espionage activity.'

It was not enough. Vavilov had named as co-conspirators only those he knew had already been tried and executed. Khvat piled on the pressure. On 11 September Vavilov confessed to being a 'wrecker' and signed a document titled 'The Wrecking in the System of the Institute of Plant Breeding That I Directed from 1920 until My Arrest on 6 August 1940'.

More interrogations followed, and more signatures on more documents. Vavilov was interrogated by the NKVD for around 900 hours altogether, with some interrogations lasting half a day or more. 'At dawn a warder would drag him back and throw him down at the cell door,' recalled his cell-mate, the artist Grigory Fillipovsky, in 1968. 'Vavilov was no longer able to stand and had to crawl on all fours to his place on the bunk. Once there his neighbours would somehow remove his boots from his swollen feet and he would lie still on his back in his strange position for several hours.'

Following the outbreak of war with Germany on 22 June 1941, NKVD investigators were ordered to close down their cases, and Khvat had to pick up the pace.

The case against Vavilov was sketchy enough, though police records of his activities began in 1931 and had by now grown to seven volumes. Evidence against Vavilov came from a motley assortment of forced confessions. Georgy Meister, who had famously lost his mind during his incarceration, had called Vavilov a 'wrecker' during interrogation but later withdrew the charges. Vavilov's old patron Gorbunov had refused to speak against him. Rather more surprising, Yakov Yakovlev, Lysenko's champion and one of Vavilov's harshest critics, had steadfastly refused to testify against Vavilov right up to his own execution. Vavilov's colleague and Lenin Academy president Alexander Muralov had testified against Vavilov, at least according to the transcript, but since the transcript was dated 7 August 1940 and Muralov had been shot in 1937, this was hardly admissible evidence.

With no decent documentation to draw upon, Khvat set about manufacturing his own. He created a commission of scientific experts under Vavilov's former employee Stepan Shundenko, who was by now back in NKVD uniform. This commission would evaluate the 'wrecking activity' of Vavilov, and here Lysenko did play a role, by vetting its members. The commission members duly signed the papers put in front of them, without discussion or any opportunity to meet.

Khvat meanwhile was orchestrating confrontations between the exhausted and broken Vavilov and the men he had been tortured into implicating: Boris Panshin, Georgy Karpechenko (who had only returned to Russia because Vavilov had persuaded him), Anton Zaporozhets (whom Vavilov hardly knew) and Lenin Academy vice-president Alexander Bondarenko.

On 8 July the Military Collegium of the USSR Supreme Court decided to hear the Vavilov case 'in a closed trial, without participation of the prosecution and defence'. The trial, on 9 July, lasted only a few minutes. On 26 July Vavilov was transferred to Butyrskaya prison for the death sentence to be carried out.

While confined to a crowded prison cell, Vavilov tried to cheer up his companions by arranging them to lecture on whatever they could think of: history, biology, the timber industry. Each of them delivered a lecture in turn. They had to speak in whispers in case the guard heard them.

Vavilov's death sentence was swiftly commuted, perhaps as early as 1 August. The plan was to move Vavilov to a *sharashka* – a specialist gulag where his scientific knowledge might be exploited. Vavilov had offered himself up for such treatment in a letter of appeal, and it is likely that Lavrenty Beria, who had charge of the *sharashka* system, was personally responsible for the petition to reduce Vavilov's sentence. (Beria's wife Nina was, like Vavilov, a former pupil of Dmitry Pryanishnikov, and this pioneering chemist, by now in his mid-seventies, was tireless in his efforts to get Vavilov released or at the very least looked after.)

The war, however, disarranged whatever plans Beria had laid. Even as his envoy was discussing with Vavilov what he could do if moved to a specialist prison, German forces were approaching the gates of Moscow.

Just as equipment was smashed to prevent it falling into enemy hands, so NKVD troops had orders to shoot their prisoners in the face of the German advance. As early as the first week of the war, 3,000 Ukrainian political prisoners were slaughtered by the NKVD. The evacuation of prisoners from Moscow started early (Beria believing, as Stalin did not, that a German invasion was imminent). Prisoners from Sukhanovo and Lefortovo Prisons were moved to an old prison in the city of Orel and executed there on 11 September, a month before Nazi troops overran that city.

Vavilov was luckier. On the night of 15 October he and tens of thousands of other prisoners were taken to the square outside Kurska railway station. He spent the night on his knees, forbidden to look up, waiting in the rain and snow to be loaded onto a train going east. (He had good company: in the crowd were the former editor of *Izvestiia* and the founder of the Marx–Lenin Institute of World Literature.)

Vavilov arrived in Saratov, the site of his early successes, on 29 October. Along with the more important prisoners, he was taken to Prison No. 1, a red-brick tsarist building on Astrakhan Street. He fell sick straight away, and once he was out of hospital, he was placed with another academician, the historian Ivan Luppol, in a cell for prisoners sentenced to death.

On 4 July 1942 his sentence was commuted to twenty years' hard labour and he was moved to the general cell block. His cellmates remember him saying as he entered: 'You see before you, talking of the past, the Academician Vavilov; now, according to the opinion of the investigators, nothing but dung.'

Also in Saratov, having fled Leningrad, were Vavilov's wife Elena and their son Yuri. They lived in one room of a brick bungalow owned by Elena's sister. Since they were the family of an enemy of the state, they had no means of earning money. Every so often Sergei Vavilov managed to get her some cash. But neither he nor they had any idea that Nikolai was imprisoned not twenty-five kilometres from where Elena and Yuri were staying. The NKVD had told Elena that her husband was still in Moscow. Out of the little she had, she used to save up food parcels for him and send them, at great expense and trouble, to the wrong city.

The disappearance of an intellectual celebrity of Nikolai Vavilov's calibre did not go unremarked. Even Winston Churchill made repeated personal appeals to Stalin, asking what had happened to Vavilov. Late in the autumn of 1942 the press attaché of the British Embassy visited Alma Ata, where the presidium of the Academy of Sciences was evacuated during the war, bringing with him diplomas for two Soviet scientists who had just been elected members of the Royal Society. One of them was Nikolai Vavilov. Perhaps this high-profile gesture would shame the Soviet authorities into revealing Nikolai Vavilov's whereabouts.

The president of the Academy at that time, the botanist Vladimir Komarov, was profoundly embarrassed. Since his appointment he had been fighting a rearguard action against Lysenko's administrative abuses within the Academy, and had even managed to smuggle some genetic principles into his books of popular science. But it was not much of a resistance, and when he was called out over Vavilov's disappearance, he trotted out the party line: of course Nikolai Vavilov was a free man!

The problem was proving it. Komarov sent the letters, diplomas and forms that Vavilov was supposed to sign to his brother Sergei, who was also an Academician. With Sergei's spurious signature, the form was then handed over to the British Embassy in Moscow. The Embassy wasn't taken in and sent the form back to Alma Ata.

Hopelessly compromised by this fiasco, Komarov wrote a letter to Stalin himself, asking for definitive information regarding the academician's whereabouts. The efforts made to find him were genuine by now, but they came too late. The NKVD traced Nikolai Vavilov at last to the records of Saratov prison hospital. He had died there of dystrophy on 26 January 1943.

Part Three
DOMINION
(1941–1953)

Stormy applause. Ovation. All rise.

Transcript of the August 1948 Session of the Lenin
Academy of Agricultural Science

Previous page: Visions of plenty: at a youth festival in 1939 young men and women in sports gear parade a papier maché harvest across Moscow's Red Square.

16 : 'Lucky stiffs'

*At various stages of the war Stalin's genius found the correct
solutions that took into account all the circumstances of the
situation: Stalin's mastery was displayed both in defence
and offence. Comrade Stalin's genius enabled him to divine
the enemy's plans, and defeat them.*[1]

    Joseph Stalin

Defeated in the Battle of Britain, in the autumn of 1940 Adolf
Hitler turned his attention eastwards. In a directive dated 18
December 1940 he declared: 'The German armed forces must be
prepared to crush Soviet Russia in a quick campaign before the
end of the war against England.'

    Stalin still imagined that Russia's best chance of avoiding war
lay in alliances (Russia signed a neutrality pact with Japan in
April 1941) and appeasement. Large shipments of Soviet material
continued to be sent to Germany, right up until the invasion –
which was late. German troops were beginning to mass on the
Soviet border by April 1941 but it was not until 22 June that three
army groups crossed onto Russian territory.

At the Volkovo cemetery, men bury victims of Leningrad's siege. Food
was scarce: police units were ordered to shoot anyone they found eating
human flesh.

Attacking on a broad front, from the Black Sea to the Baltic coast, commanders responsible for the previous year's *Blitzkrieg* to the west looked likely to repeat their earlier success. Heinz Guderian's armoured corps advanced eighty kilometres in the first day and three days after that, on 27 June, he reached Minsk, over 300 kilometres inside Russia. On 10 July he crossed the Dnieper. On 16 July he reached Smolensk. He had covered 650 kilometres in less than four weeks and was heading straight for Moscow.

By now Russia had lost around 2 million men, 3,500 tanks and over 6,000 aircraft. The state was so ill-prepared, there weren't even air-raid shelters in the Kremlin. Stalin had to work in the Moscow Metro. By 14 October law and order were breaking down. Shops were looted, empty apartments burgled. A pall of smoke hung over the city as officials burned papers.

'We are obliterating stupid business forms as a category! Look.' He dragged me to the window, out of which he passionately and with obvious satisfaction hurled the folders, which hitherto had been preserved with great care. I might have expected this, yet all I could do was rub my eyes: Could I be dreaming? They were throwing out the valuable cover vouchers and cherished reports on which our entire economic system was based!

Gennady Andreev-Khomiakov's eye-witness account in his memoir *Bitter Waters* captures the absurdity and panic of the time. With the Germans just forty kilometres from Moscow, 'secretaries, draftsmen, accountants, and typists were willingly devoting themselves to the demolition. It was as if they had been seized by the joy of destruction. Perhaps they were simply fed up with clacking away on their typewriters to produce those reams of financial reports, full of incomprehensible and boring figures.'[2]

A penetrating wind tore across the city, bringing with it a sprinkling of snow. It was cold: winter had come early. No one knew whether this was good news or bad. With luck the Germans would all freeze – but not before the ground did, and on that surface the Germans could advance as if they were driving on tarmac.

And so they would, had Hitler himself not already prevented them, insisting that his generals first disable as much of the Russian Army as they could.

Guderian, under protest, moved south towards Kiev, where a pincer movement captured another half a million men. But the delay proved crucial. By early December only a few advance detachments had managed the 200 kilometres to the suburbs of Moscow, and by then it was clear that the non-aggression pact between Russia and Japan was holding. Soviet forces rushed westwards to defend the capital. General Georgy Zhukov, a brilliant career officer in the Red Army, forced the Germans to a standstill on 5 December 1941. The weather turned out to be good news, after all: the Germans, confident in their technique of *Blitzkrieg*, had come unprepared for freezing conditions. And now the Russian winter started in earnest.

Further to the north another German army, pushing along the Baltic coast, had made spectacular progress. By August it had reached Leningrad. Unable to capture the city, the Germans besieged it, expecting it to surrender before winter took hold.

The siege of Leningrad lasted 900 days, until January 1944. Administering the city, and ensuring its heroic resistance, was Andrei Zhdanov, Stalin's right-hand man in the purges that had wracked the place only a few years before, a man with a famously weak constitution and no military experience whatsoever.[3]

Some people have greatness thrust upon them, and this was certainly the case with Zhdanov. Leaving Moscow for Leningrad in the autumn of 1941, he had fully expected to return to the capital, one way or another, before the onset of winter. In fact it would be five years before he was able to resume his post as second-in-command of the Party.

Stalin himself repeatedly accused Zhdanov of military incompetence and even insubordination. It is true he began badly, bungling both the evacuation of civilians and the business of fortifying the city. In late June men and women of working age were being ordered to dig trenches and anti-tank ditches, lay barbed wire, set mines, and build pillboxes. For three hours a day after work or school, the Leningraders laboured; pensioners and housewives toiled for eight hours a day on the lines.

What followed, as the siege took hold, was unimaginably worse – ten times worse than Hiroshima, if you count the dead. Leningrad's population dropped from well over 3 million to approximately 640,000, mostly through starvation and cold. People dropped dead on the street. Some were put on sleds and carted off to burial in immense common graves at the Piskarevskoe Cemetery. Others fell into huge snowdrifts and had to be cut out when they reappeared in the spring to stop the spread of disease. Food stocks disappeared. People ate whatever they could: shoe leather, wallpaper glue, spoiled animal fat. When the Badaev sugar stores burned down, people sold and ate the 'sweet earth' found amongst the ruins. Cats and dogs vanished. People vanished. Special police units were sent to shoot anyone who was found to have eaten human flesh.

Stalin must at some point, and for some reason, have believed in Zhdanov's military capability, but Zhdanov was soon disabused: he quickly learned to defer to his generals, Zhukov, Meretskov and Govorov, who remembered Zhdanov in their (censored) memoirs as 'polite, well behaved, and a good listener who did not pull rank'.[4] He worked out of a third-floor office in the Smolny Institute, the walls hung with pictures of Stalin, Marx and Engels. 'His asthma was much worse,' the American journalist Harrison Salisbury remembered. 'His breath came in sharp, uneven gulps. His heavy face was puffed with fatigue and only his dark eyes glowed.'

*

Leningrad was full of treasures. The Hermitage Museum, founded in 1764 by Catherine the Great, contained over 2 million paintings, sculptures, coins, articles of jewellery and other arte-facts. In just six days, three-quarters of its contents were readied for storage, in hidden vaults, in a nearby cathedral, and out of town.

Packing up Leningrad's other great storehouse – the seed and plant collection of Vavilov's Bureau of Applied Botany – presented unique challenges. For a start, its 380,000 examples of 2,500 species of grains, fruits and tubers had to be kept alive.

Collections of potato, rye and other crops were planted out at the bureau's research stations at Pavlovsk and Pushkin. The Pavlovsk station, located forty-five kilometres south-east of Leningrad, was ablaze from enemy bombing and the field containing the potato collection was under fire. Red Army soldiers helped carry the collection back to the city in military trucks to the bureau's offices in St Isaac's Square.

The plan was to evacuate the collection, one part carried by employees in their hand luggage, the other part – five tonnes of seeds – in a railway van scheduled for the town of Krasnoufimsk in the Ural mountains. Come winter the institute's director, Iogan Eikhfeld, and several staff were evacuated there. But the train car never arrived. For six months or more it was shuttled from siding to siding until all hope of sending the collection away was lost, the car was unloaded and its cargo returned to St Isaac's Square.

The largest and most important part of the collection was left in the besieged city. The institute had already lost more than thirty researchers, some in the bombing of the city, some in action, others from starvation and cold. January and February 1942 were the worst months; temperatures fell to record lows of −36 to −40 °C.

In the dark, freezing building of the bureau, the remaining workers prepared seeds for long-term preservation. They divided the collection into several duplicate parts, while bombs and shells burst around them. St Isaac's Square was a target. Its cathedral was damaged; so were many nearby buildings. It was only after the war that survivors learned why the bureau had been left unscathed: Hitler had planned to hold a victory banquet and to make a speech from the balcony of the nearby Astoria Hotel.

The Germans never did succeed in overrunning Leningrad. But the rats did. That winter, hordes of vermin swarmed into the building. All efforts to protect the collection proved fruitless. Rats started breaking into the ventilated metal boxes and devouring seeds. Vadim Lekhnovich, who was the curator of the collection during the siege of Leningrad, recalled: 'For more safety I started sealing the basement and locked it with three different locks. I also reinforced the doors with iron. However, this could not avert minor thefts.'

In the spring, thieves crashed through the blocked windows. Workers secured the windows with plywood and the seeds were transferred to a more reliable place.

Evgeny Vulf, an expert in volatile oil plants, was hit by a shell splinter and bled to death. In January, A. G. Shchukin died at his writing table, holding in his hand a packet of peanuts he had hoped to send off for a grow-out. G. K. Kreyer, head of the herb laboratory, died of starvation. O. S. Ivanov, a rice specialist, also succumbed, surrounded by several thousand packets of rice. L. M. Rodine died among packets of carefully preserved oats.

Come summer the surviving bureau staff planted out cabbages and seed potatoes in the churchyard of Saint Isaac's cathedral and at Pushkin, standing guard over the fields day and night to fight off both rats and men.

With swathes of its richest territory in German hands, Russia nonetheless managed to re-arm itself, in part using supplies arriving in convoys from the West, along the dangerous Arctic route north of Scandinavia and down to Archangel. That year alone, 20,000 tanks and 35,000 planes rolled out from factories hurriedly relocated in remote regions of eastern Russia. Russian forces took back Smolensk, Kiev and Kharkov. By the end of 1943 two-thirds of the land taken by the Germans was back in Russian hands. In February 1943 the Germans had suffered a devastating defeat in the battle of Stalingrad (now Volgograd), a major industrial centre on the Volga River. The horror of Stalingrad lasted for 199 days. Soviet losses were so great that, at times, the life expectancy of a newly arrived soldier was less than a day. The Germans, fighting from house to house and from staircase to staircase, joked bitterly that they had seized the kitchen but were still fighting for the living room. And at last, in January 1944 – a moment of huge psychological importance – Leningrad was freed.

Little enough of the city remained. On 22 September 1941 Hitler had ordered that 'St Petersburg must be erased from the face of the Earth' and 'we have no interest in saving lives of civilian population.'[5] Thousands of homes, factories, roads and railways, schools, hospitals and power plants were completely destroyed. Take a tour round the Peterhof Palace today and your guide will likely tell you about the German artillery shrapnel embedded in the trees – metal enough to draw down lightning whenever there's a storm.

The bureau's collection of useful plants survived, after a fashion. Lysenko, who had spent the war in Siberia conducting an urban gardening campaign, maintained that the whole enterprise was disordered and for a long time it was allowed to deteriorate. But the Bureau of Applied Botany survived, at least on paper, and veterans of the Leningrad siege maintained what they could. By 1969 the collection boasted 175,000 holdings, all available for the creation of new crop varieties.

Contributions to the Bureau came from all over the world. Its chief, Peter Zhukovsky (a brilliant geneticist and a former close associate of Nikolai Vavilov), once received potatoes from the Tucaman University in Argentina, thanks to a chance meeting in 1958 with a German plant collector who worked there, Heinz Brücher. Forty years later, in the late 1990s, it turned out that Brücher had been an officer in the SS. In June 1943, Heinrich Himmler had placed him in charge of a special commando unit charged with raiding Soviet agricultural experimental stations. Brücher hadn't been donating valuable varieties of potato after all: he had been returning them.[6]

To win its Great Patriotic War, the Soviet state changed its relationship with its scientists and engineers – not once, but twice. One change was already in train as war loomed in the late 1930s. The other change came later, was born of necessity, and disconcerted everyone.

The planned change was the incarceration of large numbers of specialists in a prison science network. The idea came from scientists and engineers themselves. Some time in late 1937, as they awaited sentencing, a group of specialists who were languishing in the NKVD's infamous Lubyanka prison put together a short proposal. They listed a number of specific military weapons that they could develop. In return, they asked for resources, and not to be sent to Siberia.

The bald idea was nothing new to Lavrenty Beria, then a senior NKVD official. Experiments of this sort had been tried before. In the early 1930s prisoners at the Solovki Special-Purpose Reformatory Camp had even issued their own scientific journal. What excited Beria was the idea's scaleability. The NKVD was already a powerful economic organ in its own right – ultimately, it controlled all the gold in Kolyma. If part of its gulag system could be given over to design and manufacturing, then its economic empire could expand into new territories, while its control of leading scientists and engineers would lend it huge political clout.

Beria created the Department of Special Design Bureaux in 1938. His energetic lobbying won it an entire factory plant and an initial government investment of more than 35 million roubles.

The department went through a number of name changes but it was always Beria's pet project, and shut down only when, shortly after Stalin's death, Beria was executed.

Beria's special prisons or *sharashki* solved at one stroke the two besetting difficulties of Soviet industrialisation: the political unreliability of the old guard, and the inexperience of the new. If the old specialists would work under guard, then their political unreliability hardly mattered. All you had to do was keep them incarcerated. (There was never any need to arrest and incarcerate people to feed the *sharashka* system; the arrest quotas for the NKVD were high, and there were always enough specialists already in detention.)

The aviation industry was the first to benefit: the aviation designer A. N. Tupolev and the rocket pioneer Sergei Korolev both achieved greatness and world renown (albeit indirectly and after the fact) for their work within the *sharashka* system. Tupolev's Tu-2 bomber became a sort of military design classic, remaining in service until the 1950s. Other contributions made to the war effort included engines, propellants, radio systems, guns and cannons. More than half these projects went into mass production – a superb success rate for wartime R&D. The *sharashki* were strong in communications technology, some of it quite bizarre. Cast into the gold mines of Kolyma in 1938 and 'rescued' by the *sharashka* system, the physicist and radio engineer Leon Theremin found a way of using an infrared beam to detect sound vibrations in window-glass. Beria used Theremin's eavesdropping device to spy on the British, French and US embassies in Moscow. Another Theremin invention was called the Thing. In 1945 Soviet schoolchildren presented a 'gesture of friendship' to the US Ambassador – a replica of the Great Seal of the United States carved in wood. It hung in the ambassador's residential office in Moscow, the Thing concealed there intercepting confidential conversations during the first seven years of the Cold War until it was discovered by accident in

1952. Theremin was released from the *sharashka* system in 1947, in time to receive a Stalin Prize for his work. He had enjoyed himself so much he went on working with the KGB until 1966.

Who wouldn't enjoy the life these prisons provided – given the alternative? A former *sharashka* worker, Leo Kopelev, understood the logic of such places:

They grab him by the scruff of the neck, drag him to Lubyanka, Lefortovo or Sukhanovka – confess, bastard, who did you spy on, how did you wreck, where did you sabotage? . . . They lower him once or twice into the cooler when it's freezing, when there's water in it. They'll hit him in the face, the ass, the ribs – not to kill or maim, but so that he will feel pain and shame, so that he will know that he is no longer a human being but just a nothing and that they can do whatever they want to him . . . And then, after all that, they will give him a magnanimous ten years . . . No days off. Vacation is a foreign word. Overtime is sheer pleasure; anything's better than sleeping in a cell. You chase away thought of freedom, of home – they only bring on depression and despair . . . In the sharashki they address you by your name and patronymic, feed you decently, better than many eat on the outside; you work in warmth, sleep on a straw mattress with a sheet. No worries – just make sure you use your brain, think, invent, perfect, advance science and technology.[7]

The other, unplanned and disconcerting change in the relations between government and its scientists came as a response to war itself. The Great Purge of the late 1930s had ensured that, in the coming war, there would be no White opposition to stage a putsch against Stalin's rule.[8] But the price paid was the virtual elimination of the Soviet high command. By declining to trust his experienced military men, Stalin was obliged to put his trust in the young and the untried.

This generational shift was abrupt. It used to be, to join the Party, you were put on probation and had to pass political examinations. No longer: men with no experience of the Terror were joining the Party directly from the battlefield. What mattered to them, first and foremost, was winning the war, so naturally they paid attention to the professionals around them, ideology be damned.

Those commissars for whom ideological correctness counted for more than military success were quickly put in their place. It had been the case, early in the war, that commissars were entitled to countermand orders issued by commanding officers. By October 1942 the number of military failures had risen so much that the commissars were ordered to stick to propaganda. As the conflict progressed, so party functionaries lost their status and professionals found themselves increasingly valued by the very state that had persecuted them.

'Lucky stiffs' they were called, and few made more of their luck than Alexander Luria who, in the wake of the pedology scandal, had been retraining as a medical doctor. Wartime experience brought Luria work that would establish his global reputation, when he went to treat wounded service personnel at Evacuation Hospital No. 3120, part of the Institute of Neurosurgery run by Nikolai Burdenko.

The hospital had been established in 1942 at an old sanatorium on the banks of two small lakes near the city of Chelyabinsk in the southern Urals. Arriving there, Luria brought together old colleagues from his days in Moscow and Kharkov – the cream of Moscow psychiatry – and in a few months they transformed the sanatorium into a 400-bed neurological hospital, researching and evaluating every kind of rehabilitation technique. Evacuation Hospital 3120 produced some of the most important neurology and psychology research to come out of the war.

Alongside several extraordinary clinical achievements, Luria also wrote a book that was to transform the genre of popular science when he assembled, edited and introduced the diary of one of his more badly damaged, and more superhumanly determined, patients, Lev Zazetsky.

Shot in the head, Zazetsky had woken from his coma to a shattered world. Nothing made sense to him. He found himself in a world without left and right, without cause and effect, without grammar. There was no way 'out' of Zazetsky's world – and no way 'in' either. No before or after. No with or without.

Eventually, as Zazetsky came to understand what had happened to him, he began a diary. He could write, but he couldn't read. If he studied a page, half of it would vanish. If he peered at a word, half of it would vanish. He could just about figure out words a letter at a time, but it was hard, because he could never see entire letters: only halves of letters. So he was never truly able to appreciate what he spent the rest of his life composing: *I'll Fight On*, his handwritten memoir, is 3,000 pages long.

I'll Fight On has never been published in its entirety. It is frankly unreadable: thousands of pages of fractured recollection, ripped out of time, stripped of context and causality. Luria preserved and published the clearest passages, however, and these form the kernel of one of the most widely translated and highly regarded science books of all time: *The Man With a Shattered World*.

Luria, working with patients whose injuries had left them struggling with language, began to formulate the idea of a brain divided into a hierarchy of executive levels. In Zazetsky's case, his higher executive functions were undamaged. Writing his memoir was proof enough of that. Zazetsky could plan, strategise, form intentions and pursue them. What he lacked – what the German bullet had ripped away – were certain more basic executive levels. In particular, that part of the brain that processed symbols was irreparably damaged. Zazetsky couldn't comprehend a whole word, understand a whole sentence or recall a complete memory, because to do so would require relating symbols. All he could grasp were fleeting fragments. He couldn't read, which is a perceptual activity. He could write, because it is an intentional one.

On the basis of more than 800 cases of brain dysfunction caused by military wounds, Luria began to assemble *Traumatic Aphasia*, the monograph that founded modern neuroscience. He remained Zazetsky's friend and physician, his amanuensis and his editor, till the end of his life.

Other 'lucky stiffs' had a rather more unorthodox time of it. Dmitry Fedotov's notoriously chequered career could quite easily have got him shot for the very same reasons that, in the actual course of things, brought him to high government office.

In territories occupied by the German Army, hospitals had been destroyed and patients murdered. Without enough food, soap, or fuel, the survivors fought a losing battle against dysentery, malnutrition and tuberculosis. Everywhere psychiatric patients died at an alarmingly high rate.

In January 1942 Dr Dmitry Fedotov was given the task of rebuilding the fire-damaged and vandalised Litvinov Psychiatric Hospital at Burashevo, a small village in Kalinin district. All of the hospital's 530 wartime patients had been killed.

Fedotov did what he had to do to get the hospital running again. He stole from the local government warehouse and handed out clothing, no questions asked, to his doctors and staff. He bought meat for the hospital on the black market. He restored the hospital's water supply. Most audacious of all, he ran high-voltage cables to the hospital from a local power station and charged local farmers to piggy-back off the line. In 1945 Fedotov joined the Party. He was an exceptionally lively example of that generation for whom Party membership was an act of simple patriotism, rather than an expression of political conviction.

In 1946 the Ministry of Health[9] brought Fedotov to Moscow to answer for his entrepreneurism. 'How could you allow yourself such things?' the minister exploded, reviewing the electricity scheme. Others were either condescending (he was inexperienced) or harsh ('out of control . . . outside the law'). Fedotov earned a 'stern demerit with a warning' and was sent back to Kalinin.

In his absence, however, the conversations continued. Say what you like about his unorthodox methods, Fedotov was a man of action, a practical man who got hold of what he needed when he needed it. In the poverty and ruin of the early post-war years, with half the country in ruins and the population living on the edge of famine, were men like Fedotov really to be discouraged? By 1947 the official mood toward specialists and entrepreneurs had warmed considerably, and Fedotov found himself in Moscow once again – and being put in charge of all the hospitals in the Soviet Union.

Fedotov joined a new elite of administrators and scientific workers who, from March 1946, enjoyed high wages, bonuses, and priority access to housing, food and goods. Members of academies dined in special restaurants originally meant only for Party officials. They used special government hospitals and sanatoriums. Specialists were so few in number, and in such high demand, that a doctor of sciences heading a laboratory in the Academy of Sciences now earned nearly twice as much as an official in the Central Committee.

Even as Fedotov was taking stock of Soviet psychiatry ('catastrophic', in his own estimation, with numerous cases of hospital workers violently beating patients), Sergei Gurevich, a hygienist, was arguing that psychiatrists should help design new apartment blocks, sunny, uncluttered and quiet, to help people overcome the effects of 'grief for the killed and the lost, deprivation, illness, [and] evacuation'. Each apartment ought to have a view of 'river banks, forest, meadows . . . an apartment without a piece of land to go with it is flawed,' he wrote.[10] You could be forgiven for thinking that the purges of the late 1930s had never happened, such was the speed with which Russia's mandarin class reasserted itself.

Meanwhile thousands of Soviet citizens were still living in dugouts and struggling to get enough to eat.

17 : 'Can I go to the reactor?'

Nuts!

    Physicist Enrico Fermi, responding to the suggestion
    that nuclear fission might become an issue of national
    security[1]

'I am born a Russian,' Nikolai Timofeev-Ressovsky explained, declining the offer of German citizenship, 'and I do not see any way to change that fact.'

It was 1938, and no less a figure than Bernhard Rust, minister of science and education, was lobbying for Timofeev-Ressovsky and his family to stay in Germany. Timofeev-Ressovsky's high standing had come about in a bizarre way. In May 1935 the Nazis had insisted that Oskar Vogt resign as director of his own institute. In March 1937 a new director was appointed: the neuropathologist Hugo Spatz, a staunch Nazi supporter. Spatz said that he did not want to deal with Russian scientists. In fact he wanted to dismiss Timofeev-Ressovsky then and there, judging his department of genetics 'an alien body' in the institute. Oskar Vogt, however, managed to convince the Ministry of Science and

Yulii Khariton sits beside a copy of his brainchild: the first Soviet atomic bomb, detonated on 29 August 1949. This photograph was taken in 1992.

Education that Timofeev-Ressovsky's work was important. To keep Spatz happy, the genetics department was taken out of the institute and reorganised as an institute in its own right, directly under the supervision of the Kaiser Wilhelm Society. So it was that Timofeev-Ressovsky, far from being thrown out of his job, found himself completely free to do what he wanted. And his work, studying the basics of genetics, was an asset the minister was determined to hold on to.

Rust need not have feared losing the celebrated geneticist. Timofeev-Ressovsky had already overcome temptation. In the mid-1930s colleagues from the United States had offered him a position at the Carnegie Institution – but Hermann Muller had spoken so discouragingly of life and science in the United States that, even after the war, Timofeev-Ressovsky showed not the slightest interest in emigrating there.[2]

It helped that the Kaiser Wilhelm Institute for Brain Research was located in the suburbs, where the Nazi presence was not so overbearing. The political pressures on scientists were anyway remarkably slight. You didn't have to be a Nazi Party member to get a grant; in fact membership didn't make funding more likely.

Timofeev-Ressovsky's position was further helped by his collaboration with the physicist Karl Zimmer, who worked for Auer, a metallurgy company involved in uranium production. Though Timofeev-Ressovsky had nothing to do with this business, the friendship lent him an aura of untouchability. So did his being on first-name terms with Werner Heisenberg, the man in charge of the development of the German atom bomb, at another institute just down the road. 'No one followed him,' Auer's research director Nikolaus Riehl recalled in 1985. 'The Nazis left him alone. He lived there and we all supported him. We had very close contact. We were all good friends . . .' This friendship circle extended even to the local Gruppenführer, who ensured that anyone denouncing Timofeev-Ressovsky's staff was roundly ignored. (Later in the war, and with food in short supply,

Timofeev-Ressovsky repaid the favour by making him a regular guest at experimental-rabbit dinners.)[3]

When war began, Vogt's institute became a modest intellectual safehouse, successfully sheltering Jews and political refugees for the duration of the conflict. The Kaiser Wilhelm Institutes in Berlin were classified as important to the war effort, and this qualified them to apply for labourers. Timofeev-Ressovsky diverted several prisoners to his genetics department, wildly inflating their importance and qualifications. Forged documents magically transformed displaced persons and prisoners of war into 'guest-workers'. The anti-Nazi resistance found work for them where they could – usually on distant farms – and while they waited, they hid in Timofeev-Ressovsky's lab.

As the war developed, and he found his freedom of movement restricted – there were to be no more family vacations on the Baltic coast – Timofeev-Ressovsky made minor compromises. He delivered gloomy and discouraging lectures to SS doctors on the technicalities of mutation research. His respirator studies, using isotopic markers to test the filters, were certainly of military significance.[4]

But even minor compromises were too much for Dmitry, the Timofeev-Ressovskys' eldest son, whose hatred of the Nazis grew ever stronger as he associated with Russian groups in Berlin. Much against his father's advice (their arguments were terrible), Dmitry involved himself in the work of an underground resistance group, providing support to prisoners of war. Because he knew French, he was chosen to shelter two French pilots from arrest. The group was betrayed and Dmitry was arrested by the Gestapo on 30 June 1943.

Nikolai Timofeev-Ressovsky tried everything in order to free his son. He lobbied leading SS officers, politically influential colleagues, and his superiors in the Kaiser Wilhelm Society. Heisenberg himself appealed on behalf of his friend. Nothing worked. Dmitry was sent to the concentration camp at

Mauthausen and died – so rumour had it – during an uprising of the prisoners just before the Americans came. (To her dying day, Elena Timofeev-Ressovsky nursed a wild hope that her son had somehow survived.)

The tide of the war was turning. In 1944 the Brain Institute became a clinic for pilots with head and brain injuries. Anoxia from high-altitude flying became a focus for research, and the institute's modest 'pantheon of brains' acquired new and gruesome holdings: brains from at least 500 patients murdered as part of the Nazis' euthanasia programme.

Soon all the ordinary food was gone. In Berlin people were told to gather herbs, mushrooms and snails, make coffee from acorns and bake bread from rapeseed. Staff at the institute got off relatively lightly, subsisting on cornmeal and molasses meant for their experimental fruit flies. (For some reason this stuff kept arriving by the truckload.) There were rabbits, too, meant for radiation experiments. Since you couldn't eat an irradiated rabbit, the experimenters lowered the dosage. Eventually Timofeev-Ressovsky decided not to expose the rabbits to any radiation at all. Science could wait, and there was always a hot meal for you at the Timofeev-Ressovskys' apartment.

In the first half of February 1945, as Soviet forces neared Berlin, Hugo Spatz decided to transfer the staff of the brain research institute to Göttingen in the south-west of Germany. Timofeev-Ressovsky stayed put, saying that it was his responsibility to try to save his staff members from harm, and he could best explain the nature of the institute to the commanders of the advancing troops. His plan, as he explained it to Nikolaus Riehl, was to return to the USSR and to put his department at the disposal of the Soviet Union.

Riehl was appalled: 'I said to him, "Timofeev-Ressovsky, you are subject to treason. It can go very badly with you." He said, "No. Nothing will happen." I said, "I am formally in order being a German citizen, but you are a Russian citizen." He repeated,

"No, nothing will happen to me." . . . Timofeev-Ressovsky said that the fact that his son was lost in itself should be regarded as justification for staying in Germany. That was, of course, not the case.'[5]

The Soviet Army reached Buch on 21 April 1945. Timofeev-Ressovsky's colleague I. B. Panshin, himself a Russian, recollects: 'I and many employees of the Institute were in Timofeev-Ressovsky's house. There was sporadic shooting. I saw our soldiers from the window. I went to them with a self-made white flag of surrender.' On the whole, everything took place peacefully. 'One of our first actions was sending a telegram to Stalin stating that it was necessary for us to serve in the Soviet Union and that the Institute with the staff and the equipment would be very valuable for Russia.'[6]

When the Soviet troops arrived, Timofeev-Ressovsky was promptly arrested – and just as promptly released – by NKVD General Avrami Pavlovich Zavenyagin. Recognising the potential value of Timofeev-Ressovsky's research in radiobiology, Zavenyagin took the institute under his wing, staff and all, and saw to it that Timofeev-Ressovsky's work continued while he tried to sort out the institute's transfer to the Soviet Union.

Not everyone was so keen to welcome Timofeev-Ressovsky home, however. The first sign of trouble came when two physicists flew in from Moscow dressed in the uniforms of NKVD colonels, and one of them, Lev Artsimovich, flatly refused to shake Timofeev-Ressovsky's hand. The physicists were part of a commission organised by the Academy of Sciences, and it is one of the great unexplained ironies of the time that, while the NKVD were working to keep Timofeev-Ressovsky free and productive, it was a denunciation from the Academy of Sciences, of all places, that, on the night of 12–13 September, got Timofeev-Ressovsky re-arrested.

From October 1945, all trace of Timofeev-Ressovsky evaporated. Zavenyagin attempted to trace his charge, only to find that his own people had lost track of the paperwork. Some other division of the NKVD had got hold of Timofeev-Ressovsky and Zavenyagin had no means to know which.

Meanwhile the Soviet investigators to whom Timofeev-Ressovsky had been assigned began assembling the case against him. Not only were they unaware of his importance; they neither could nor wanted to understand his work, and put hostile interpretations on everything they learned.

From Zimmer they learned what work Timofeev-Ressovksy's institute had been assigned. This catalogue of projects – from the highly technical (investigating the biological effect of neutron radiation) to the frankly mad (using X-rays as a weapon against enemy aircraft) bore little relation to the actual work Timofeev-Ressovsky had conducted. But even this was vulnerable to hostile interpretation. A close colleague, Hans-Joachim Born, inadvertently incriminated his friend when he described radiation experiments on animals, on voluntary human subjects (Born himself was one of those), and on two human corpses. This grim-sounding project merely used safe dosages of thorium X to map blood circulation disorders – a research technique common-place even in the 1920s. Nevertheless, Born's testimony was used to imply that Timofeev-Ressovsky had experimented with Soviet prisoners of war – a rumour so persistent it turned up in the academic literature as late as 1988.[7]

Having found ample material to conclude that Timofeev-Ressovsky's institute had supported the German war effort, on 4 July 1946 the Soviet authorities convicted Timofeev-Ressovsky of having 'betrayed the Motherland by going over to the side of the enemy', and sentenced him to ten years' imprisonment in labour camps.

On the way to Karaganda, one of the most terrible camps of the gulag, Timofeev-Ressovsky was lodged in a common cell where, twice a day, water was poured from a bucket without a spout into the prisoners' teapot, usually spilling on the floor.

'There was nothing at the Lubyanka . . . which so offended him as this spilling on the floor,' recalled one inmate:

He considered it striking evidence of the lack of professional pride on the part of the jailers, and of all of us in our chosen work. He multiplied the 27 years of the Lubyanka's existence as a prison by 730 times (twice for each day of the year), and then by 111 cells – and he would seethe for a long time because it was easier to spill boiling water on the floor 2,188,000 times and then come and wipe it up with a rag the same number of times than to make pails with spouts.

That inmate was Alexander Solzhenitsyn, and his memoir, *The Gulag Archipelago*, became the founding document of the Soviet dissident movement. Solzhenitsyn also described the way Timofeev-Ressovsky, with his characteristic energy, organised a scientific and technical society, with lectures by members who included physicists, electronic engineers and a chemist.

From August to November 1946, Timofeev-Ressovsky was kept in Karaganda. There, too, he organised a scientific society to save himself and his fellow scientists from despair. By now he was developing dystrophy and hovered on the verge of death. In the end, he recalled, 'I remembered only that the name of my wife was Lyolka [a nickname], but I forgot her full name. I forgot the names of my sons. I forgot everything. I forgot my last name. I remembered only that Nikolai was my first name.'[8] Neither friends nor family were able to learn where he was or even whether he was alive.

It was the French physicist, Nobel Prize winner and resistance worker Frédéric Joliot-Curie who brought Timofeev-Ressovsky at last to light, when he appealed directly to Lavrenty Beria, the Minister of Internal Affairs.[9] Where Zavenyagin's enquiries had stalled, Beria's impatience and terrifying reputation brought results. By then Timofeev-Ressovsky was close to death from starvation and nearly blind.

He was immediately shipped back to Moscow in a crowded prison wagon, forced to stand crushed together with others most of the time, and arrived unconscious, in a carriage crammed with the dying and the already dead. Officers carried him from the prisoner transport in their arms.

In the Central Prison Hospital, strenuous efforts saved his life, but his vision, although improving, remained permanently impaired. He was never again able to use a microscope and could read only with a large magnifying glass and a brightly illuminated page. It took him months to recuperate to the point where he was strong enough to travel. At last, in the spring of 1947, he left Moscow, bound for a secret destination in the Urals.

He found himself deposited on the shore of a blue lake. There were dark green stands of fir. There were deserted sandy beaches. He was very weak. 'I could hardly climb a stair.' Cottages, lab buildings and warehouses made up the village of Sungul. 'When I put a foot on the next step, I had no strength to pull the second foot.' When he got to the door assigned to him he looked back and saw that the lake was dotted with islands. To the west, there was a magnificent view of the mountains.

Object 0211 was a *sharashka*: a prison village built on a small, wooded peninsula. It was barricaded on the land side by a barbed-wire fence. There were guard posts, and dogs patrolled the shore. None of that mattered. Here, at last, was a haven. A place to work. A place that understood him. Someone had even left flowers on his doorstep, ready for his arrival.

The only puzzle was, why had Timofeev-Ressovsky been so favoured? What made him so important? What work was he expected to perform?

Leo Szilard had spotted it first. The Hungarian physicist moved to Britain in 1933 and in September that year it occurred to him that if a nuclear reaction produced neutrons, and those neutrons then caused further nuclear reactions, then he knew of no pressing reason why process would ever stop. Szilard's dream, or nightmare, was the nuclear chain reaction. The fear that nuclear chain reactions, once started, would proliferate through all available matter, remained a slim but besetting possibility right up until the test of the first atom bomb by the United States on 16 July 1945.[10] For Szilard, in 1933, the prospect did not even seem particularly slim. The more likely possibility – that nuclear chain reactions might set off an explosion of as yet unimaginable power – was worrying enough. In any event, he was sufficiently exercised by the idea that in 1939 he wrote to Enrico Fermi saying that the United States should from now on conduct its fission research in secret.

'Nuts!' Fermi replied.

It was a French team, in the end, who put flesh on Szilard's anxieties, announcing in *Nature* on 7 April 1939 that nuclear chain reactions were real. Each fission would generate three or four neutrons, each of which would trigger further fissions.

Physicists would later look back on the 1930s – with some incredulity – as a golden age: a period of naive speculation, collaboration and friendly competition. The possibility of fission, far from being a dread secret, fast became the hottest topic in physics, and a new weapon in the discipline's insatiable quest for funding. In 1939 Heisenberg used the possibility of making an atom bomb as leverage to get funds for reactor research.

At the end of April 1940 William Laurence, who followed nuclear research in Germany for the *New York Times*, learned that a large part of the Kaiser Wilhelm Institute for Physics was being turned over to uranium research. The Germans, he concluded, were working on an atom bomb. On Sunday, 5 May 1940, the *New York Times* broke the story on its front page: 'Vast Power Source In Atomic Energy Opened by Science'. According to Laurence, 'every German scientist in this field, physicists, chemists and engineers . . . have been ordered to drop all other researches and devote themselves to this work alone'.

The US government assumed (rightly) that Laurence was exaggerating. Still, the article was serious enough that it caught the attention of George Vernadsky, émigré son of the geologist and environmental visionary Vladimir Vernadsky, and now a professor of Russian history at Yale. George clipped the article and sent it to his father with a note that read, 'Don't be late!'

Vernadsky, by then in his eighties, received his son's letter while staying at a sanatorium on the Uzkoe estate, near Moscow – a monument of aristocratic culture whose facilities included a functioning church (Easter was celebrated; May Day was not). Though his failing health did not permit him much of a direct role in the development of the Russian nuclear programme, he acted a deal more promptly than the American authorities did. He assembled a Uranium Commission for the Academy of Sciences, bringing on board distinguished scientists Igor Kurchatov, Sergei Vavilov, Dmitry Scherbakov and Peter Kapitsa. Equally important, he argued that the field's practical problems should be given priority. It was all very well for physicists to bandy theories back and forth, but they would get nowhere unless they actually started separating isotope 235 from uranium. In his diary for 16 May 1941 Vernadsky recorded a conversation with one of the Academy vice-presidents:

I pointed out to him that now obstruction was caused by the physicists (Ioffe, [Sergei] Vavilov – I did not name names). They are directing their efforts to the study of the atomic nucleus and its theory and here ... much of importance is being done – but life requires the mining-chemical direction.[11]

Progress was steady, if unhurried. A story in *Pravda* on 22 June 1941 described the construction in Leningrad of 'a building which looks like a planetarium ... This is the first powerful cyclotron laboratory in the Soviet Union for splitting the atomic nucleus.'[12] It never operated. German armies invaded the Soviet Union that very day, forcing the physicists to abandon their work and hurriedly evacuate much of Ioffe's Physico-Technical Institute to Kazan by train.

Here, and for a long while, the story was paused, as scientists switched from nuclear physics to more pressing problems. Yakov Zeldovich and Yulii Khariton, who had been conducting experiments on uranium chain reactions, moved over to the development of chemical explosives. Igor Kurchatov, whose own interest in chain reactions led him to invent nuclear reactors, for the moment was working to protect ships against magnetic mines. Some of his colleagues laid physics aside entirely and joined the army.[13]

It was left to one young physicist, Georgy Flyorov, to notice just how quickly the hottest topic in physics was becoming a matter of global security.

Flyorov's own work in the field was considerable: in 1939, with Konstantin Petrzhak, he had demonstrated spontaneous fission – a form of radioactive decay found only in very heavy chemical elements. Now he was a lieutenant, studying to be an aviation technician at a training airfield near the city of Voronezh. In late 1941, at a local university library, it occurred to him to see how work on fission was progressing. He leafed through the latest available English-language physics journals. Pique at finding no references to himself quickly turned to disquiet. His wasn't the only name missing from the conversation. In fact, there was no conversation. His casual literature trawl revealed *no* papers on nuclear fission. This could mean only one thing: the allies had classified the problem as a military project.

Flyorov wrote to Kurchatov, urging a return to uranium research. He wrote to Sergei Kaftanov, who supervised scientific work at the State Defence Committee. He may even have written to Stalin. He travelled to Kazan, where most academic institutes had been evacuated, to discuss the matter with as many senior colleagues as he could find. He sketched out the workings of a crude atomic bomb. But Ioffe and Kapitsa, in particular, were sceptical. A bomb, if possible at all, would require uranium they did not have and ten years they could not afford.

As the war dragged on, however, and became primarily a war of attrition, people began to suspect that an atomic device might prove decisive. Neither the Russians nor the Western Allies knew much about the German effort: what if they got the bomb first? Britain was first to respond seriously to the threat, and news of its 'Tube Alloys' project reached Moscow in October 1941 thanks to Klaus Fuchs, a communist refugee from Nazi Germany who had started work on the atomic project in Britain.

Fuchs proved to be a gift that kept on giving: in 1943 he arrived in Los Alamos as a member of the British team working on the Manhattan Project, and carried on sending information to the Soviets. A pupil of Nobel laureate Max Born, Fuchs was a brilliant mathematical physicist, likeable, polite and sincere. (When his new Russian contact offered him $1,500 for his trouble, Russia's most valuable spy held the envelope containing it 'as if it were an unclean thing' and flatly declined the offer.)

Fuchs understood that the war could not be won without the Red Army. It made sense to him, therefore, to keep the Russians informed of any technological or military advance. The idea of sharing the bomb with the Russians did not seem at all outlandish to his colleagues at Los Alamos. The consensus there was that Russian physicists were quite capable of building their own bomb, so the sooner they were made partners in controlling this fearsome technology the better. Physicist Martin Deutsch, an Austrian émigré and MIT professor, remembers that such discussions at Los Alamos were frequent and quite open; only Fuchs seemed lost for words.[14]

The Germans had stocks of heavy water and uranium. They had brilliant physicists, and excellent facilities. Nevertheless, for one reason or another – some say sabotage, others say dumb luck – the German atomic project stalled. In the USA and Britain, meanwhile, industrial facilities were being built for atom bomb construction. In 1943 the Soviets launched their own project –

not to contribute to the war with Germany, but more as a hedge against the atomic future.

In 1943, with the worst of the siege over, Leningrad's Radium Institute managed to dig up and ship out some vital parts of its cyclotron on two goods wagons and reassemble it in the newly established 'Laboratory Number 2' in north-west Moscow. This was Kurchatov's nuclear facility, created in secret by the Academy of Sciences in buildings originally intended for the Institute of Experimental Medicine. Here, Kurchatov and others worked on constructing and operating an experimental atomic pile – and for that they needed, and needed urgently, supplies of uranium.

The hunt for enemy holdings of uranium began in Vienna, just four months after the Red Army fought their way over the German border and prior to the Soviet occupation of Berlin. Forty physicists were recruited to the effort – roughly half the workers of Laboratory No. 2. Their haul was tremendous: along with reams of valuable technical paperwork, they traced and seized nearly 340 kilograms of metallic uranium.

The achievements of the Berlin search team under Avrami Zavenyagin,[15] however, were of a wholly different order.

Briefed on their mission only once their plane was in the air, Zavenyagin's team arrived in Berlin on 3 May and took up residence in Berlin-Friedrichschagen, in a heavily guarded building big enough to house not only the team members, but also some of the German scientists recovered by the group.

The next day, at the Kaiser Wilhelm Institute of Physics, the Russian team acquired a complete description of the German uranium project. This was paperwork quite as important as any captured in Strasbourg by the ALSOS mission – America's own effort to impound the enemy's advanced technology.

Then there was the uranium itself, found hidden among barrels of lead in a tanning plant. More than a hundred tonnes of uranium oxide were sent to Moscow – enough to power an experimental nuclear pile capable of sustaining a chain reaction.

Uranium was vital to the Soviet atomic effort. In southern and south-western Saxony, in Germany's Soviet zone, silver and cobalt mines were stripped to provide equipment for new uranium mines. The silver, cobalt, bismuth, nickel and other ores dug out along with the uranium were simply thrown on waste piles.

Equally important, and in short supply, were the skills required to purify uranium. Georgy Flyorov and his colleague Lev Artsimovich found themselves bundled in with Zavenyagin's 'fishing expedition'. In the uniforms of NKVD colonels they went to pick up Karl Zimmer, the research radiologist who had collaborated with Max Delbrück and Nikolai Timofeev-Ressovsky on research into the size and nature of the gene.

With Zimmer in tow, they visited Zimmer's close colleague Nikolaus Riehl, inventor of the fluorescent lamp and, more to the point, the research director of Auer. Auer specialised in the production of special materials and in 1939, figuring uranium was a 'special material' of the future, Zimmer and Riehl had got Auer involved in the German bomb project.

Flyorov and Artsimovich asked Riehl to join them for a 'few days' of discussion in Berlin. The Russian-born (and, it so happened, half-Jewish) Riehl offered no objections. Word had gone round that the Soviets were promising one- and two-year employment contracts. For Riehl, as for virtually everyone else the Soviets approached, such a business trip, with its chance to send regular food parcels to relatives in destitute Germany, was very welcome; certainly it overrode any political qualms.

Riehl quickly discovered how important he was to the Russian effort. Zavenyagin's group wouldn't let him go home. They detained him in Berlin-Friedrichschagen until 9 June, when he was flown to Moscow, along with his family and some of his staff.

In Moscow, Riehl was handed the task of purifying uranium metal. He was, he supposed, a captive. But having survived the Nazi regime, this was a captivity he found it hard to resent. He was treated like royalty. 'In other respects it was really terrible, the living standard, terrible,' he recalled. 'But not for us. We had everything.' More important, he was treated with civility: 'We never experienced any direct expressions of hate toward us Germans, or even felt it. Teenagers sometimes made minor derogatory remarks in our early days . . . but were quickly called to order by older people.'[16]

On 16 July 1945 at 5.30 a.m., the United States tested an atomic bomb in the desert at Alamogordo, New Mexico. The news was telegraphed to Harry Truman, who'd succeeded Roosevelt as US president: 'Babies satisfactorily born.'

The next day, the Potsdam conference began, at which Churchill, Truman and Stalin were to discuss the post-war settlement.

On 24 July, says Truman's official interpreter Charles Bohlen, the president told him that he would 'stroll over to Stalin and nonchalantly inform him' about the bomb. Truman walked around the conference table and sidled up to Stalin. He 'casually mentioned' that the United States had 'a new weapon of unusual destructive force'.

Stalin managed not to show the hit: 'All he said was that he was glad to hear it and hoped that we make "good use of it against the Japanese".'

In private, Stalin was rattled. He told Molotov, his minister of foreign affairs, 'We'll have to have a talk with Kurchatov today about speeding up our work.' He still had no idea how far behind the Soviet Union had fallen in the race for the bomb. The obliteration of Hiroshima and Nagasaki, less than a month later, left him in no doubt: he had been trumped.

'Stalin was really enraged,' recalled Sergei Kruglov, Stalin's minister of internal affairs:

That was the first time during the war that he lost control of himself . . . What he perceived was the collapse of his dream of expansion of socialist revolution throughout all Europe, the dream that had seemed so real after the capitulation of Germany and was now invalidated by the 'carelessness' of our atomic scientists with Kurchatov at the top.[17]

Construction of a Soviet uranium plant in Elektrostal, in Moscow's Noginsky district, began immediately, using equipment brought from Riehl's old company, Auer. 'On one occasion Zavenyagin visited us in the tiny munitions laboratory where we were first located,' Riehl recalled. 'He asked the staff of Russian workmen, who encircled him respectfully, from where the various pieces of equipment had come. The response was uniform. Each had been liberated as war tribute from Germany. Just as this exercise was finished, a rat suddenly ran by. He said harshly, "That clearly is ours."'

By the end of the summer of 1946, three tonnes of uranium ore were arriving at Laboratory No. 2 every week. Nikolaus Riehl and his colleagues worked conscientiously and quickly. The effort won Riehl one of the batch of Stalin Prizes granted in November 1949. He was also made a Hero of Socialist Labour, with the Golden Star, 'which, when worn on my breast, stirred looks of wonder on the faces of many top-ranking politicians'. And the rewards kept on coming: a Lenin Prize, a dacha in the woods west of Moscow, and best of all the opportunity to reassemble around him, as far as he could, the intellectual and professional circle that he had enjoyed in Berlin during the war.

One day Zavenyagin, the Atomic Minister in Russia, called me and said, 'Do you know a Dr. Zimmer and a Dr. Born?' I said, 'Yes, I know both of them very well.' 'Do you wish them?' 'Yes.' I said, 'Of course.' Then it became clear to me that the poor gentlemen were captives. I had not known it previously. I had thought they were walking about Berlin as free men and envied them . . . I tried to arrange the situation so that Zimmer and Born would not actually be working directly in the uranium activity but in ways that would be closely related to it. That was, of course, somewhat of a swindle, since we had no need for a geneticist and also had relatively little to do with radiochemistry. However, I wanted to keep them busy.[18]

This, then, was the reason that Nikolai Timofeev-Ressovsky had been saved from the gulag and – as best the doctors could manage it – restored to health. Object 0211 (sometimes referred to as Laboratory B) was a fascinating, many-sided institute exploring radiation biology, dosimetry and radiochemistry. Timofeev-Ressovsky was there to work.

He found himself among radiologists, chemists and botanists gathered from all over the Soviet Union and those territories under Soviet control. (One young geneticist, Nikolai Viktorovich Luchnik, arrived out of nowhere, caught on the wrong side of the lines at the end of the war. He had managed to blag his way out of the work camps by passing himself off as a worker on the Nazi atom bomb project.) Technically, Timofeev-Ressovsky remained a criminal (there were criminal prisoners among the service staff, including a rather likeable murderer), but he was housed very well, and treated as well as the Germans – a group which, to his delight, turned out to include his old friends from Berlin, Buch. Better still, he was given leave to write to his wife in Germany.

Since the end of the war, Elena had been working at Berlin University. In 1948, she left with her younger son Andrei for Object 0211. The couple, reunited, once more worked hand-in-glove – only now, Elena had also to be the purblind Nikolai's amanuensis.

In September 1950 there was another important arrival at Object 0211: Nikolaus Riehl himself. Now that he had solved how to purify sufficient quantities of the uranium arriving from Elektrostal, the authorities didn't know quite what to do with him. It was Zavenyagin, in the end, who offered Riehl the chance to direct operations at the *sharashka*.

The smallest of all the laboratories under Zavenyagin's care, 0211 was a pure research effort, investigating the effects of the radiological weapons that were being developed in the USSR until 1954. The atmosphere was set by the scientists, and life was good, though the weather was, as Riehl put it, 'intensely continental'. (In winter, the temperature regularly dropped below –40 °C.)

Left to set their own research agenda, the Timofeev-Ressovskys and their German colleagues set about trying to understand how radioactivity impacted living organisms and ecosystems. Their experiments were simple affairs, done in slab boxes and rain barrels. The barrels were filled with boxes of earth; a mixture of radioisotopes was poured in from one end, and the components could be measured as they came out the other.

This new field of study, radiation ecology, revealed how radioactive isotopes spread, accumulated and migrated through experimental and natural biological systems. It was conducted very much in the spirit of Vladimir Vernadsky's major environmental work. When, much later, Timofeev-Ressovsky finally got the chance to publish, he acknowledged the debt: he called this field 'Vernadskology'.

Object 0211's gaoler was Alexander Konstantinovich Uralets, an NKVD colonel who knew nothing whatsoever about genetics, and very little about science in general. He was an economist. Uralets turned out to have a lively mind, however, and the more he learned about the work of the scientists in his care, the more he wanted to learn from them. Timofeev-Ressovsky took him under his wing, and Uralets proved an able and conscientious student. When genetics once again came under attack, he proved an extremely loyal ally. Do what you would usually do, he would say: just don't publish. As a consequence (and at considerable personal risk), Uralets's 'jail' became virtually the only place in the USSR where geneticists were *not* harried or repressed.

Despite enjoying privileged living conditions, Russian scientists and engineers hit the walls of their little world quite easily. Beria, whose suspicion of his own intelligence-gathering had delayed the development of the Soviet bomb by years,[19] infected the staff of the NKVD with the same malady. His men were not above asking scientists handling plutonium hemispheres daft questions like 'Why do you think it is plutonium?' At the very end of 1946, a few days after Kurchatov's pioneering reactor[20] went critical, generating 100 watts of electricity (enough to power a decent light bulb), Beria himself visited and was given control of the three cadmium cooling rods. Raising and lowering these through the lattice of uranium lumps, each embedded in a graphite sphere, controlled the rate of the chain-reaction. The Geiger counter screamed as Beria raised the rods, but Beria was unimpressed: 'Is that all?' he demanded, suspiciously. 'Nothing more? Can I go to the reactor?' Kurchatov stopped him.

Beria, a notorious sadist, was a natural manager of *sharashki*. More than most, he thrived on the entrepreneurial possibilities of terror. The rumour went that he had a system: those who were marked to be shot in case of failure would receive the title of Hero of Socialist Labour should they succeed; those who would have received maximum prison terms would be given the Order of Lenin, and so on. He told one of his managers 'You're a good worker but if you'd served six years in the camps, you'd work even better.'[21]

He was far less effective at managing his free personnel – figures often highly regarded in their own fields, who had taken full personal advantage of the relaxed ideological atmosphere of the war years. His run-ins with the physicist and inveterate letter-writer Peter Kapitsa were particularly revealing: the old police chief had no reputation for being polite or tolerant, and Kapitsa had a reputation for leaping fiercely to his own defence. When Kapitsa joined the bomb project, tensions developed quickly.[22]

Just before the war, Peter Kapitsa had developed a new method for producing liquid oxygen for industry – work that had earned him his first Stalin Prize. Not one to take success lying down, Kapitsa made plans for factories, and planned to transport the liquid oxygen throughout the country in specially designed railway tanks. He lobbied hard and, as usual, his letters took no prisoners. To Molotov he wrote: 'all this time, I have worked as a mule driver, and I have been denied both a stick and a switch. I think that in one way or another, I should be granted official power to direct all the processes of industrial inculcation.'

Within a month, he had been put in charge of a new Chief Directorate for Oxygen. This high bureaucratic post got him listened to far more than his scientific achievements ever had, and at the end of the war his achievements as an administrator earned him the country's highest civil title, Hero of Socialist Labour, and a brief involvement in the Soviet bomb project.

The brevity was largely down to Lavrenty Beria. Under Beria's supervision the Soviet bomb was evolving by slow, expensive, deliberately painstaking and redundant steps, as it slavishly reproduced the American effort. Kapitsa could not understand why Beria persistently ignored the Soviet Union's great advantage – their certain knowledge that an atomic device would work. 'We want to retry everything the Americans did rather than try to follow our own path. We are forgetting that to follow the American path is beyond our means, and will take a long time.'[23]

It seemed to Kapitsa that he himself could do a better job if only he was given more power over the project. On 3 October 1945 he wrote to Stalin in terms that just a few years before would have got him shot:

There was a time when the Patriarch stood alongside the Emperor; the Church was then the repository of culture. The Church is becoming obsolete and the patriarchs have had their day, but the country cannot manage without leaders in the sphere of ideas . . . it is time for comrades like Comrade Beria to begin to learn respect for scientists.

After Beria saw the letter he turned up at Kapitsa's institute carrying a double-barrelled shotgun. It was a gift, a peace offering – but the men could not work together.

Kapitsa was a member of both the Special Committee gathered for the bomb project and its Technical Council. In his letter to

Stalin he had offered to resign from both posts – a gesture that he probably meant rhetorically. In December, Stalin called Kapitsa's bluff, met his 'request' and released him from his duties.

Beria, still smarting from Kapitsa's patrician criticisms, did not let the matter lie. In 1946 he tried to have Kapitsa arrested, but Stalin told him: 'I will remove him for you, but don't you touch him.' Stalin was as good as his word, and eventually signed the official decision that said that Kapitsa had failed to fulfil government orders for the production of gaseous oxygen.

Having lost almost all his official positions except the Academy membership, Kapitsa retreated to his dacha outside Moscow. There he arranged a garage laboratory, ironically nicknamed the 'Hut for Physical Problems', published a few academic publications, and otherwise disappeared from public view.

Stalin had raised Kapitsa; now Stalin had cut him down. It was a pattern that took no one by surprise – a pattern repeated in small ways across the science and engineering base of the country. In the *sharashka* system you were invisible, but safe. Though physically a captive, you had freedom of thought. Outside the *sharashki*, you earned good money, had the ear of the state, and had all the privileges of a bureaucratic career. But it was a precarious business.

18 : 'How did anyone dare insult Comrade Lysenko?'

I received a report that a human head and the soles of the
feet had been found under a little bridge near Vasilkov,
a town outside Kiev. Apparently the corpse had been eaten.[1]
Nikita Khrushchev, recollecting the 1946–7 famine

In mid-June 1945 the Soviet government flew a large party of foreign scientists to Moscow. Some 122 delegates from eighteen countries, including the United States, Britain, France and Canada, found themselves being chauffeured round the war-torn and devastated capital in comfortable cars. Their rooms were luxurious, their meals decadent. Until the end of the month, in Moscow and in Leningrad, a virtually bankrupt Soviet government treated representatives of world science to every luxury.

On the last day the Kremlin held a special banquet for its guests. Eric Ashby, a botanist sent to Moscow as Counsellor of the Australian Legation, recalls:

When all were assembled the music stopped and the curtains of the great Bolshoi stage parted to reveal, brilliantly lit by floodlights, an impressive and unusual spectacle.

The August 1948 Session of the Lenin Academy of Agricultural
Sciences, held in the club of the Ministry of Agriculture in Moscow.

At the front, behind a long table covered with a red cloth, sat the Council of the Academy of Sciences. They were backed by ranks of other distinguished scholars. The impressive depth of the Bolshoi stage was occupied by a huge bank of hydrangeas. Among the heaped flowers stood a bust of Lenin three times life size; behind Lenin, in a setting of red curtains, hung a picture about twelve feet high by eight feet wide of Stalin in the uniform of a Marshal.[2]

The occasion – the 220th anniversary of the Academy of Sciences – produced a lot of promises. Peter Kapitsa declared in his speech that there was really no such thing as Soviet science or British science; there was only one science, dedicated to the betterment of humankind. Molotov – to many a guest's bemusement – concurred, promising the 'most favourable conditions' for the development of science and technology and for 'closer ties of Soviet science with world science'.

There was a motive for this extraordinary thaw in relations. The Soviet government badly needed international support if it was going to recover from the war. International contacts were now considered a national asset. What better way to arrange a wide pro-Soviet propaganda campaign in the West than have science's chattering classes take an interest in the Soviet project?

Eric Ashby, a critical but not unfriendly observer, got a book out of the experience, *Scientist in Russia*. In it he noticed that a visiting scientist had to cultivate some 'sales resistance' when he visited an agricultural research institute, 'for he will hear stories of perennial wheat with prodigious harvests, bacterial treatment of seeds which doubles the yield, new potatoes for the Arctic, and new sheep for deserts, which do not stand up to international standards of criticism'.

Ashby was particularly curious to find out where, on the spectrum between iconoclast and plain fraud, the notorious agronomist Trofim Lysenko sat. With his friend Julian Huxley, he arranged for Lysenko to give a special talk at the Academy's jubilee, translated for the benefit of English-speakers. From this lecture they learned, to their mounting embarrassment, Lysenko's views on sexual reproduction: how the best oocyte chooses the best spermatozoon, and everything occurs as a 'love-based marriage' during which one cell 'eats' another. Both visitors shuddered at Lysenko's description of sex cells 'eating' each other. Huxley hid his discomfort by fussing with his glasses. Afterwards they spoke privately to Lysenko, in an attempt to make sense of what they had just heard. To their dismay, they found that the translator had been extremely proficient, accurately rendering some of the maddest material they had ever encountered. 'Fertilisation is mutual digestion,' Lysenko explained,

The egg digests the sperm, and at the same time the sperm digests the egg. They assimilate each other. This assimilation is not complete. We know in our persons what happens when it is so. We belch. Segregation is Nature's belching. Unassimilated material is belched out.

Huxley and Ashby left the sumptuous lecture-hall in stunned silence. When they gained the street their eyes met – and, according to the geneticist Raissa Berg, who was with them, 'these two tall restrained men simultaneously put their hands on each other's shoulders and burst out laughing'.[3]

Ashby devoted a considerable portion of his account of his year-long visit to Lysenko, and noted that biologists put little stock in his queer ideas. His anomalous status greatly puzzled Western commentators, none more so than Robert Cook, writing in the *Journal of Heredity* in 1949:

In a country as great as Russia, with such an impressive body of first-class scientists, who are familiar with science in the rest of the world and are contributing substantially to it, the 'new genetics' is a strange anomaly. It is well past its zenith but it still flourishes in uneasy truce beside the 'old genetics'. Lysenko and his school are clearly a deep embarrassment to bona-fide biologists; yet the school goes on, and Lysenko was made a Hero of Socialist Labour, the Soviet equivalent of an Order of Merit, in June 1945. How can the Academy tolerate such a departure from its catholic standards? And how can Lysenko pose as leader of genetics when he is patently unfamiliar with most of the advances in the subject over the last twenty-five years?[4]

The question was a good one, and it was being asked in Russia, not just within the Academy but in the corridors of government. The more the Soviets engaged with world science, the more Lysenko proved a political embarrassment. Around the talented geneticist Anton Zhebrak – a Party member and president of the Belorussian Academy of Sciences – an anti-Lysenko campaign began to form. In May 1945 Zhebrak went to San Francisco, to the founding conference of the United Nations. A few weeks before, he had an audience with Molotov, the minister of foreign affairs, and one of the subjects of their conversation was the situation in Soviet genetics. It was presumably with Molotov's permission that Zhebrak, once in the United States, set about spreading the word among the genetics community there: Lysenko was out. As the evolutionary biologist Isador Lerner reported to Hermann Muller, 'It will not be too long before Lysenko has enough rope to hang himself. In the present situation, the support of American geneticists is tremendously important.'[5]

Critics of Lysenko in the West needed little encouragement. Many blamed him personally for the disappearance of his mentor and rival, the much-revered Nikolai Vavilov. Publication in Russia of Lysenko's book *Heredity and Its Variability* in 1943 had coincided unhappily with the release of definite information that Vavilov was dead, and had inspired several critical articles in British and American scientific journals.[6]

This kind of criticism, however – which condemned, either by implication or quite stridently, the way Russia treated its elites – was unlikely to be of much practical use to living Russian geneticists. The chance for a more targeted, scientifically rigorous campaign against Lysenko arose early in the spring of 1945 when the publishing company McGraw-Hill sent the Russian edition of Lysenko's book to Leslie Dunn. They were thinking about commissioning an English translation.

Dunn, leaping at the opportunity, immediately recruited the Russian ex-pat geneticist Theodosius Dobzhansky to read and review the book, and wrote back to the publisher:

Notwithstanding the fact that most geneticists here believe that [Lysenko's] views are erroneous, he is a person of such importance and the question at issue is so important that it would probably be of much service both to American and Russian science to have this book available in English, even though the views expressed prove to be wrong.

McGraw-Hill backed out of the scheme in the end, but Dunn and Dobzhansky decided to publish a translation under their own auspices. As Dunn wrote to Lerner:

We believe the best way to deal with Lysenko's influence is to make known his ideas and evidence in the form in which he himself has published them. We have no doubt that the judgment of Americans will be adverse and that this will strengthen the hands of those in the Soviet Union who oppose him.

Dobzhansky started the translation. It was, he told Dunn, 'one of the most unpleasant tasks I had in my whole life, and surely I would never undertake a thing like that for money – it can be done only for a "cause"'.[7]

Heredity and Its Variability was published by King's Crown Press in 1946. A storm of embarrassment and hilarity ensued. Fellow travellers, apostates and lifelong critics of the Soviet project tussled in the columns of most major newspapers. In Britain, Hermann Muller and George Bernard Shaw gave readers of *The Listener* the greatest intellectual punch-up that polite publication ever ran.[8]

With Soviet pride no longer firmly nailed to the Lysenkoist standard, real work could begin again, and Soviet biologists once more found themselves part of an international community. At the Academy of Sciences, an 'International Publishing House' was established. The Academy's Bulletin ran special columns entitled

'On the Pages of Foreign Scientific Periodicals' and 'The Western Press on Soviet Science'. The exchange of genetics literature and even of *Drosophila* stocks was revived. Large numbers of reprints, journals and books were sent to Russian geneticists, and research papers of Soviet geneticists found their way into Western periodicals. Alexander Serebrovksy, Anton Zhebrak and Koltsov's successor Nikolai Dubinin used American journals to announce a new dawn in Soviet genetics, and worked hard to make that claim a reality, chasing foreign contacts, and grabbing at the suggestion from Western colleagues that they host the next international congress of genetics in the Soviet Union.

Lysenko and his colleagues could only fulminate in the popular press, but even here they found their currency devalued. The Party apparatus, normally so sensitive to adverse headlines and angry letters, had stopped listening. When opponents of Lysenko tried to get Prezent dismissed as Leningrad University's head of Darwinism, the Party cell of Lysenko's Lenin Academy issued a memorandum 'on the controversy in genetics', claiming (quite correctly) that geneticists were trying to undermine Lysenko's authority. They asked no less a figure than Andrei Zhdanov to take 'appropriate measures'. But Zhdanov – returned from Leningrad a hero, now Stalin's right-hand man and tipped to be his successor – never replied.

Andrei Zhdanov's rise to the brink of absolute power was, even by Russian standards, a curiously gloomy one. Barely into his fifties, ill-health had already reduced him to an old man. His own conservatism, his hatred of cultural pretensions (indeed, his hatred of large swathes of culture) made him a natural if regrettable companion for Stalin, a self-isolated man whose own grip on reality was, by the war's end, beginning to slip.

Together, like two crabby professors, Stalin and Zhdanov set about making over post-war Russia in their own nineteenth-century image. Writers from Akhmatova to Zoshchenko (once

Stalin's favourite author) were condemned for 'defaming Soviet reality'. Shostakovich and Prokofiev were blamed for composing 'formalist' music that ordinary people found 'incomprehensible'.[9]

It is hard to say whether Zhdanov's mounting exhaustion made him more reactionary, or whether his reactionary temperament drove him to exhaustion. The one fed the other. Dmitry Shepilov, a senior Party official, remembered that Zhdanov 'had to occupy himself with the most varied questions: the re-evacuation of factories and the abolition of bread ration-cards, the growth of fishing and the development of book-printing, the production of cement and the increase of scientists' wages, the sowing of flax, and television'.

Then there were the ideological questions: Zhdanov seems to have had an unlimited appetite for those, and rather than waiting for his staff to come up with the answers he insisted on conjuring them up himself by re-reading classic Marxist and Leninist texts. Admirable attention to detail, perhaps, but not a survival strategy for a powerful man in ill health who was anyway beset with the obligation to attend Stalin's notorious dinners – all-night affairs of numbing tedium, and Stalin's favourite method of breaking the spirit of anyone who got close to him. These 'informal' evenings at Stalin's dacha surely drove Zhdanov to drink, though his doctors had told him not to touch the stuff.

With Stalin now in virtual retirement, Zhdanov had no chance to sleep in. He had control of the Party. Staff whispered about 'our crown prince'. Bureaucrats would bow to him and scrape their way in reverse out of his presence: 'and in bowing himself out, he backed into the door, nervously trying to find the doorknob with his hand', remembered the Yugoslav ambassador.[10]

Zhdanov handled the pressure of Stalin's favour poorly. Emotionally and physically unfit for the role for which he was being groomed, he grew increasingly touchy and paranoid. Of course some form of 'cold war' between former Second World

War allies was inevitable, given the nature, power and potential of atomic energy. But the tone of that conflict within Russia – the extreme isolationism, the extreme chauvinism, and the undoing of all the benefits the end of the war had ushered in – was set a good two years before Russia ever acquired its bomb, when Zhdanov found he had been left out of a conversation about the treatment of cancer.

The affair had its roots in the work of two doctors: Nina Kliueva and her husband Grigory Roskin. Roskin, who had once been a pupil of Nikolai Koltsov, had been studying *Trypanosoma cruzi* – the tropical parasitic microorganism that causes Chagas disease. When he found that the parasite had a particular appetite for tumour cells, he and his wife Nina, a microbiologist, set about extracting the protein responsible for this anti-cancer effect. In March 1946 Kliueva and Roskin published their clinical results. Their tumour suppressor 'Preparation KR', or 'cruzin', was harmless for human beings and in several cases caused tumours to dissolve.

It was a remarkable claim, the science was solid, media interest was high; even Western newspapers ran the story.

What Roskin and Kliueva needed now, if they were going to develop their work, was foreign equipment for their laboratory. Immediately after the war, and with a savage famine to contend with, obviously the government had no spare funds, but the minister of health, Georgy Miterev, invited US ambassador Walter Bedell Smith to visit Kliueva and Roskin in an attempt to awaken American interest.

Smith visited on 20 June 1946: '"K" and "R" turned out to be wife and husband, charming, modest, and obviously devoted to each other and to their work.' Smith proposed the Soviets and Americans collaborate on an institute for cancer research: if Kliueva and Roskin could deliver the ideas, America would deliver the equipment.

Georgy Miterev happily signed off on the idea, and on 6 July

1946 his ministry officially conveyed the good news to Kliueva and Roskin.

But though both the Foreign Affairs Ministry and the Ministry of State Security had cleared Smith's visit, the Central Committee had somehow never been informed. On 3 August 1946, Andrei Zhdanov received the report on Smith's visit, and exploded.

The report confirmed many of his and Stalin's worst suspicions about the subversive ways of foreign intelligence. (Smith's later career rather proved the point: in 1950 he became director of the CIA.) Furthermore, Stalin and Zhdanov saw in the report an example of government bureaucrats bypassing the Party as if it had no authority in such matters. Zhdanov saw to it that the joint institute never took shape. He took no steps against the researchers themselves though, and managed to persuade Stalin and the three deputy premiers to issue a secret Council of Ministers resolution to support their laboratory.

The connection with the United States, however, once forged, was not so easily broken. In autumn 1946 Vasily Parin, secretary of the newly organised Academy of Medical Sciences, and quite unaware of any problem, accepted an American invitation to tour hospitals and cancer centres in the USA. As he was packing he dropped into his luggage a manuscript copy of a book, already published and freely available in Russia, called *The Biotherapy of Malignant Tumours* – by Kliueva and Roskin. He also threw in some samples of cruzin, though before handing them over to his interested guests he made sure, once he reached America, to get permission from the foreign affairs minister, Vyacheslav Molotov, who was also visiting.

When Zhdanov heard all this he responded with fury, and on 28 January 1947 he summoned Kliueva to the Kremlin for interrogation. In a Politburo meeting held on the very day the American propaganda station Voice of America began its broadcasts, 17 February 1947, Stalin announced the renewal of the spiritual struggle with the West. All this pervasive 'kowtowing to the West' had to stop, once and for all! He picked out the KR Affair as typical. Parin (who had only just got back home) was arrested as soon as the meeting was over. He was later condemned to twenty-five years' imprisonment as an American spy.

The question then remained, what to do with the couple whose sterling medical research had triggered the crisis. Specialists now operated much closer to power; they were not to be thrown away without consequence and, anyway, they were now of a generation that had grown up under Soviet rule. They were the state's own children, and disciplining them required tact as well as authority.

In late March 1947, 'honour courts' appeared in ministries and other central agencies, including the Academy of Sciences. Honour courts were a throwback to tsarist times, when they had dealt with unbecoming conduct in the military – cheating at cards and the like. The court, consisting of the commander and a few other elected members, would conduct its investigation and announce its verdict at a public meeting of the entire corps. There were no official punishments: it was assumed that those judged guilty would be shamed into leaving their unit and sometimes the military altogether. Honour courts were also convened in pre-revolutionary Russian universities to deal with cases of plagiarism.

Resurrecting them now, the Politburo announced that the courts were to be 'a new and effective form of the reeducation of the intelligentsia'.[11]

Zhdanov directed the trial of Kliueva and Roskin. He even wrote the script, for this was truly a 'show trial'. Between 5 and 7 June 1947, in a large Moscow theatre, before an audience of more than a thousand specially invited senior bureaucrats, Roskin and Kliueva, those two 'murderers in white coats', were accused of treason against the Motherland. 'Clean yourselves of the disgrace and shame that you have inflicted upon yourselves by your unworthy deeds,' cried the biochemist Boris Zbarsky, 'you have never been patriots.'

The show concluded with a 'public reprimand' for Kliueva and Roskin – and that was all. There were no press campaigns, no hostile publicity, no sackings, no internal exile, and no lasting consequences. Just a month later, on 12 July 1947, ministries were ordered 'to help Professor Kliueva with materials and equipment to build her laboratory'.

But the purpose of the show trial had been fulfilled: it had re-established the authority of the Party, which had been so loosened by the massive and indiscriminate expansion of membership during the war. Disclosing information regarded as a state secret was now 'punishable by confinement in a reformatory labour camp for a term of from eight to twenty years'. The Academy of Sciences issued a special instruction, 'On Principles of Scientific Publications', and all scientific journals of the Academy ceased translating their abstracts and tables of contents into English and other Western languages. The column 'On the Pages of Foreign Scientific Periodicals' vanished from the Academy's bulletin. Soviet scientists were told to resign from all foreign scientific societies, and foreign visits were curtailed.

The 'patriotic' campaign triggered by the KR affair ended the post-war thaw in Soviet science, and it could not have happened at a worse time. International contacts were necessary to allow Soviet science to 'catch up with and overtake' Western science – and in agronomy they desperately needed to.

Spanning 1946–7, the third Soviet-era famine was not just the result of bad weather or war-damaged infrastructure. The Party and state, desperate for foreign exchange, had resorted to the very policy the tsarist authorities had followed in 1897: selling grain on the international market, instead of using it to offset the famine at home. Now, though, the consequences were far worse. Even regions not visited by drought, starved. Virtually the whole Soviet countryside suffered, some 100 million people. Two million died. In 1947, one in three children perished in infancy.

While the newspapers censored reports of the famine (which remained largely hidden from the West for twenty years, when Khrushchev published his memoirs), it was clear enough that the country was in crisis, and that something would have to be done.

On New Year's Eve of 1946, Stalin summoned Trofim Lysenko to the Kremlin.

The subject of their conversation was so-called branching wheat – a variety of wheat that produces an unusually large number of seeds per plant. There was very little else good you could say about it, though. Its outsize grains were frequently disease-ridden. Ancient Egyptians had tried to grow it and gave it up as a bad job, and ever since it had been used almost exclusively for stock and poultry feed. It might have been possible to make some rather grim pasta out of it – its nearest relative is durum wheat. For bread though, it was useless; it had no gluten to speak of and only half the protein of other varieties.

The press photograph that emerged from Lysenko's meeting with Stalin (their only personal exchange) circulated in newspapers, books and posters for years. Indeed, the photograph was the real purpose of the meeting. Branching wheat was virtually inedible, but it was certainly photogenic, and the message was clear – that Stalin and his favourite barefoot scientist were again on top of the situation, poised and ready to pull the nation from the brink of catastrophe.

If that interpretation is too cynical, then the alternative is actually worse: that Stalin, the old-fashioned Michurinist, who dreamt of one day plucking fruit from Arctic lemon trees, genuinely thought that branching wheat would help to overcome the crisis – and ran roughshod over the expert opinion available to him in order to promote this latest hobby horse.

Shortly after his meeting with Stalin, 200 kilograms of seeds arrived at Lysenko's door, to aid his further study of branching wheat. Even Lysenko balked at this one. He had never had much to guide him but the practical wisdom picked up from his father, and while this did him little good in the normal course of things, it surely told him that branching wheat was so much diseased rubbish. He kept Stalin abreast of his work, and he kept promising better results tomorrow, but he made no attempt to conceal the abject failure of his trials.

If the meeting with Stalin did not provide Lysenko with a magical cure-all for Russia's famine, it at least saved his career. From this point on, he countered critics and hostile bureaucrats by pointing to his important personal correspondence with the nation's leader.

The recent and widely circulated photograph of him in Stalin's company was especially timely, since the very next month, February 1947, the Western propaganda channel Voice of America started broadcasting; and one of its very first voices belonged to Trofim's brother Pavel Lysenko. Pavel, who had escaped to the United States after the war, was speaking none too flatteringly of his homeland, and this made Trofim a family member of an enemy of the people, and therefore liable to arrest.[12]

The protections offered him by one meeting, and one photograph, would, he knew, not last him long. Lysenko knew he was in trouble, and that the trouble had been long in coming and was impossible to avoid. The number of members of all other Soviet academies had risen after the war. The Lenin Academy under Lysenko, however, had grown moribund. A ministerial report in November 1946 stated that the Lenin Academy was on the point of 'disintegration'. It had become exclusively preoccupied with problems invented by Lysenko and had neglected everything else. Even Nikolai Tsitsin, the vice-president of the Lenin Academy, 'in fact does not work in the academy and does not attend its plenary meetings, because of his disagreements with academician Lysenko over organisational matters and principles'.[13]

For twelve years Lysenko had held off new elections, waiting in vain for friendly faces to dominate his academy's candidate lists. By the end of the war, on the contrary, fewer than half of his academicians were still alive. By 1947 arrests and natural deaths had reduced the number of academicians to just seventeen. Try as he might to procrastinate even further – pointing Benediktov to his important work on branching wheat and so on – elections had to happen for the Lenin Academy to survive, and new elections, along with a whole raft of other measures, would amount to a wholesale reorganisation.

Andrei Zhdanov was deaf to Lysenko's entreaties, so Lysenko now shifted his allegiance to the Council of Ministers. Here the situation was more favourable to him, because in charge of agriculture was Georgy Malenkov, Zhdanov's life-long rival for political power.

Malenkov's star had declined since the end of the war. He had been one of the top five most powerful men in the Soviet Union, supervising military aircraft production and the development of nuclear weapons. Now Stalin was cutting his wartime political aides down to size. (The demotion of Marshal Zhukov and the arrest of Marshal Alexander Novikov of the Air Force, both of whom had dared to criticise some of Stalin's military decisions in private, were among the more notorious examples of Stalin's 'reorganisation'.) In May 1946 Malenkov had been removed from the Party secretariat and sent off to supervise the grain harvest in Central Asia. For several months he and his family had been awaiting his arrest – but it never came. Now, as Zhdanov's power reached its apogee, and his health began once more to deteriorate, it seemed that the see-saw of Stalin's political favour was about to lift Malenkov again.

The autumn of 1947 saw Andrei Zhdanov leaving Moscow for Sochi, a resort town on the Black Sea. At fifty-one, exhaustion and hypertension had finally overtaken him.

Lysenko took advantage of Zhdanov's absence, writing directly to Stalin in October concerning branching wheat. Almost immediately, however, he slid into 'theoretical' concerns. 'I dare state', he wrote, 'that Mendelism–Morganism, Weismannist neo-Darwinism is a bourgeois metaphysical science of living bodies, of living nature developed in Western capitalist countries not for agricultural purposes but for reactionary eugenics, racism and other purposes.'[14]

Three days later, Stalin dashed off a personal response: 'Regarding the theoretical tenets in biology, I think that Michurin's tenet is the only scientific tenet,' he wrote. 'The future belongs to Michurin.'[15]

For some while, Lysenko and his confederates had been citing Ivan Vladimirovich Michurin as an intellectual influence. Michurin, born in 1855, had been a talented plantsman, a sort of Russian equivalent of the self-schooled American Luther Burbank. Ignored under the tsarist administration, he had been championed as a homegrown talent by the Bolsheviks, who liked to think that they had unearthed all manner of geniuses their predecessors had been too hidebound or prejudiced to appreciate. Michurin was, in other words, one of those latecomers to the intellectual party who shoehorned a lifetime's valuable practical experience into an odd vessel and named it 'theory'. This naturally appealed to Lysenko.

Stalin's own interest in Michurin was rather more wide-ranging. Part of his attraction to the Michurin story was personal: Stalin considered him an upholder of Lamarck. But there was a political angle also to Stalin's regard.

The late 1940s and the early 1950s saw the publication of countless 'lives', describing the manifest destiny of the Russian people through stories of the nation's 'forefathers'. This bit of national myth-making was exceptional in both its fulsomeness and its monotony. Whether you were Michurin, or Tsiolkowsky the rocket pioneer, or Pavlov the physiologist, or one of any number of others, your story followed the same pattern. You were Russian (or, in those few instances where the historical record could not be got around, as Russian as made no odds). You spent your life in splendid isolation, founding an entire field of intellectual endeavour. Not even God was on your side, because from the very beginning you were a convinced materialist. You sympathised with socialism, and when finally the opportunity presented itself, you worked fruitfully for the common good, and this made up somewhat for a lifetime spent being vilified, abused, mistreated or, worse, just plain ignored by every foreigner who crossed your path or stumbled on your work.

From a fashionable love of the homegrown underdog, Lysenko's respect for Michurin had evolved into a full-grown myth in which Mendel, Weismann, and finally Thomas Hunt Morgan had concocted, across the span of generations, a fiendish method of withholding practical knowledge of nature from the proletariat – a conspiracy supported and encouraged by a superstition-ridden church and craven, wealthy capitalists.

This was not exactly the political use to which Stalin hoped to put Michurin – in these Cold War years Michurin's Russianness was more important to Stalin than his proletarian credentials – but it would serve. Michurin was a founding father, supported in his last years by the Bolshevik state, elected an honorary member of the Academy of Sciences, and safely dead since 1935. He was a hero in exactly the mould required by Stalin and Zhdanov. Their fusty and stentorious tales about Russia's isolated greatness were not just the maunderings of old men (though they were certainly that): they were concerted national campaigns, meant to shore up their nation for the big chill to come. As yet without a bomb of their own (it was still more than a year away) stories like Michurin's were more than folk tales. They were ammunition.

The first to spot storm-clouds on the horizon was the zoologist Ivan Schmalhausen. In early January 1948 he wrote two letters to Andrei Zhdanov, complaining about the way Lysenko's latest theoretical pronouncement had been covered in the *Literary Gazette*, under the editorship of the philosopher Mark Mitin.

Lysenko had decided that animals of the same species did not compete in any way that influenced their evolution. (Like most of his ideas, this one came to him frayed and third-hand – on this occasion, from the nineteenth-century zoologist Nikolai Nodzhin, who had Darwin down as a 'bourgeois naturalist' who relied on 'Malthusian assumptions'.) When critics took Lysenko to task for this latest bit of baseless thinking, Mitin had published the most violent and dangerous political smears against them. Schmalhausen was appalled.

So, rather unexpectedly, was Nikolai Tsitsin. Tsitsin had for years been a card-carrying Lysenkoist and had chaired the 'Court of Honour' of the Academy of Sciences which condemned Anton Zhebrak, the man who had told the American scientific community that Lysenko was on his way out. Shortly afterwards Zhebrak had been dismissed as president of the Belorussian Academy of Sciences.

However odious some of Tsitsin's actions appear, he was a conflicted man, more a victim of circumstances than their author: 'a very cultured, modest person', according to the ecologist Sergei Zonn. 'He found himself in a bind. He was squeezed between his public position of authority and his inner beliefs.'[16]

Tsitsin's early association with Lysenko was sincere enough: both men had been born the same year, and both had shown early talent as plant breeders. Tsitsin had crossed wheat and couch grass, making perennial varieties of wheat. They had turned out to be useless for agriculture, but that took nothing away from Tsitsin's scientific work, earned him a reputation as an expert in interspecies hybrids – and, given that he clearly knew what he was doing, probably put a brake on his enthusiasm for Lysenko.

Not only had Tsitsin been playing second fiddle to Lysenko as the Lenin Academy disintegrated around them, he was also chairman of the Government Commission for Seed Testing, the whole purpose of which was to separate genuine from bogus advances in agricultural research. If Tsitsin had not had reservations about Lysenko's work before, he surely developed them now that he was obliged to read the results from about a thousand eighty-hectare testing stations.

Written on 5 February 1948, Tsitsin's thirty-page letter to Stalin about the uselessness of Lysenko's work in general, and branched wheat in particular, is a masterpiece of exasperated invective. 'Many representatives of biological science in recent years have been living in an atmosphere in which they feel shut in,' he complained,

an atmosphere of fear of one-sided criticism and of biased exposition of many questions that have already been solved by them. An organised discussion could identify much that is new and of value both for theory and for practice and could enable us to assess what is of value, to discard that which is not or which is actively harmful. A discussion might allow us to find a common line in tackling the problems of our agriculture . . . I ask you, Comrade Stalin, to permit us to hold such a discussion at the nearest future time.[17]

The same day, Tsitsin sent an extended scientific critique of Lysenko to Andrei Zhdanov. Since Zhdanov's contempt for Lysenko was well-known, Tsitsin wrote more directly. He lambasted Lysenko's theories, 'on the basis of which, by the way, during the course of twenty years, not one acceptable variety has been produced, notwithstanding the numerous promises and loud assurances'.[18] But what really exercised him was the way Lysenko was 'surrounding himself with a clique of unscrupulous individuals'. Lysenko had transformed the Lenin Academy into a vacuous bureaucracy, excluding all scientists except his own yes-men. 'I may boldly state not fearing to exaggerate matters that the situation now is such that the normal development of biological and agricultural sciences . . . has become impossible without the intervention and serious assistance of our Party and government.'[19]

The need to have it out with Lysenko, once and for all, was becoming painfully apparent, and it wasn't just scientists themselves who were saying so. Lysenko's cock-and-bull theories were starting to interfere with the work of Stalin's own inner circle.

When the Ministry of Agriculture wanted to breed a new variety of rubber-bearing plant, kok-sagyz, invented by the well-known cytogeneticist Mikhail Navashin, Lysenko used every means to stop 'this genetic monster' from being put into production – this in spite of the fact that its great champion was Yuri Zhdanov, Andrei's son and suitor to Stalin's daughter Svetlana.

In February 1948 Yuri Zhdanov sent a long memorandum to Stalin, accusing Lysenko of 'sabotage': 'From the very beginning, instead of objectively studying the new breed . . . [Lysenko] has created a poisonous atmosphere of hostility and mistrust.'[20]

Yuri, though only twenty-nine, was already head of the Central Committee's science section. Nepotism undoubtedly had a role in this: Stalin liked Yuri. Increasingly lonely, he found that Yuri and Svetlana were the only people with whom he got along. At the same time, Yuri was no fool. He had a degree in chemistry from Moscow, and a brief internship in genetics had made him a convinced Mendelian. In a speech entitled 'On Issues of Contemporary Darwinism', delivered on 10 April at the Moscow Polytechnic Museum, Yuri made it crystal clear exactly what he thought of Lysenko's opposition to genetics.

Yes, chromosomal genetics lacked practical results, and generally suffered from a divorce of theory from practice. Once these rote criticisms were out of the way, though, Zhdanov turned to the alternative offered by Lysenko – and launched into a series of devastating criticisms. Lysenko had rejected the use of hybrid maize. He had not fulfilled the promise of new useful varieties of cereals in two to three years. He had suppressed other schools of thought in ways that were intellectually and morally bankrupt. And Party philosophers who should have known better had backed him.

As they sat in Mark Mitin's office, eavesdropping on the speech through the public address system, it would have been hard to say who was the more outraged: the philosopher Mitin or his guest, Trofim Lysenko. When a man of Yuri Zhdanov's standing spoke such things, some *apparatchik* was bound to turn it into an official directive. Lysenko needed to act, and act fast.

It took him a week to formulate his complaint, addressing his letter both to Stalin and Yuri's father Andrei Zhdanov. A month went by without a response, so Lysenko upped the ante. On 11 May he wrote to the Minister of Agriculture, Ivan Benediktov, offering his resignation as President of the Lenin Academy.

It was a clever ploy. Because the post of Lenin Academy president was in the *nomenklatura* of the Politburo, Benediktov could neither accept nor reject his resignation. All he could do was pass the news on to Malenkov (Lysenko's ally) and to Stalin himself. Lysenko had ensured that Yuri Zhdanov's speech could not be glossed over: one way or another, Stalin was going to have to express an opinion.

In late May or early June, the text of Yuri Zhdanov's lecture finally reached Stalin's desk. Stalin read the transcript carefully. 'Ha-ha-ha,' he wrote in the margin. 'Nonsense.' In one place: 'Get out!'

The amateur Lamarckian gave no quarter to Yuri's new-fangled genetics. At one point Yuri had written,

We Communists are by nature more sympathetic to a doctrine that establishes the possibility of the reconstruction ... of the organic world, without waiting for sudden, incomprehensible, accidental changes of some mysterious hereditary plasma. It is this aspect in the neo-Lamarckian doctrine that was emphasised and valued by Comrade Stalin in 'Anarchism or Socialism?'

Stalin boldly underlined in pencil the words 'It is this aspect' and commented in the margin: 'Not only "this aspect", mister.'[21]

On 28 May Stalin called Malenkov, both Zhdanovs and other Party leaders into his Kremlin office. The government's chief propagandist Dmitry Shepilov was there: 'Pipe in hand and puffing on it frequently, Stalin paced the room from end to end, repeating practically the same phrases over and over: "How did anyone dare insult Comrade Lysenko?" "Who dared to raise his hand to vilify Comrade Lysenko?"'

Shepilov had in fact approved Yuri Zhdanov's speech. He defended it now, and quite openly criticised Lysenko, the trouble-making underachiever who was now, to top it all, disrupting rubber plant production. Shepilov had long since come to the conclusion that Lysenko's theories 'were a joke among real scientists throughout the world, including those who were well-disposed to the Soviet Union'.

Stalin paid him no mind. 'He went up to his desk, took a cigarette, shook the tobacco out into his pipe and slowly walked along the table where everyone was sitting. Then he said very quietly, but I could hear the ominous tone in his voice: "No, it can't be left like this. There has to be a special commission to look into it. The guilty must be punished as an example. Not Yuri Zhdanov, he's still young and inexperienced. It's necessary to punish the 'fathers'."'[22]

From the very start, and at the highest level, Stalin's decision to reject genetics, and rehabilitate Lysenko, evoked either bemusement or dismay. Historians have had no easy time of it either. Super-human efforts have been made to show the strategy, the *Realpolitik* behind Stalin's decision. These efforts are not so much wrong as insufficient. Yes, Lysenko was Michurin's great champion, and was identified as such in the public mind. An attack on Lysenko was therefore an attack on one of the nation's founding fathers. But it is hard to imagine that the nation's psyche would have been irredeemably scarred had Lysenko – or even Michurin, for that matter – been quietly dropped from the news cycle.

And, yes, it is true that Lysenko represented home-grown science, while Mendelian genetics prided itself on being international. In the developing chill of the Cold War, it was necessary to champion Lysenko, the patriot, and vilify genetics as a nest of 'cosmopolitanism'. But here again, this is not nearly enough of an explanation. Genetics was not alone in being the brainchild of an international community. Nuclear physics was an international, cosmopolitan effort – but no one now felt a burning need to contradict Einstein. The state took physics seriously, and wanted to lead the field by any means necessary, including the use of prison camps for intellectuals. It did not attempt to replace physics with some antiquated, homespun variant.

The sheer peculiarity of Stalin's decision raises a possibility of the sort historians try very hard to avoid. Is it possible, even likely, that the foibles and prejudices of one man, and one man alone, had enormous consequences for the whole of the science base of the Soviet Union? Did Stalin singlehandedly destroy genetic science in his own country, for no better reason than that he was a keen gardener with ambitions for lemon trees and set views about how plants grew?

Unlikely as this sounds, there's evidence for it. Stalin, while in some ways a very modern ruler, was also the last in a long line of European philosopher kings. He shared with his fellow Bolsheviks the idea that they had to be philosophers in order to deserve their mandate. And by 1946, when Stalin was sixty-seven and exhausted from the war, this conviction was manifesting itself more and more as a determined effort to reshape reality in his own image. He schooled the USSR's most prominent philosopher, Georgy Aleksandrov, on Hegel's role in the history of Marxism. He told the composer Dmitry Shostakovich how to change the orchestration for the new national anthem. He set the celebrated war poet Konstantin Simonov the task of writing a play about the Kliueva and Roskin case, treated him to an hour of literary criticism, and then rewrote the closing scenes himself. Eisenstein and his scriptwriter on *Ivan the Terrible Part Two* were treated to a film-making masterclass by Stalin and Zhdanov both. 1950 saw Stalin negotiating a pact with the People's Republic of China, discussing how to invade South Korea with Kim Il Sung, writing a combative article about linguistics, and meeting with economists three times to discuss a textbook.

As for the rout in genetics itself, it had Stalin's personal thumbprint all over it. In spite of the fact that President Harry Truman had dispatched B-29 bombers to Europe, and the ambassadors of France, Britain and the USA were all awaiting a lengthy discussion of the Berlin crisis, Stalin still found time to edit speeches by Lysenko extensively, checking them for scientific mistakes as well as political gaffes. No one then or since has ever been able to explain why Stalin was more concerned with the evils of genetics than with the collapse in trust between wartime allies that was even then threatening to bring the USSR and NATO to the brink of war.

Stalin's 'special committee' examining the state of Soviet genetics was a fudge. Andrei Zhdanov and Georgy Malenkov had to co-chair it. The report they wrote, and rewrote, and rewrote, was titled 'On the Situation in Soviet Biology'. Zhdanov repeatedly removed references to his son's *faux pas* in trying to reconcile the two trends in biology. Finally, the report declared that 'the discussion of this question is considered finished'.

And so it would have been, if the Party had simply banished genetics by means of a simple resolution. It had got rid of pedology by that method, after all, in 1936.

But it was Stalin's conceit to play the matter out in public, in as 'scientific' a manner as possible, by means of a special session of the Lenin Academy. On 15 July the Council of Ministers changed the *nomenklatura* so that Lysenko's allies would fill the Lenin Academy come the scheduled elections. Next, Lysenko was told to prepare a report on the controversy in genetics. Eight days later, on 23 July, Lysenko sent 'On the Situation in Soviet Biological Science' to Stalin, and at 10 p.m. on the 27th, Stalin and

Malenkov sat down with Lysenko in the Kremlin to discuss the speech in private.[23]

Lysenko's initial address had separated biology into two camps along class lines. The followers of Mendel were bourgeois. Michurinists came from the proletariat. This was classic Lysenko, but it would not serve the new conditions. Stalin took out all references to class, and replaced these with code-words that opposed 'idealist' (Western) with 'materialist' (homegrown Soviet) science. (Of course there was no time to brief every ally, and come the session, which ran from 31 July to 7 August, Lysenko's supporters continued to fling at their opponents class-war insults that were already more than a decade old.)

Stalin not only changed the political emphasis of Lysenko's opening speech; he actually made it *less political*, much more like a piece of genuine and objective science. He took out an egregious reference to himself, and deleted the word 'Soviet' in the title so that the paper became something of more international consequence: a view 'On the Situation in Biological Science'.

Ensuring the 'scientific' character of the Lenin Academy's August session had its risks. In particular, the Lenin Academy was not the only authority that could decide biological questions. If members in the Academy of Sciences got to hear about the session, they might throw the whole scenario into disarray. So Lysenko kept the session as secret as he could; most of his critics did not attend the session, because they did not know about it.

On Saturday, the first day, Lysenko delivered 'On the Situation in Biological Science'. It was, clearly, a declaration of war, and on the next day, Sunday, an excursion to Lysenko's model farm in the Lenin Hills near Moscow gave participants a chance to absorb it.

The agricultural economist and statesman Pavel Lobanov, who was presiding, expected no counter-attack when he invited formal geneticists to take the floor and respond to Lysenko's paper. But Lysenko's secret preparation had not been watertight, and one man stood to speak. Iosif Rapoport had learned about the meeting only by chance and at the last moment. An acquaintance had given him his ticket, but he had still had difficulty getting into the building. Journalist Mark Popovsky recorded the sensation that greeted Rapoport's appearance:

... with his black curls and boyish expression. He looked very handsome in his military tunic without shoulder boards but with rows of medal ribbons on his chest. Even the black bandage over an empty eye socket did not disfigure him, but lent a keener expression to his pale nervous face. His address too – he was a candidate of science at the time – was nervous but firm in tone.[24]

In 1941 Iosif Rapoport had volunteered for the Moscow People's Militia – a force made up of volunteers who were too old or were otherwise unqualified for the draft – and had returned to the front even after he was wounded and lost an eye. Made an officer, he had become a legend among his men, and was nominated three times for the title Hero of the Soviet Union. He had been rejected each time because he could never stop criticising his superiors – a flaw in his character which now came splendidly to light as, with fine sarcasm, he stripped the emperor Lysenko of every shred of intellectual clothing. 'The transformation of animals and plants', he concluded, witheringly, 'cannot be achieved merely as a result of our wish.' The stenographers were so frightened they edited the sharpest of Rapoport's comments on the fly (although they did allow him to scream 'Obscurantists!' in the middle of another, Lysenkoist speech).

Open opposition to Lysenko on matters of fact was unfortunate, but of no lasting consequence. What really mattered was that the Central Committee's involvement in the session be kept under wraps. This was supposed to be a scientific discussion, not a political one.

On Wednesday, however, Boris Zavadovsky, a Party member of very good standing, founding member of VARNITSO and a celebrated scourge of bourgeois biology, stood up to ask a question. More in sorrow than in anger, he wanted to know why he had not been invited to the meeting. Why had he had to hear about it by accident? Why, above all, were Lysenko's colleagues bad-mouthing his Party reputation in their speeches?

There was, Zavadovsky reckoned, something unconstitutional in all this, and old as he was, sick as he was – his life reduced to a tour of government sanatoriums – Zavadovsky demanded to know why Lysenko was not responding to his critics with due diligence and self-criticism. 'Hence, as I understand it, this conference is obviously taking place not in accordance with, or at least without the participation of, the Central Committee's Science Department.'[25]

Lysenko and Prezent, appalled that Zavadovsky had stumbled on the real nature of the session, heckled him constantly in a desperate bid to shut him up. But forms were there to be followed: Zavadovsky was allowed seven extra minutes. He said in his closing remarks:

To whom and by whom was it necessary to include me among the Weismannists and formal geneticists? Only because I have constantly come forward and will continue to come forward to point out the mistakes of Comrade Lysenko's works, only because I have constantly pointed out that Comrade Lysenko while being an innovator in one sphere has in other spheres become a heavy brake on many necessary and useful trends. I have frequently said this both at sessions of the Academy and in front of Comrade Lysenko. I do not conceal it. But is that a reason to defame me and stick labels on me?[26]

With the room stuffed with Lysenko's supporters, turning the tables on the session was out of the question. Still, what few opponents there were took strength and made their voices heard. Peter Zhukovsky, a member of the Lenin Academy, refused to be interrupted by Lysenko and even went so far as to declare that 'one should worship' at Mendel's grave. Zhukovsky could not understand why there was a 'veritable craze' at the meeting to defame Mendel when all he had done was make some very sound observations about the laws of heredity in peas. Mendel had never written about evolution, nor had he excluded external factors from influencing heredity. There was no need to assume that Mendelism and Michurinism were in contradiction. Zhukovsky called for unity: 'We are all Soviet citizens, and we are all patriots. Some of us went personally and others sent their sons to the front. We all fought for our country, and should we really allow things to reach the point where people refuse to greet Professor Zhukovsky when they meet him?'[27]

His speech won applause, though of course he had missed the point: unity was the very last thing the session had been convened to establish.

With the cat out of the bag, and voices – few in number, but difficult to silence – questioning the legitimacy of the August session, the Central Committee had no choice now but to act. Stealth and subtlety were out of the question. The decision to extinguish Mendelian genetics in the Soviet Union would have to be made openly, and Stalin's personal role be revealed.

A letter from Yuri Zhdanov to Stalin, apologising for his criticisms of Lysenko, was prepared for publication in *Pravda*, and on the evening of 6 August, the night before the last day of the session, Stalin met Lysenko again and dictated the opening paragraph of Lysenko's closing speech.

'Before I pass on to my concluding remarks,' Lysenko announced, the next day, 'I consider it my duty to make the following statement. The question is asked in one of the notes handed to me, What is the attitude of the Central Committee of the Party to my report? I answer: The Central Committee of the Party has examined my report and approved it.'

This was, and remains, says the evolutionary biologist Stephen Jay Gould, 'the most chilling passage in all the literature of 20th century science'.[28]

The transcript concludes: 'Stormy applause. Ovation. All rise.'

19 : Higher nervous activity

Imagine that someone wants to know you better.

He takes a picture of your skull, and if this skull contains some thoughts, the negative will reveal them as black blots, or snakelike spirals, or some other unattractive form.

If he wishes, he can try to photograph your conscience, and the negative will also show all the excrescences and blots.

In a word, every person will be seen through now, and however thick and impenetrable your skin might be, the new light makes it transparent like glass.[1]

Maxim Gorky, 1896

Within a week of the August Session, letters began appearing in *Pravda*. One by one, intellectuals with a stake in Mendelian genetics backed down, apologised and experienced Damascene conversions to 'Michurinism'. Given the memories these people carried of the Terror, their actions now seem only politic.

More culpable were the actions of Academician Alexander Oparin, a food scientist best known for some remarkable

Boris Yefimov's cartoon to accompany the article 'Fly Lovers, Man Haters' in the popular journal *A Little Light* (1949) casts genetics in a fascist guise.

hypotheses about the origin of life. Oparin knew perfectly well what genetics involved. Nevertheless he chose this moment to publish a full-page letter in *Pravda* railing against 'fenced-in pontiffs toying with fruit-flies' – and he called on the Presidium of the Academy of Sciences to renounce genetics.

And so it did. Between 24 and 26 August the Presidium of the Academy of Sciences convened for one of those self-conscious sessions of 'criticism and self-criticism' so beloved of its Bolshevik paymasters. The Academy's president, Nikolai Vavilov's brother Sergei, bit his tongue and took upon himself the 'blame' for the Academy's support of genetics, and reproached his colleagues for their neutrality in trying to balance the claims of 'the two directions in biology'.

The meeting dismissed the Stalin Prize laureate Leon Orbeli as head of the Biological Sciences Section, though, significantly, it put Oparin in his place instead of the obvious candidate, Trofim Lysenko. (Oparin at least had a scientific track record.) Oparin promised that all experimenters in natural science would fundamentally revise their work and cease their 'fawning and servility before foreign pseudo-science'.

For working geneticists, disarming in the face of political disapproval would not be anywhere near enough. Over the coming weeks they saw their *Drosophila* stocks destroyed, and their literature taken from the libraries. Courses and textbooks were rewritten to reflect the victory of Michurinism.[2] Hundreds of decrees were issued, filling noticeboards and agendas in ministries, bureaux, boards, publishers, universities, institutes, experiment stations, editorial boards ... In 1949 even *Soviet Frontier Guard* published an article on the reactionary nature of the Mendelism–Morganism passing across Russia's borders. The suppression continued for years. 'In 1951,' remembers Raissa Berg, a geneticist,

India ink was used to cover the names of the most famous geneticists in all books where they were mentioned ... I witnessed these procedures in the library of the Geographical Society. My father had died in 1950. Deprived of any hope of continuing with population studies, I had begun to study the scientific life of my father, and I was using the archives and the library of the Geographical Society. At first, I could not even understand what was happening.[3]

A week after the session, the sackings began. In research and teaching establishments, scientific journals and elsewhere, Lysenko's cohort replaced the 'reactionary supporters of Mendelism–Morganism'. Thousands were removed from their teaching and research positions. They took whatever work they could: as translators, accountants. Zoia Nikoro, who'd once assisted Sergei Chetverikov, found work as a ballroom pianist at a club. Most ended up working on dreary agrostations far from the capital, their years of experience wasted.

Nikolai Dubinin, a main target of Lysenko's attacks during the session, was pushed out of Koltsov's institute and the whole institute was abolished. The botanist and pioneering ecologist Vladimir Sukachev took him in, giving him a job at the Institute of Forestry. Dubinin spent the next five years bird-watching in the Ural mountains.

Pushed out, his department dissolved, the wheat specialist Anton Zhebrak offered a wishy-washy self-criticism in *Pravda*, got a job at the Moscow Pharmaceutical Institute, and promptly created a secret little centre of genetics of his own. Dmitry Sabinin, a professor of plant physiology at Moscow University, who used to make fun of Lysenko in lectures, was exiled from Moscow altogether and wandered jobless from place to place until, in 1951, he shot himself.

On 24 September 1948, Hermann Muller, now back in the USA and working at Indiana University,[4] publicly resigned from the Soviet Academy of Sciences. So too did virtually every other foreign member. The actions of the leaders of the Soviet Academy were 'disgraceful', Muller said; he compared them to Nazis. He wrote to Leslie Dunn in 1949: 'Let me say that I do not think there is the slightest chance for reconciliation with the Soviet authorities over this matter – it has been tried for some thirteen years now – and I think all which is left is to call a spade a spade.'[5]

The Michurinist campaign of 1948 appears on the face of it to have got far out of hand, swallowing fields that had nothing remotely to do with genetics or even biology. A member of an editorial board of a learned journal 'struggled against the Morganism' of the editor-in-chief; textbook authors attacked rival textbooks for their 'idealist content' (sometimes they even laid into their own co-authors); researchers noted the 'anti-Michurinist tendencies' or the 'servility to the West' of their colleagues. Leaders of fields as far apart as physics and linguistics gathered to discuss how to revamp their work 'in light of decisions of the meeting of the Lenin Academy of Agricultural Sciences'. Chemists found 'Mendels' and 'Morgans' in the prominent Western scientists Linus Pauling and Christopher Ingold, and their own 'Michurin' in a founding father of Russian chemistry, Alexander Butlerov. They even coined a label, 'Ingoldist–Paulingist,' to demonise their opponents. None of this had anything to do with political conviction. It had to do with self-preservation.

In truth, this great quantity of noise in fields other than genetics hid a marked lack of real action. Leaders in all disciplines, survivors of the Great Purge, had learned their parts well, and used the Michurinist campaign to pursue their own agendas: declaring their loyalty, sticking the knife into the handful of people they didn't like, and laying a layer of protective political cant over whatever work they happened to be engaged in at the time – work that rarely had anything to do with genetics.

By staging your own purge, you could leave business as usual free to operate beneath a gauze of insincere political hysteria. Psychologists and hygienists, pedagogues and specialists in physical education preserved their fields (and their tenure, and their incomes) by staging shows of conformity. The academies of

Estonia, Latvia, Lithuania and other republics passed resolutions underlining the necessity of discharging all 'Mendelians' from their posts. They didn't sack anyone. The central academies, such as the Academy of Sciences, made much of the urgent need to dismiss people who, funnily enough, had already been dismissed by order of the Central Committee. If the Central Committee hadn't booted you out of your job, you stood a good chance of staying in your post. At worst you might be moved between your academy's various bases and branches around the country.

In medicine, too, the August Session served as an alibi for institutional feuding, rather than as a rallying point for true believers – and in so doing, it brought into the open conflicts that had been simmering since the mid-1930s.

The nation's vast medical bureaucracy – bureaucracies, rather, because two gigantic organisations vied for dominance in public health – had become top-heavy, with too many old men holding too many senior posts, and too many younger men and women fighting for scraps. The trouble had started in 1934 when the government, pursuing a vision of nationwide medical care, had established an All-Union Institute for Experimental Medicine. Rather like the Lenin Academy in agriculture, this was an umbrella organisation controlling several institutes. In 1944 it expanded again, becoming a truly gargantuan Academy of Medical Sciences.

Yet despite its size and its vast and complex bureaucracy, the Academy was only the second biggest medical organisation in the Soviet Union. The biggest was the Ministry of Health, which between 1936 and 1937 had been extended across the whole Soviet Union. These two behemoths duplicated each others' efforts for years without too much friction; after all, if you were a senior administrator of one, you most likely held a high official position in the other. In the straitened circumstances that pertained after the war, however, this doubling of effort could not be permitted to continue. Fifteen years of discord followed.

Like maggots trapped in a tin, medical disciplines fought each other for survival, the larger disciplines consuming the smaller ones until only one was left. That discipline was physiology. Since 1936, when the decree on abuses in pedology had put competing disciplines out of work, physiology had been spreading like knotweed into psychology, psychiatry and pedagogy, until it became the theoretical and experimental basis for the whole of Soviet medicine. The commissariats of health, agriculture, and public education, the Academy of Sciences, the Committee for Higher Education, and even the Red Army boasted physiology departments.

'Removing reactionary idealist biological concepts from medical science' was a heaven-sent chance to stir up a bloated and top-heavy discipline, and shake off the old guard. The chief trouble was how to join the fray. What connection was physiology supposed to have with genetics?

The Academy of Medical Sciences invited Lysenko himself to deliver a report at their meeting in September 1948, but even he found himself at a loss, declaring at the meeting: 'There is no direct connection between Michurinist teaching and medical science.'

The physicians did, though, identify a worthy opponent for their 'Soviet progressive materialist medical science'. They picked on the only woman in the medical academy, Lina Shtern, the sour Old Bolshevik (Pavlov wasn't alone in not being able to stand her) who had discovered the 'blood–brain barrier', and had dared to acquire an international reputation for her work.

Shtern belonged to that generation of women who had had to travel to Switzerland in order to get a higher education. A long-time revolutionary sympathiser, in 1923 Shtern had accepted a Soviet invitation to return home just as many of her generation were doing everything they could to get out. Shtern became a Distinguished Scientist of the USSR in 1934. The government even gave her a car. Five years later she became a full member of the Academy of Sciences, and its first female member. Election to the Academy of Medical Sciences soon followed. Shtern was a leading member of the government-sponsored Jewish Anti-Fascist Committee during the war, and it was this association that did her the most damage, as the state's campaign against international contacts – against 'cosmopolitanism' in general – inevitably began playing to Russia's small but persistent vein of anti-semitism.

After being denounced in the meeting of the Academy of Medical Sciences, Shtern's fall was precipitous. A couple of weeks later, at a joint session of the Academy of Sciences and the Moscow Society of Physiology, Shtern was accused of anti-scientific ideas, of undermining the ideas of Ivan Pavlov, of disloyalty, and of contacts with the West. She was fired from the Institute of Physiology.[6]

In Shtern, the medical community had found its cosmopolitan scapegoat: its Nikolai Koltsov, its Nikolai Vavilov. What it lacked was a contemporary hero: a medical Trofim Lysenko. Olga Lepeshinskaya, an Old Bolshevik and a personal friend of Stalin, threw her hat into the ring, with a homegrown vision of biology that certainly passed on partisan grounds:

What happiness! At last, the dialectical materialists have triumphed, the idealists are paralysed and are being liquidated as the kulaks were once liquidated. To prevent their obstruction of the forward motion of science and their propaganda of idealism . . . it is necessary to remove them from all leading posts and to exercise a special vigilance toward repentants, because, perhaps, among the sincere repentants there are some wolves in sheep's clothing, trying to save themselves from liquidation.[7]

Praising herself as a 'materialist and innovator', Lepeshinskaya dubbed her opponents (meaning cytologists, histologists, and morphologists) 'idealists and reactionaries'.

Her gambit failed, and for a while the medical community made do without a contemporary hero. They had, after all, a founding father, and one far more celebrated and established than Michurin would ever be. They had Ivan Pavlov.

How had Pavlov, the state's most celebrated opponent, become its greatest intellectual hero? Age and vanity no doubt played their part. For Pavlov, the 1930s were years of unparalleled privilege and prosperity. When he complained that traffic outside the Institute of Experimental Medicine was disturbing his dogs, the street was moved. When he fell ill and a doctor prescribed champagne, it was imported the next day from Helsinki. When he needed more space for his Physiological Institute at the Academy of Sciences, the president was ousted from his flat. Pavlov was chauffeured in a Lincoln, and his larder was full of imported food.

In 1934 his eighty-fifth birthday was marked with even more funding for his Institute of Experimental Genetics of Higher Nervous Activity at Koltushi – a science village supplied by its own collective farm ('the Magnitostroi or Dneprostroi of the medical sciences', the press explained). Here, Pavlov studied conditional reflexes in two apes, named Roza and Rafael, and hoped this work would lay the ground for eugenics; 'the science of the development of an improved human type', he told the newspapers, even as anti-eugenic campaigns were destroying the careers and lives of Nikolai Koltsov and medical geneticist Solomon Levit.

But there was a darker side, too, to Pavlov's reconciliation with the authorities. Simply, Pavlov feared for his son Vsevolod's safety, for he had been an officer in the White army.

By the mid-1930s Pavlov had withdrawn his disapproval of the Soviet regime. Reconciliation turned to positive approval just before he died when, in August 1935 at a large reception in the Kremlin, he thanked the government for its generous support and modestly wondered whether he would justify it. Molotov called out: 'We are sure that you will, no doubt about it!' Pavlov returned the compliment lavishly:

As you know, I am an experimenter from head to foot. My whole life has consisted of experiments. Our government is also an experimenter, only of an incomparably higher category. I passionately want to live, to see the victorious completion of this historic social experiment.[8]

Ivan Pavlov died of pneumonia on 27 February 1936, at which point his public reputation fell into the hands of the government, who refashioned it to their own taste. On the 100th anniversary of his birth, in September 1949, a public campaign was launched to establish Pavlov as a founding father of Soviet science. A front-page editorial in *Pravda* ran 'A Great Son of the Russian People', instructing readers that Pavlov was 'close to the heart of every Soviet person' and had been personally supported by Lenin and Stalin. He had founded 'a new epoch in physiology' and continually struggled against 'reactionary, idealist, false theories . . . of the bourgeoisie in the United States, England and other capitalist countries'. Apparently Pavlov's work 'expanded especially quickly after the Revolution' and he had 'selflessly served his socialist motherland and his people'.[9]

With no obvious link to be made between the medical sciences and the controversies shredding agriculture, the September meeting of the Academy of Medical Sciences set about examining Pavlov's legacy in genetics. Perhaps, through the work of their founding father, some politically convenient link could be made.

In his report to the meeting, the academy's vice-president, Ivan Razenkov, who also had happened to be a Pavlov pupil, recalled the experiments his colleague Nikolai Studentsov had conducted on mice:

In a whole series of investigations, it was proved that typical characteristics of higher nervous activity could be radically changed under the influence of external factors artificially created in an experimental environment. This Pavlovian position wholly corresponded to his anti-Morganist orientation on the question of the inheritance of acquired activity. The belief in the possibility of such transmission through heredity had led him to his personal dispute with Morgan and to his arranging of special experiments.[10]

In other words, Pavlov had overseen work that proved that acquired behaviours could be inherited, and this had put him at loggerheads with the international genetics community. A nice story – and a complete reversal of what had actually happened.

At once, a dissenting voice arose. Leon Orbeli, recently dismissed as head of biology at the Academy of Sciences, was an Armenian physiologist who had known Pavlov better than anyone else in the hall: he had joined the Military Medical Academy at seventeen and he and Pavlov had trained together. He had been Pavlov's loyal lieutenant throughout the great man's life, and on Pavlov's death had effectively inherited his empire. He was scientific director of three Pavlovian institutes, overseeing the work of twenty-eight laboratories in the Leningrad area. He ran the colossal All-Union Institute of Experimental Medicine, which boasted a staff of around 3,000. He was head of the Military Medical Academy, president of the All-Union Physiology Society, and a member of numerous governmental commissions and committees. And he remembered everything about Studentsov's experiments with mice.

To Razenkov's deep embarrassment, Leon Orbeli proceeded to recount in detail the history of Studentsov's experiments and Pavlov's later forays into the inheritance of acquired behaviour. Not only had these later experiments proved that acquired behaviours were not inherited, said Orbeli, Pavlov did not even think the studies were particularly important.

Orbeli was far too significant a bureaucrat to be attacked head-on, and Lysenko's response to Orbeli's opposition was – for Lysenko – positively diplomatic: he proposed that Orbeli organise collaborative research: 'Let's show the inheritance of conditioned reflexes, their transformation into unconditioned ones, in all wild birds, mice, rats.' In the meantime, Orbeli was expected to review the scientific personnel of the Koltushi Institute, and sack its geneticists.

Orbeli did not dismiss anyone; he simply removed and stored the bust of Mendel which had been set up by Pavlov in 1934 in front of the building. (However, the situation at Koltushi remained so tense that R. A. Masing, a talented geneticist and a former pupil of Yuri Filipchenko, stopped working with *Drosophila* altogether and, without Orbeli's knowledge, set them free.[11])

The collaborative work that Lysenko proposed never happened. Instead, Orbeli used Lysenko's ideas as rhetorical cover, including 'the inheritance of conditioned reflexes' in the research plans of his institutes. He set his young assistant Viktor Fedorov the task of 'studying the hereditary transmission of acquired features in mice' but when Fedorov began to outline how he would do his research, Orbeli interrupted him: 'We will talk about it later. What we need now is a *title* for the plan. I think it could be entitled "The Fixation of the Changes in Functional Characteristics of the Nervous System".' The 'corrected' plan was sent to the Academy of Medical Sciences and approved. It was the only study Orbeli's institutes ever conducted into 'the inheritance of acquired behavioural features', though that formula appeared in virtually every plan and report Orbeli signed. Needless to say, Fedorov's work showed that the phenomenon did not exist.

Orbeli's rise to pre-eminence in the field of physiology had not been without controversy. Pavlov had clearly tipped him as his successor, but Pavlov's family weren't very keen on him for some reason, and protested when he assumed (in an entirely above-board way) the running of Pavlov's institutes. There was resentment, too, from a couple of Pavlov's less illustrious colleagues, who disliked Orbeli's approach to physiology, not least because it involved them having to think.

Rather than expect his colleagues slavishly to work through research projects set by Pavlov himself, Orbeli expected physiology to surpass Pavlov and develop in new directions.

This had been Orbeli's attitude while Pavlov had been alive (the two had enjoyed magnificent arguments) and the great man's death had done nothing to dampen Orbeli's ambitions for their field.

Pavlov's old hangers-on now tried to challenge Orbeli's position. They informed Party agencies of his 'misconduct' and 'monopoly', and accused him of 'deviations' from Pavlov's legacy. They found a willing ear in Yuri Zhdanov who, having dodged a bullet over the Lysenko affair, was determined, as he wrote to Stalin, to 'work hard to correct previous mistakes'. Yuri came to the conclusion that Pavlov's scientific legacy was not being properly developed by Soviet physiologists, and that his successor, Leon Orbeli, had concentrated too much institutional power in his own hands. He wrote to Stalin: 'It is necessary to liquidate Orbeli's monopoly of the development of Pavlov's science and to subject his mistakes to criticism.'[12]

Yuri was speaking Stalin's language at last. Stalin wrote back positively, offering advice on how to carry out the struggle most effectively:

By falsely appointing himself as Pavlov's most important student, Orbeli did everything he possibly could to disgracefully silence Pavlov with provisos and ambiguities. His cowardly and disguised raids against Pavlov [were meant] to dethrone and slander him ... The sooner Orbeli is denounced and the more soundly his monopoly is liquidated, the better. [13]

The blunt instrument Stalin and Yuri Zhdanov chose for this task – and he was none too happy about it either – was Konstantin Mikhailovich Bykov, the new director of the institute once run by the disgraced Lina Shtern. Bykov, a former pupil of the great man, was an orthodox Pavlovian, politically pliant, and seemed ideally suited to the task: 'We are guilty of confining Bykov's luminous science to the shadows,' Stalin opined, 'while a random meteor like Shtern can occupy leadership positions. It is time to be done with such abomination.'[14]

Bykov had other ideas. In fact he agreed with Sergei Vavilov that Orbeli and his Pavlov Institute were doing important work. 'It's true,' Stalin once admitted to Yuri Zhdanov, '[Bykov] is a little timid and doesn't like to mix it up.'

More than that, Bykov was at heart a peacemaker with a love of European culture and a creeping dread of Moscow politics. He was reported as saying, of a trip to Bucharest,

The influence of French and German culture is still felt there. They still have not wasted this influence and it benefits them. Among them I felt like a human being. That pleasant environment was so beneficial that I felt healthy and did not even worry about my bowels. That's sanity! I returned here and the depression, exhaustion, irritation and troubling atmosphere began again.[15]

Happily for Stalin and Zhdanov – and luckily for Bykov – Bykov did not have to do very much.

Yuri Zhdanov arranged a Scientific Session for June 1950, when the Academy of Sciences and the Academy of Medical Sciences could thrash out between them how to become more 'Pavlovian'. The joint meeting was in many ways a more tightly scripted copy of the fateful August 1948 Lenin Academy meeting. It was even held in the same venue – Moscow's House of Scientists. The Central Committee's Science Department wrote the script, making clear who was righteous and who was guilty. Yuri Zhdanov directed the show.

The session began on 28 June. There were more than a thousand participants and guests from over fifty cities and all the Soviet republics' academies of science. The cost of the session came to nearly half a million roubles, including train fares, hotel rooms (guests stayed in the elite hotels Moskva, Grand and Europa), buses, stenographers' fees, flowers and posters. *Pravda* ran daily coverage.

Bykov showed up with a speech which Stalin reviewed and edited personally. Armed with a manuscript that bore Stalin's imprimatur, Bykov declared there were in the Soviet Union genuine Pavlovians, and opponents, such as Orbeli. Bykov told his colleagues that Stalin urged the congregation to criticise and self-criticise.[16]

Orbeli's fate hung in the balance for some days, with Pavlov's old pupils as likely to demand Bykov's resignation from his many posts as Orbeli's. In the end it was Orbeli's own response to savage criticism that spoiled his chances. He refused to jump through the hoop offered – to concede that he was a good scientist whose own interests had led him to skew the field he administered towards his own concerns. Instead he faced down his accusers in forthright terms:

If predetermined individuals are selected for rather severe criticism, then, if it is to be a free scientific discussion, it would be extremely important to let them know beforehand what they are to be accused of and criticised for. Even when criminals are involved, they are given an indictment to read, so that they can defend themselves or say something in their defence. In the present case that wasn't done, and we, the accused, find ourselves in a difficult situation . . .[17]

Orbeli's defence scared even his own side. When close colleagues were called upon to speak, they claimed to be taken aback by his lack of genuine self-criticism.

On 4 July, the last day of the session, Orbeli spoke again, claiming that he 'immediately understood the erroneous and dissatisfactory nature' of his first speech, blamed his lack of experience in this style of debate, and promised to open up the field to more 'self-criticism'.

It was not enough. Orbeli was dismissed from all his posts at both academies. Later resolutions fired him from his editorial posts at the *Physiology Journal of the USSR*, *Achievements in Modern Biology*, *Herald of the Academy of Sciences* and the Soviet science magazine *Priroda* [*Nature*].

With Orbeli marginalised, there was no one now to challenge Pavlov's ideas; a founding father had been saved from cosmopolitan corruption; and the once vibrant science of physiology ossified overnight, becoming just one more sounding board for the political perspicacity of Old Bolsheviks in general, and Joseph Stalin in particular.

As part of the propaganda effort organised in 1948 to distinguish Russian science from Western science, Stalin saw to it personally that Lysenko and Michurin were celebrated in story and song. Poems referring to 'the eternal glory of the academician Lysenko' who 'walks the Michurin path' became fashionable.[18]

The Michurinist propaganda campaign reached its high point with the celebration of the fiftieth birthday of 'the famous heir to the Michurinist doctrine T. D. Lysenko'. Lysenko's portraits were hung in all scientific institutions. Art stores sold busts and bas-reliefs. Monuments were erected. Central and local newspapers offered congratulations to the latest recipient of the Order of Lenin, awarded 'for outstanding public service in the development of progressive Soviet science'.

Absurdity piled on absurdity. On 22 May 1950 Alexander Oparin, as head of the Academy of Sciences' biology department, invited Olga Lepeshinskaya to receive her Stalin Prize. Now in her dotage, Lepeshinskaya – who had once recommended soda baths as a rejuvenation treatment – was by this time completely entranced by the mystical concept of the 'vital substance', and had recruited her extended family to work in her 'laboratory', pounding beetroot seeds in a pestle to demonstrate that any section of a plant ovule could germinate. Now she claimed success in actually filming living cells emerge from non-cellular materials. Lysenko hailed Lepeshinskaya, declaring that her demonstration that 'cells need not be formed from other cells' should be taken as the basis for a new and eminently workable theory of species formation. (Actually, she had filmed the death and decomposition of cells, then ran the film backwards through the projector.)

No one else said a word. 'In later years, they did their best to wash themselves clean of the dirt,' writes the physician and contemporary witness Yakov Rapoport, 'but the fact remains that there was not one Giordano Bruno among them.'[19] And after all, what did it matter? Lepeshinskaya always had been a lousy scientist, but she would make (and did make) a splendid myth. The subject of poetry. The heroine of countless plays. In school and university textbooks she was hailed as the author of the greatest biological discovery of all time.

It did not take long for Stalin to usurp credit for Lepeshinskaya's incredible and absurd discovery. As Stalin's cult of personality expanded into the natural realm, Lysenko celebrated 'Stalin's teaching about gradual, concealed, unnoticeable quantitative changes leading to rapid, radical qualitative changes, [permitting] Soviet biologists to discover in plants ... the transformation of one species into another'.[20] Fraudulent reports of species transformation in Lysenko's journal *Agrobiology* included wheat turning into rye, cultivated into wild oats and barley, cabbages into swede and rape, sunflowers into strangleweed, and pines into firs. All these miracles were credited to Stalin's teaching.

 И ЗАСУХУ
ПОБЕДИМ!

20 : 'The death agony was horrible'

I realised that one cannot demand heroism from everybody.
Yakov Rapoport, *The Doctors' Plot*

In December 1948 a new colour film, *Michurin*, was shown in all the country's movie theatres and even abroad. Its music was composed by Dmitry Shostakovich. Its hero was played by one of the most popular actors of the day, Grigory Belov. The screen-writer and director was one of the Soviet Union's most famous film directors, the Ukrainian Alexander Dovzhenko.

'The orders for filming *Michurin* at Mosfilm have been issued,' Dovzhenko wrote in his diary in 1946:

Leaving Mosfilm after making *Aerograd* eleven years ago, I swore to myself, 'Praise the Lord that He has freed me from this hideous den of iniquity! Never again, under any circumstances, will I return to this house of cretins where everything is twisted and perverted.' And here I am again.

Aerograd (1935), a semi-science-fictional film about building an 'air city' on the border with Japan, had been such a hit, the authorities had let Dovzhenko off the leash a little: he had been able to return to the Ukraine for his film *Shchors*, about the Ukrainian Civil War. But his good fortune did not last: Stalin castigated him for the 'nationalist bias' of a subsequent documentary *Ukraine in Flames*, and Dovzhenko was summoned to Moscow for a severe reprimand by no less a figure than Lavrenty Beria.

Filming *Michurin* in Moscow was the only way Dovzhenko could see of clinging on to his career. A film on Michurin seemed 'a cozy retreat' at first: 'The subject doesn't seem to go with my "nationalism",' he reflected, 'after all, it's Russian.'

Dovzhenko was already familiar with the material. Years before, he had written a play about Ivan Michurin, and the chance now to make a film of his play, in sumptuous colour (his first colour film!) and with a virtually unlimited budget, attracted him. The script needed some work, of course. The play was weak stuff – a poetic paean to nature – and Mosfilm wanted to see some sharper dialogue that reflected the real issues and brought out the global scientific importance of this humble old plantsman. And there would have to be new characters, of course – predatory foreigners, obfuscatory opponents – to bring out Michurin's credentials as a homegrown genius and a man of the people . . .

The film took four years to make, and Dovzhenko hated every minute of it.

I pulled that film out of the barren ground. I suffered, exhausted by heart attacks and the insults of dull bureaucratism. Then, after superhuman efforts, when the film finally began to show signs of life, and pleased even trained snobs, I found myself in some mystical zone where the film was to be judged by the Supreme Arts Council. Then the Minister ran around with it someplace, and they showed it to the Great Leader, the greatest mortal since time began, and the Supreme Being proscribed my work . . . Now the film agency have me on the rack again. I spend days sitting at my desk. I have to throw away everything that I have written, turn against everything I was enthusiastic about, which is composed of many delicate elements, and create a hybrid – an old poem about his work and a new story about selection. And my heart bleeds. I often get up from my desk after a day's labour and look over what I have written, and I see how mournfully little of it there is. And I am exhausted as if I had been dragging heavy rocks around all day long.[1]

Shostakovich, meanwhile, was having an utterly miserable time with the music. There is no written record of what Shostakovich thought of the movie itself, but his score, being very dreadful, speaks for itself. As he wrote in a letter to his good friend Isaak Glikman on 12 December 1948: 'Physically, I feel quite low ... I suffer from frequent headaches and besides that I feel constantly nervous, or, to put it simply, I feel like throwing up.' As well he might: after being condemned in 1948 for a second time by the Communist Party, after his music was banned and he was fired from his teaching posts, Shostakovich was having to compose film music as his only means of earning money. This time round, apparently, Stalin had personally chosen Shostakovich to write the score. The pressure was well-nigh unbearable.

The plot of *Michurin*[2] goes something like this. Near a railway station in the province of Ryazan stands an orchard, looked after by a man who repairs watches for a living. He has a season ticket, and he sells little pamphlets to his fellow passengers as he travels back and forth along the railway line that passes the bottom of his garden. The pamphlet describes the kinds of apples his orchard grows. They're not the usual kind. Everyone admires his orchard. It's the pride of the town.

One day the old man decides to select new varieties of apple that will grow just about anywhere – maybe even in the severe dry cold of northern Russia. He buys a plot on the riverbank, where the soil is all gravel and silt and appears utterly sterile. He digs up his trees and his wife helps him carry them down to the river, and bites her lip as she does so. The neighbours think the old man's gone out of his mind.

Michurin faces down every conceivable obstacle, frustration and disappointment in pursuit of a crazy dream: that one day, fruit gardens will bloom all over his cold, tough country. The work takes many decades of fanatically focused effort. And then his wife dies: the only person who ever truly believed in him. She dies, and Michurin steps into the night, into a howling wind and drifting autumn leaves and the horror of death. This is, rightly, one of Dovzhenko's most celebrated cinematic sequences, and just about the last good scene in the movie.

Because, from this point on, everyone starts shouting. Katashev, a fictional figure, is talking to Michurin about genes, chromosomes and mutations. Michurin answers: 'It is time for biology to get off the pedestal! It should speak the language of the people, and not get lost in the fog!' Two pot-bellied capitalist professors from the United States tempt him with bags of gold and promises of glory in America. Michurin waves a violet-scented lily under their noses. (Whenever his ideas are challenged, he reaches into a pocket or a desk drawer and – *eh, presto!* – produces a specimen that deflates the opposition.) He tells them his flower is a hybrid of a violet and a lily. 'That's the trouble with the Mendelians, they can't explain hybrids!' The capitalists withdraw, muttering and cursing.

A priest wags a finger, shouting: 'Do not turn God's garden into a brothel!' The old gardener ignores him. He focuses even deeper on his work.

At last the tsar is overthrown, and simple folk come to power. They understand Michurin. They understand everything. They come up with brilliant suggestions, and Michurin makes several important discoveries. In a long, lyrical passage, Michurin stands on a ladder in the orchard waving his arms like an orchestra conductor.

This is the moral. Be pure. Be impatient. Be focused. Be mono-syllabic. Do your work. Fulfil your crazy dream. Do not despair: the natural world will yield.

'All those who interested themselves even a little in the national situation saw the difficult situation in agriculture,'

Nikita Khrushchev reported later, in his famous 'de-Stalinisation speech' of February 1956,

But Stalin never even noted it ... He knew the country and agriculture only from films. And these films had dressed up and beautified the existing situation in agriculture. Many films so pictured life on collective farms that the tables were bending from the weight of turkeys and geese. Evidently, Stalin thought that it was actually so.[3]

The grain harvests of 1947 and 1948, while better than the disastrous harvest of 1946, nonetheless failed to match pre-revolutionary levels, and Stalin's love of commanding, cinematic gestures extended now beyond the studio doors of Mosfilm and into the natural world itself. On 20 October 1948, Stalin made his grandest cinematic gesture yet, signing off the Stalin Plan for the Great Transformation of Nature.

More properly known as the 'Plan for Shelter Belt Plantings, Grass Crop Rotation, and the Construction of Ponds and Reservoirs to Secure High Yields and Stable Harvests in Steppe and Forest-Steppe Regions of the European Part of the USSR', the Stalin Plan was much more than a leader's whim. Its recommendations had been put together at a conference of agronomists and forestry experts, who themselves were drawing from over half a century of ecological planning. As early as 1892 the influential soil scientist Vasily Dokuchaev, appointed by Tsar Alexander III to explain the devastating drought the year before, had concluded that the steppes of Russia had been damaged by centuries of desultory agricultural development. Dokuchaev had urged a series of water conservation measures, including the planting of enormous forest belts throughout the south, hoping to recreate a time when southern and central Russia were united under one canopy of trees.

The Stalin Plan revived Dokuchaev's ambitious vision. The thirty-year projections of the Stalin Plan foresaw the creation, by

1965, of nearly 6 million hectares of new forest – an area greater than that of all the forests of Western Europe. These new forests would halt the arid winds from the steppes of Kazakhstan and the deserts of Central Asia, cool and dampen the climate of southern Russia, and eliminate the periodic droughts that afflicted the steppe. Additionally there would be seven forest belts, each one thousands of kilometres long, running north to south across the dry steppes of the Volga basin.

The lion's share of the afforestation work fell to the ministry of agriculture. No one there knew much about forestry and to make matters worse they were deluged by contradictory instructions. One manager from Kursk province recalled being asked, out of nowhere, and without advice or assistance, to plant 'five hectares of field protective belts, four hectares of forests on gullies, create a 1.5 hectare forest nursery, build two ponds, grow 50,000 seedlings and prepare ninety kilograms of acorns'.[4]

To these long-discussed planting programmes, the architects of the Stalin Plan added their own technocratic and utopian schemes. The biggest and most far-reaching of these was an ambitious programme to transform the major rivers in the USSR – the Volga, Don, Dnieper, Kama, and Svir – into a single system of canals, dams, reservoirs, irrigation systems and hydroelectric power stations. At the heart of this scheme was the mechanisation of the Volga, the largest river in Europe, 3,700 kilometres long from its source near Tver to its mouth at Astrakhan on the Caspian Sea.

Most of the Volga's flow is generated by snowmelt; left to its own devices, the river would run most powerfully in late autumn and spring. In the summer growing season, when agriculture needed it most, the untamed Volga ran at barely a fifth of its full capacity. Before the Second World War Soviet engineers had set out to build dams, reservoirs and other huge structures to regulate the flow more evenly through the year, to keep channels deep enough for navigation, and to open the way with canals to the Don and other rivers. The war had virtually eradicated these efforts: German armies had destroyed bridges, factories and hydroelectric stations, including the iconic Dneiper station. Eager to eradicate evidence of the invasion, the government now brought the Volga to a virtual standstill with twelve major hydroelectric stations, each inundating thousands of square kilometres of land, displacing around half a million people and destroying homes, churches, schools and farmland. Six cities were submerged by the Rybinsk reservoir alone – once the largest artificial lake in the world.

The symbolic importance of the Stalin Plan was immense. Now that the war against Germany had been won, the Soviet people were once again called upon to revive and prosecute the campaign for which the revolution had been fought – that epic struggle for a new, more rational, more humane environment. Hundreds of authors tried to outdo each other in celebrating the power of Soviet science to transform the planet. The image of thousands of kilometres of sturdy oaks breaking the strength of the parched eastern winds cropped up endlessly in films and newsreels. In newspapers and booklets, children stuffed themselves with fruits and berries while strolling through desert landscapes turned into oases. The propaganda blitz even included a patriotic oratorio composed by Dmitry Shostakovich, *Song of the Forests* – a work which consciously strives and only narrowly fails to be worse than his music for *Michurin*.

Isaak Prezent's article 'The Refashioning of Living Nature' captures the spirit of the campaign. 'Bourgeois professors assure us that nature will not tolerate human interference and will avenge itself with natural disasters for intrusion into its regularities,' Prezent began. Needless to say those obscurantists were mistaken. Citing Lysenko's 'discovery' that planting grain around young trees prevents them from being smothered by weeds (it doesn't, by the way), Prezent declared:

Field and forest – protecting plantings – trees and bread – what a wonderful idea about cooperation and struggle in the green kingdom of plants. And what about the construction of ponds and reservoirs! Truly, never in the history of the world was there ever and could there ever be such a huge scale of hydroconstruction! . . . Soviet biologists are joyously creating new kinds of life, are renewing and enriching living nature, and together with our entire people are building Communism.[5]

Lysenko's involvement in the Stalin Plan came very late. He had not been remotely involved in the development of shelter belts in the 1930s and 1940s, and indeed had published nothing before 1948 about tree biology. Appointed to the council of the Main Shelterbelt Administration, Lysenko lost no time in formulating Promethean promises. He believed that all plants possessed a quality called 'self-thinning', which allows them to work together in fighting against weeds during their early years and then to pool their energy for the benefit of one shoot, to which the other shoots sacrifice themselves.[6]

Lysenko's forestry was built around this idea. To plant an oak, for instance, a central hole was dug, and four auxiliary holes dug around it, creating a 'nest' in the shape of a plus sign. Lysenko claimed that this formation would allow the oak seedlings to defend one another from weeds most effectively.

The main practical advantages of the nest method were simplicity and speed. 'Only three man-days are needed for the hand-planting of one hectare of oak forest using the nest-method,' his booklet explained. Workers could accomplish in hours what ordinarily would require a decade or more.

Forests sown by this method did not last long. By September 1951, all the nested forests in the Urals had died. By 1952 fully half of the 'nest method' forests had died, and the first two forest belts were near-total losses.

But while the vast scale of the Plan invited errors and losses on a heroic scale, it did afford shelter for some real science – science that might otherwise have been expunged as 'anti-Michurinist'. For example, while Lysenko was busy confusing the Main Shelter Belt Administration, another quite separate administrative body, the Comprehensive Scientific Expedition for Problems of Field-Protective Silviculture, was busy studying local conditions and giving genuine technical guidance to workers. Leading the expedition was the ecologist Vladimir Nikolaevich Sukachev, whose Forest Institute, founded in 1942, was by now virtually the only bulwark left against Lysenko.

Sukachev turned the whole Expedition into a refuge for geneticists and tried to save as many persecuted biologists as he could. Sergei Vavilov, president of the Academy of Sciences, who was secretly intervening where he could to save genetics, kept Sukachev informed about whom he should hire for the project.

The Stalin Plan afforded some shelter for scientists considered politically unreliable by the state. A far greater level of protection was afforded to physicists who, at around the same time, were nearing completion of the first Soviet bomb. They enjoyed privileges, of course. Kurchatov got an elegant double-storeyed mansion with marble fireplaces, wooden panelling and a sweeping central staircase. Italian craftsmen were imported in 1946 to finish the interior. Much more important, physicists on the project were accorded a certain amount of intellectual tolerance.

Far from having to pass through the flaming hoop of their own 'Michurinist session', Kurchatov and his fellows were entirely exempted from political education.[7] 'Do not bother our physicists with political seminars,' Stalin had said. 'Let them use all their time for their professional work.' Stalin even went so far as to tick off Kurchatov for not being demanding enough: 'If the baby doesn't cry, the mother doesn't know what he needs. Ask for anything you need. There will be no refusals.'[8]

There was no guarantee that this topsy-turvy state of affairs would continue. A telling anecdote from the time has Stalin saying to Beria, 'Leave them in peace. We can always shoot them later.' Still, the physics community was now the closest thing the Stalinist regime had to civil society – a point not lost on the physicists themselves.

Yulii Khariton directed the final construction of bombs at a secret installation deep in the forests of the Volga. His laboratory used the buildings of a wartime munitions factory on the premises of what had been, before the revolution, one of the holiest Orthodox shrines, the monastery of Sarov. The name Sarov disappeared from official maps and other Soviet documents and for the next forty years, internal documents referred to the place as Post Office Box Arzamas-16.

Arzamas-16 grew into a town of 80,000 inhabitants and was a paradise compared with half-starved Moscow. Politburo member Lazar Kaganovich complained that the atomic cities were like 'health resorts'.

The scattered resources of a devastated country were mobilised to realise the bomb. According to a CIA estimate, between 330,000 and 460,000 people were employed in the effort. The physicist and dissident Andrei Sakharov began working with thermonuclear weapons in 1948 and later described how, in the years 1950–3, he watched through the window of his office at Arzamas-16 as columns of prisoners marched by under armed guard. One rebellion by manual labourers there was suppressed with live rounds: every rebel was killed. Many prison labourers were never set free; instead they were labelled especially dangerous and carted off to the gold mines of Kolyma.

By the summer of 1949, a nuclear device was ready for testing at Semipalatinsk in Kazakhstan.

The area was barren, at least from a human point of view: a stony, sandy expanse grown over here and there with feather-grass and wormwood. There were no houses, no trees – even birds were rare. By early morning, the heat had already grown oppressive, and by the middle of the day the haze was generating mirages of mysterious mountains and lakes.

Near the tower housing the device, wooden buildings had been constructed. Railway locomotives and carriages, tanks and artillery pieces were scattered about the area. Animals were placed in open pens and covered houses nearby, so that the effects of irradiation could be studied.

Lavrenty Beria arrived. Suspicious of his scientists even now, he had with him two Russian observers who had witnessed the American nuclear tests at Bikini in 1946 to confirm the authenticity of the blast. With a complete lack of ceremony, the order was given to detonate the device. Kurchatov went to the door and opened it. It would take thirty seconds for the shock wave to reach the command post. When the hand on the clock reached zero, the doorway was flooded with light.

For a moment or so it dimmed, then grew brighter very quickly. A white fireball engulfed the tower and rushed upwards, changing colour as it went. The blast wave at the base swept up everything in its path: houses, machines, stones, logs, girders. The mushroom cloud reached a height of about eight kilometres before losing its outlines and turning into a torn heap of clouds.

Beria asked his observers, 'Haven't we slipped up? Doesn't Kurchatov humbug us?' When he was told that the test had been successful, he ordered a telephone call put through to Stalin. It was two hours earlier in Moscow and Stalin's secretary warned him that Stalin was still asleep. 'It's urgent, wake him up,' Beria insisted.

Stalin came to the phone. 'What do you want?' he muttered, 'Why are you calling?'

'Everything went right,' Beria announced.

'I know already', said Stalin, and hung up the phone.

Beria went wild with anger. 'Who told him?' he cried. 'You are letting me down! Even here you spy on me! I'll grind you to dust!'[9]

*

Beria was right to be afraid. Stalin's mind was weakening and, as his grip on reality loosened, so he became ever more of a danger to those around him. 'It became more and more obvious that Stalin was weakening mentally as well as physically,' Khrushchev recalled in his memoirs. 'This was especially clear from his eclipses of mind and losses of memory.' Stalin's morbid suspicions destroyed the family members of his closest supporters. Molotov's wife Polina was arrested in 1949. Members of Stalin's own family were arrested and sent to camps. In 1951 Stalin confessed to Khrushchev, at that time one of his closest advisors, 'I'm finished, I trust no one, not even myself.'[10]

Stalin's paranoia and formidable survival instinct pushed him into pronouncements that were more and more peculiar. Unable to trust even himself, it came to Joseph Stalin that people were, or ought to be, completely readable from first to last. All it needed was an entirely verbal theory of mind. 'There is nothing in the human being which cannot be verbalised,' he asserted, in 1949. 'What a person hides from himself he hides from society. There is nothing in the Soviet society that is not expressed in words. There are no naked thoughts. There exists nothing at all except words.'[11] For Stalin, at the end of his life, even a person's most inner world was readable – because if it wasn't, then it couldn't possibly exist.

The paranoid fantasies of an old man – even an old man with his finger on the nuclear button – would have had little purchase on history. What made Stalin's final mental manoeuvres extraordinary was the way they played out in the real world. Stalin's last rewritings of the historical and political record were extraordinarily effective. They were, in a terrible and malign way, his masterwork. Yes, Stalin lost touch with reality – but he took the entire country along for the ride.

These final and terrible manipulations had begun with the purge of colleagues who, in Stalin's view, had attained too much power and importance by the end of the Second World

War. Andrei Zhdanov was one such victim. Already his health was failing, and as the time of the Lenin Academy's Autumn Session neared, the Politburo had given him permission to leave for a two-month holiday in order to recover. Once Zhdanov left Moscow his political career went into a predictable decline. During July, several of his protégées lost their posts. Zhdanov's health, meanwhile, came under the care of unfamiliar doctors who advised not total rest, but exercise and harmful massages. Only one of them – Lydia Timashuk, a cardiographer – diagnosed Zhdanov's heart attacks correctly, but her diagnoses were over-ruled and Zhdanov was told to take long walks in the park. Timashuk, appalled, wrote personally to Stalin, denouncing her colleagues. Stalin received the letter, filed it, and did nothing. Andrei Zhdanov died at the end of August. At his grand funeral Stalin helped bear the coffin and spent the rest of the day getting blind drunk.

After Andrei Zhdanov's death, a new purge swept through Leningrad, and the bases of power and patronage he had managed to establish there. Around seventy senior officials were arrested and convicted of various offences, along with family members, in rigged trials. The purges spread from patrons to clients, from the centres of power in Leningrad to outlying acquaintances in regions as far flung as Novgorod and Crimea.[12]

It took Stalin around three years to transform the inciting incident of this purge – Andrei Zhdanov's unexpected and suspicious death – into a national Jewish conspiracy and an anti-semitic witch-hunt.

Anti-semitic feeling was not new to Russia and its client republics, although the Bolsheviks had always paid lip service, at least, to the duties they owed their Jewish minority. Now campaigns against 'stateless cosmopolitanism' – campaigns that had been conceived largely by Andrei Zhdanov – were stirring up considerable anti-Jewish feeling, and even the government-funded wartime Jewish Anti-Fascist Committee came under fire.

In January 1949 – once its leader, the actor Solomon Mikhoels, was killed in a highly suspicious car crash – the committee's members were arrested and charged with spying for the United States. (The newspapers went even further, accusing committee member Lina Shtern and others of conspiring to create a Jewish state in the Crimea.)

All those arrested were tortured. All 'confessed'. Semyon Ignatiev, the new minister of state security, offered to sentence 'all Jewish nationalists, American spies . . . excluding Shtern, to be shot. Shtern is to be exiled to a remote area for ten years.'[13]

On 12 August 1952, the sentence was carried out. Thirteen of the fifteen were secretly executed and their families deported to remote parts of Siberia and Kazakhstan. One died in prison. Shtern was sentenced to five years' exile in Kazakhstan.[14]

Soon afterwards the anti-cosmopolitan campaign lost all political trappings and became openly a campaign against Soviet Jewry. The campaign was indiscriminate. It hardly mattered what your political history had been. Abram Ioffe, the founder of Soviet physics, came unstuck when he was caught calling himself a Russian on official paperwork. On 5 July 1951 he wrote to the personnel office of the Leningrad Physico-Technical Institute: 'In view of the disagreement in my documents on the question of my nationality, I request to define my nationality according to the origin of my parents as Jewish.' He was then relieved of his position as director of the institute.

Another highly placed victim was Isaak Prezent, following an anonymous denunciation pointing up his Jewishness. For years Prezent had been making a nuisance of himself with the female students[15] and few were sorry to see the back of him. 'Do you know who just came to me?' announced Grigory Roskin to his laboratory staff one day. The man who with his wife had developed the 'cruzin' anti-cancer treatment, and been denounced for his trouble, had been made blisteringly angry by the visit of this 'strange, short man'. 'Bloody Prezent!' Roskin exclaimed. 'He asked me to give him cruzin. I didn't give it to him.' (Prezent died of cancer in 1969, a few days after being expelled from the Lenin Academy.)[16]

The 'Doctors' Plot' was unveiled on 13 January 1953 by *Pravda*, in an article heavily edited by Stalin and under the title 'Ignoble Spies and Killers under the Mask of Professor–Doctors'. 'Some time ago,' the TASS news agency reported, 'the bodies of State Security uncovered a group of terrorist doctors who set themselves the task of cutting short the lives of prominent public figures in the Soviet Union by administering harmful treatments.'

Andrei Zhdanov was listed as one of the victims. Much was made of this, though Zhdanov had been virtually forgotten – the first anniversary of his death had merited no more than an article on *Pravda*'s page three.

On 20 January 1953, Lydia Timashuk – the woman who had accused her fellow physicians of mishandling Zhdanov's illness – was invited to the Kremlin to receive Stalin's personal thanks for her 'great courage'. The next day a bemused Timashuk was awarded the Order of Lenin for trying to save the life of a Soviet leader in the teeth of a Jewish conspiracy. Her letters to the Politburo and to Stalin were unearthed. Clearly the 'conspiracy' had hidden them from their intended recipients. Within days, Timashuk was cast as the heroine of the Doctors' Plot. Poems were dedicated to her. She was compared to Joan of Arc.

Nine doctors, six of them Jewish, were arrested. Among them was Yakov Rapoport, the young physician who had responded so sarcastically to Olga Lepeshinskaya's 'ground-breaking' discoveries, and was now one of Stalin's personal physicians.

Natalya Rapoport, Yakov's daughter, remembers the hatreds unleashed by the plot: 'People refused to be treated by Jewish doctors. The air reeked of pogrom ... There were rumours that, for the sake of "protecting" the others – the innocent Jews – from all this mass hatred, camps were being set up for them in Siberia.'[17]

The Doctors' Plot was a masterpiece of political manipulation, providing Stalin with an alibi for Zhdanov's death while busying the Soviet public with yet another conspiracy scare. At the same time, there is the most dreadful emptiness about the whole affair. Anti-semitism was not a strong force in Soviet life. Once Stalin died, the entire campaign petered out in a matter of days. Why then had he been so set on stirring hatreds that could bring neither him nor the state any advantage? The direct victims of the affair – the doctors accused of murdering high officials of the Party – were just doctors. They wielded no especial influence. They had no power. They were entirely replaceable. A phone call would have done it. Why, then, the manufacture of a national scandal? Stalin's physical and cognitive decline had reduced his world to a few rooms and the ministrations of a handful of doctors. Stalin no more trusted them than he trusted anyone else, so he had done what he had always done: he had set out to destroy them. And he had used the weapons he had always used: spell-binding, nation-swallowing terror.

Death beat Stalin to the draw. The leader's last hours were extra-ordinarily painful. Following a stroke, Stalin's doctors struggled in panic to keep him alive. They tried everything. They even bled him with leeches. 'The death agony was horrible,' his daughter Svetlana later said,

he literally choked to death as we watched. At what seemed like the very last moment he suddenly opened his eyes and cast a glance over everyone in the room. It was a terrible glance, insane or perhaps angry and full of fear and death.[18]

At 9.50 p.m. on 5 March 1953 the reign of one of the greatest mass murderers in history came to an end.

21 : Succession

*Every theory of the world that is at all powerful and covers
a large domain of phenomena carries immanent within
itself its own caricature.*[1]
  Richard Levins and Richard Lewontin, 1985

Upon his death, Joseph Stalin's natural successor, deputy premier
Lavrenty Beria, set about dismantling the case against the
Kremlin doctors. He announced that their confessions had been
obtained through torture, and he threw out the charges. Various
reforms were proposed – but Beria's bid for self-reinvention
would not play. The idea of the man who had overseen the Soviet
atomic project using that knowledge as leverage to seize power
galvanised Khrushchev and other highly placed leaders into a
conspiracy. Beria was arrested on 26 June 1953. Denounced as an
agent of international imperialism, he was tried in secret and, on
23 December, shot.

  When Khrushchev was made General Secretary of the Party,
the Academy of Sciences did what they thought was necessary
and elected him as an Honorary Member. They were in for a

Full circle: Nikolai Timofeev-Ressovsky (left) and the mathematician
Alexey Liapunov conduct 'Vernadskological' studies of radioisotopes at
an experimental station near Lake Miassovo, 1957.

surprise: Khrushchev refused to accept the honour. He had, he said, no pretensions of being a scientist.

Many expressed delight at Khrushchev's modesty, and welcomed his bid to replace the forced marriage of academia and government with a cordial working relationship. The new leader's gesture did not convince everyone, however. The geneticist Raissa Berg commented: 'This means that he will annihilate the Academy of Sciences.'

The problem was that Khrushchev was as much of a sucker for homespun folk wisdom as the rest of his generation, and took considerable pride in those little bits of folk wisdom that had come down to him personally. He caused a small sensation when he told a group of agricultural specialists that he lacked their expert knowledge, and they had to tell him when he was wrong. But when a couple of them took him at his word and sent him a memo criticising his wild effort to spread maize cultivation into unsuitable regions, he scolded them angrily. What on earth were they on about? He himself had grown maize successfully in his dacha garden near Moscow![2]

With hindsight, it is easy to see that it would be only a matter of time before Khrushchev fell under the sway of Trofim Lysenko, completely and unquestioningly, in a way Stalin never had.

To begin with, Stalin's death ushered in a certain pluralism. Shtern returned from exile. Vladimir Engelhardt, a friend of genetics, replaced Oparin as Secretary of the Biology Division of the Academy of Sciences. Orbeli turned up at a meeting of the Physiology Society with an ovation that lasted minutes.

Still, Lysenko's allies were secure in their high bureaucratic positions at ministries, academies and universities.

The biggest threat to Lysenko's position was not the reopening of old fault lines in biology – he had long since learned how to profit from them – so much as the way his old enemies were regrouping under a new banner entirely: one he did not understand and did not know how to respond to.

To Lysenko's considerable alarm, genetic research was coming back into fashion, but this time it was being conducted, not in biological institutions (which were controlled by the Lysenkoists), but under the roofs of physical and chemical research institutes.

Three years after Stalin's death, Soviet genetics was effectively reestablished, albeit under some strange-looking camouflage, as 'radio-biology', or 'radiation bio-physics', or sometimes 'physico-chemical biology'. Having earned certain privileges in the Cold War, Russia's physics community was now coming to the rescue of genetics – a field that physicists had held in fascinated esteem ever since Niels Bohr began dreaming up ideas of 'quantum biology'. At the very least, genetics promised to explain living processes by chemical and even physical laws.

Igor Kurchatov's Institute of Atomic Energy – the institute at the very heart of Soviet atomic bomb development – soon included half a dozen separate genetics laboratories. Restored as director of his Institute of Physical Problems in January 1955, Peter Kapitsa, the pioneer of low-temperature physics, sponsored a major event on molecular genetics, where Igor Tamm and geneticist Nikolai Timofeev-Ressovsky ('conditionally released from prison because of great successes in scientific research work') spoke to a packed audience. It was an occasion unthinkable at any biology institute.

A year later, in the spring of 1956, genetics returned to Moscow State University, when the rumour went round that Timofeev-Ressovsky, that infamous traitor and Nazi collaborator, was coming to speak at the biology department. The young radio-biologist Gennady Polikarpov remembered that the head of his department, Professor Boris Tarussov, used to show Timofeev-Ressovsky around the university's labs 'without permission and against the sound categoric prohibition, issued by the Dean of the Biology department, to "slap the doors of the department in front of this Mendelist–Morganist–Weismannist"'.[3]

Now it was time to come clean: Polikarpov remembered standing with half a dozen other postgraduates to see Timofeev-

Ressovsky safely into the department through a side door. The main entrance was on the alert and impossible to enter – the dean had issued orders not to let Timofeev-Ressovsky set foot in the building. Timofeev-Ressovsky arrived, if not in secret, then discreetly, in a hat and loose coat. The students ushered him in. Polikarpov's colleague Vladimir Ivanovich Korogodin takes up the story:

I found out where the lecture was going to be and went there ... The lecture hall was overcrowded. Then a man came out from behind the scene. He was thickset, with mane-like hair. He looked over the audience. Then he put off his jacket and hung it on the back of the chair. The next moment he started speaking ... and his speech was so vivid that we realised we had never heard such a lecture before ... we returned to the department room feeling enchanted and bewildered.[4]

Timofeev-Ressovsky's freedom turned out to be rather restricted. Denied the opportunity to live and work in the capital, he returned to the Urals and worked in Sverdlovsk, where he organised a biophysics lab at the Ural Division of the Academy of Sciences. He also founded an experimental station and summer school at nearby Lake Miassovo in the Ilmen National Park with an agenda ranging 'from astronomy to gastronomy'.

This school, with more than a hundred participants – some released from his old *sharashka*, Object 0211 – re-awoke the tradition of Mendelian genetics. The social climate at the station was amazingly informal. For all the work that was done there, there seemed to be no such thing as a working day. 'While doing research work you should not be aggressively serious,' Timofeev-Ressovsky would say. The days were filled with lab work, excursions around the reserve, discussions, meals, lectures (sometimes conducted in the lake itself – the summer heat could be punishing) and practical jokes. 'Those who worked in Miassovo became invisibly marked with a special token, like Muslims after they visited Mecca,' Korogodin recalled. This vibrant atmosphere of intellectual holiday was a job of work to maintain, and a lot of it fell to Timofeev-Ressovsky's wife Elena, whose indefatigable letter-writing kept the summer school running smoothly and out of trouble. She had extraordinary tact. 'Quite an absurd gentleman!' she once exclaimed, out of earshot of her target. It was the nearest she ever got to being rude.[5]

To couch genetics as a branch of physics involved a certain amount of special pleading. As the 1950s wore on, however, a new discipline emerged that promised to refashion genetics completely, turning it from a branch of natural history into a science of how information is replicated and exchanged.

It was Alexei Andreevich Liapunov – a Moscow graduate who had studied under the celebrated mathematician Nikolai Luzin – who introduced cybernetics to the Soviet Union. His encyclopedic list of interests had drawn him into a long-term friendship with several leading Soviet geneticists. Towards the end of the 1940s Liapunov had organised a home study group for his two school-age daughters and their friends, to teach them the fundamentals of genetics. Consequently when his daughters entered the Biology Department of Moscow University in 1954, they were practically the only ones to know what classic genetics looked like. Liapunov's study circle expanded as students from Moscow came to hear his ever-more-impressive list of guest speakers. Dubinin, Sakharov, Timofeev-Ressovsky, Zavadovsky, and Zhebrak all spoke to the group.

Running the group was a risky business for Liapunov, a Party member who was working at the time on classified projects.[6] Lysenko's colleagues in Moscow University were incensed by all the 'young Morganists' running around, spreading tales of 'formalist genetics'. 'Students gather at Professor Liapunov's home,' one professor fumed, 'they chatter about Lysenko, about his behaviour, and so on.' Adding insult to injury were the songs the students cooked up and sang in the lab. These apparently 'discredited Lysenko and the Michurinist approach' in the most ribald fashion.

But though attempts were made to expel Liapunov's daughters from the Young Communist League – which would surely have got them sacked from the university – these failed. Liapunov was a loyal Party member, working responsibly in an area far from Lysenko's zone of influence. Lysenko could only watch as Liapunov's geneticist friends were taken up and nurtured by patrons who had nothing whatsoever to do with agrobiology. Indeed, these patrons came from the military, where computation and cybernetics were emerging as key disciplines of the Cold War.

The geneticists regarded cybernetics with a certain amount of bemusement. 'Liapushka is a sweetie,' Timofeev-Ressovsky remarked. 'With great enthusiasm, he would carry you along the wrong path; later, upon discovering his blunder, he would carry you along another path with no less enthusiasm than before, and this new path would be equally wrong.' Timofeev-Ressovsky teased his friend mercilessly, using the term 'cybernetics' in their private letters in the sense of 'cock-up'. He once described his having put a letter in the wrong envelope as a 'complete cybernetics'. Nor was he unaware of the broader problem with cybernetics – that it was one of those ideas that people adopted without thinking through the consequences: 'Anybody who met with mathematicians more or less regularly', Timofeev-Ressovsky half-joked, 'can easily imagine the catastrophic picture of the world at the moment when our life and activity are mathematised.'

Nevertheless, the conjunction of genetics and cybernetics was a tantalising one. 'Perhaps it would be better [for cyberneticians] to refrain from solving all global problems,' Timofeev-Ressovsky argued, 'but instead jointly [with biologists] to attempt the construction of mathematical and computer models of simplified ecological systems.'

In a private letter to Liapunov in October 1957 he wrote:

I sat down and wrote for you, cyberneticians, [an article titled] 'Microevolution.' I tried, on the one hand, to cover everything essential, and on the other, to be brief. It came down to 33 paragraphs in an aphoristic-axiomatic style. It came out not bad, I think, quite original and different from other writings on evolution. All basic definitions seem to be sufficiently brief and rigorous.[7]

Timofeev-Ressovsky classified nature into four nested 'levels of organisation' – the cell, the organism, the population, and the ecology – and Liapunov, in turn, interpreted these systems as cybernetic 'control systems', each with its own mechanisms of information exchange. According to Liapunov, they were well on their way to modelling life *as information*.

Genetics was, therefore, an information science:

On close examination, it turns out that what is transmitted from the parents to their offspring by inheritance is hereditary information. The task of genetics is to study the structure and methods of material coding of this information and the forms of its expression in a new organism in the process of individual development.[8]

Cybernetics – the study of the world as information – was familiar to Soviet philosophers through the writings of Alexander Bogdanov. Bogdanov's works had long been proscribed, however, while in the West, through the writings of Norbert Wiener, cybernetics was making inroads in all manner of fields, including the social sciences. The irony – that models of rational, scientific government and social control were even now being built by Americans, using a science Lenin himself had suppressed – was lost on no educated person, and for one person in particular it provided a Damascene moment.

At the turn of the 1930s Ernst Kolman had left a career as a professional revolutionary and spy to become one of the most doctrinaire philosophers of the Stalin era. While minding the delegation to the 1931 International Congress for History of Science he had kept a close eye on Boris Hessen; in the pages of *Under the Banner of Marxism* he had accused the geneticist Solomon Levit of fascist sympathies countless times; and he had been a not-at-all-impartial judge of the debate between Lysenko and the genetics community in 1939. Come the war, he resumed his more overtly political activities, helping to establish Communist Party control over the Czechoslovak scientific community. The reward for his tirelessness on the state's behalf was a predictable one: in 1948 he was summoned back to the USSR and spent the next three years in the Lubyanka prison.

The experience was salutary. Of all the self-serving and self-justifying memoirs written by Stalin's team after the great leader's death, Kolman's is the most interesting, and in its way the most convincing. He could certainly turn a phrase: 'True, scientific disciplines were sometimes abused for reactionary aims,' he wrote, recalling his work as a Party philosopher, 'but instead of removing such tumours with a scalpel, we struck the healthy body with a cudgel and threw it afterwards in the cesspit.'[9]

In the summer of 1953, half a year after he was released from the Lubyanka, the rehabilitated Kolman was holidaying in a little village on the Black Sea. One evening as he was taking a walk he passed by the cottage belonging to an old friend of his, the psychologist Victor Kolbanovsky, and heard the tapping of a typewriter. Kolbanovsky welcomed him in and said that he was writing an article about cybernetics.

It was an attack, of course. Cybernetics had entered Soviet public discourse in 1951 as an eloquent example of America's fetishisation of technology. 'Contemporary technocrats' in the US were conspiring 'to step in, "scientifically" explain, and "fix" the "malfunctioning" of society ... with exact mathematical formulas'.

Kolman, who had spent the last three years in prison, was immediately captivated by this new science, which investigated the processes of control in technical automata, in living bodies and in society. It reawakened all his old hopes. In his youth he had edited *Life and Technology in the Future*, a collection of futuristic fantasies, and even as he was arrested he was in the middle of a book about mathematical logic in which he claimed that logical operations might one day be performed by machines. Cybernetics was all his intellectual Christmases come at once, and he ticked off his friend: 'Victor, you are a pure philosopher without any knowledge of mathematics and of foreign languages. You have not read a single line of cybernetics. How can you judge it? How can you think that American businessmen would spend millions on faked electronic computers?'

Kolbanovsky went ahead and published his polemic, signing it with a pseudonym 'Materialist'. It appeared in a leading Soviet philosophical monthly magazine. Kolman, meanwhile, set out to learn as much as he could about cybernetics. But in the Lenin Library in Moscow, he found that the discipline's founding work, Norbert Wiener's *Cybernetics: or Control and Communication in the Animal and the Machine* (1948) was locked away, a prohibited book. Kolman protested to one of the secretaries of the Central Committee – and to his surprise, was given access to the book.

In November 1954 the Party's institute of social sciences invited Kolman to deliver a lecture on the philosophical problems of the natural sciences. Kolman chose cybernetics as his theme, and disconcerted everyone by speaking in its favour. 'They all pounced on me as a "technophile", and as an "admirer of bourgeois fashion",' Kolman recalled later. 'The discussion continued for two weeks. Only one of those in attendance, a young postgraduate, did not fear to join me.'[10]

It took Kolman a long while to get his lecture into print, but he was stubborn, and the effort was worth it. In 1958 the Academy of Sciences finally organised its own scientific council to explore the possibilities for cybernetics. Axel Berg, a former tsarist naval officer, was put in charge of it, and promptly invited Kolman aboard.

The first Soviet stored-program digital computer, the MESM, was completed in December 1951 by a small group of twelve designers and fifteen technicians led by Sergei Lebedev, director of the Institute of Electrical Engineering in Kiev. The MESM was the first operating stored-program computer in continental Europe.

The nuclear weapons researchers (led by Igor Kurchatov) and the designers of ballistic missiles and spacecraft (supervised by Sergei Korolev) used up almost all the resources of the first Soviet digital computers. The cosmonaut Georgy Grechko has recalled his experience of working on the BESM (MESM's successor) at the Academy Computation Centre in the mid-1950s: 'Kurchatov's people used it in the daytime and during the night Korolev's people. And for all the rest of Soviet science: maybe five minutes for the Institute of Theoretical Astronomy, maybe half an hour for the chemical industry.'

By the turn of the 1960s, cybernetics had taken on all the characteristics of the Bolsheviks' longed-for 'one science'. There was hardly a discipline cybernetics did not promise to subsume. As 'physiological cybernetics', Nikolai Bernstein's studies of locomotion were widely applied, even finding their way into the Soviets' nascent space programme, where they were used to assess how reliably cosmonauts would be able to move around in conditions of weightlessness. Fields as various as structural linguistics ('cybernetic linguistics'), experiment design ('chemical cybernetics') and legal studies ('legal cybernetics') found shelter under this new institutional umbrella – and there is no doubt that these fields found cybernetic metaphors useful.

The further Soviet society departed from Stalinism, the more radical the cybernetic project became, until its practitioners were seriously advocating the 'cybernetisation' of the entire science enterprise. Speaking at Moscow University in November 1961 Axel Berg looked forward to developing methods and tools 'for controlling the entire national economy ... to ensure the optimal regime of government'.

This was Bogdanov's dream revived. And not just Bogdanov's dream: scientific government was every Marxist's grail.

In 1953 no one expected Trofim Lysenko to continue. To the death of his greatest patron, Joseph Stalin, there was added the matter of the hornbeam tree, and an intellectual scandal so outrageous, no scientist with a conscience could ever have survived it.

The story broke in the November–December 1953 issue of the *Botanical Journal*, when A. A. Rukhkian (who later made his name as a sheep breeder) examined one of Lysenko's most celebrated examples of species change, reported in Lysenko's journal *Agrobiology* in 1952: the case of the hornbeam tree that had been persuaded to turn into a hazelnut. Rukhkian wrote that the branch everyone was getting so excited about had actually been grafted into the fork of the hornbeam. He even managed to track down the man who had made the graft in 1923.

Rukhkian's approach was thoroughly decent: one case of fraud by a third party, he said, hardly discredited all Lysenko's work on species formation. But Lysenko, true to form, spurned the graceful exit being offered him. Catching wind of the article before it was published, Lysenko wrote to the editors saying that his colleague S. K. Karapetian's original report was entirely correct. The hornbeam had been changed into a hazelnut, and that was that.

The *Botanical Journal* disagreed. They published Rukhkian's exposé, Lysenko's letter – and photographs showing clear evidence of the graft. Never mind Lysenko's theories of species formation: now his personal integrity was threatened.

For a while, things only got worse for him. In this brave new world of measurement and control, ushered in by the new science of cybernetics, Lysenko's monolithic public profile was proving hard to maintain. Such was the state's appetite for information, it was rushing foreign scientific and technical literature into translation. The small Academy Institute of Scientific Information was transformed, staffed with hundreds of researchers and equipped to provide translation and dissemination of the most recent reports. 'Technical attachés' and 'agricultural attachés' were assigned to Soviet embassies to gather relevant information. On 2 February 1956, *Izvestiia* ran a report by a member of a Soviet agricultural delegation to the USA. It was an eye-opener: 'In the USA and Canada we did not find a single farmer who planted maize with his own seed . . . Our trip to the USA and Canada has convinced us that we are poorly acquainted with scientific achievements in foreign countries.'[11]

Lysenko stepped down from the presidency of the Lenin Academy in April 1956. Across the world, newspapers hailed his downfall. A headline in the *New York Times* proclaimed: 'Lysenko, Stalin's Protégé, Out as Soviets' Scientific Chieftain'.

At last, Soviet agricultural and biological science could breathe again – could, perhaps, begin to catch up with the astonishing advances made in other sciences. The Soviet Union was, after all, a nuclear power. Historically it had invested more into scientific research and development than any other nation. It had a higher proportion of scientists and engineers than any other nation. The hegemony of Lysenko had been a terrible error and an accident, to be sure, but that was all it was. An error. An accident.

But the institutional weaknesses that had made Lysenko's rise possible in the first place still pertained. The massive centralisation of the state around its leader, and its leader's obligation to be a plain-speaking man of the people, played straight into Lysenko's hands, even at this late hour.

The leader of this new, more open, ostensibly more rational Soviet Union, Nikita Khrushchev, had happily admitted that he knew nothing about science, and this pronouncement, if comforting to the laity, had its shadow side. It meant that Khrushchev judged rival agricultural proposals intuitively, showing interest only in quick fixes and immediate returns, and associating savings with losses. How else to explain Khrushchev's disastrous 'virgin lands' campaign? This was a plan, launched in 1953, to plough and sow 13 million hectares of previously uncultivated and dangerously dry land on the right bank of the Volga, in the northern Caucasus, western Siberia, and northern Kazakhstan. The crop of choice was hybrid maize. Lysenko, the man who had kept Soviet fields free of hybrid maize for a generation, acted as though nothing had happened and in 1956 humbly offered the Twentieth Party Congress a 'square-cluster' method of planting the stuff which would, he said 'accumulate and preserve moisture' in the soil.

Lysenko's 'quick fix' mentality appealed to Khrushchev. Here was someone he could work with – which is to say, someone who would not contradict him. Khrushchev's celebrated bull-headedness and the warmth and loyalty he showed to his friends were of a piece. In May 1962, for example, Khrushchev went on a press junket to Lenin Hills, where Lysenko conducted his research. This trip happened to coincide with Lysenko being censured by an Academy commission into molecular biology. The commission was promptly dissolved, and all its materials sequestered by the Central Executive Committee. To insulate Khrushchev from embarrassment, Lysenko was given space in *Pravda* and *Izvestiia* for an article (handed down by the Central Executive Committee and not to be edited) reaffirming his ever more absurd beliefs.

The following year, two Lysenkoist scientists were up for the Lenin Prize: Alexander Samsonovich Musiyko, director of the Plant Breeding and Genetics Institute, and Vasily Remeslo, a leading wheat breeder. The prize committee, by secret ballot, rejected them. Normally the committee's decisions were rubber-stamped, but not this time. On 13 April, a week before the announcement, Khrushchev blew his top and insisted the committee reconsider.

A year passed, and still the two Lysenkoite candidates were being repeatedly rejected. In 1964, Lysenko added fuel to the fire by adding Nikolai Ivanovich Nuzhdin as a candidate. This sent the Academy into uproar. Nuzhdin had worked at Vavilov's institute under Hermann Muller and was widely considered to have contributed to the arrests of both Vavilov and Timofeev-Ressovsky. Andrei Sakharov spoke for the large group – uniting mathematicians, physicists, chemists, astronomers and cyberneticists – who opposed the nomination, blaming Nuzhdin directly 'for the shameful backwardness of Soviet biology and of genetics in particular, for the dissemination of pseudoscientific views, for adventurism, for the degradation of learning, and for the defamation, firing, arrest, even death, of many genuine scientists'.[12] With Nuzhdin not only rejected but publicly unmasked, Lysenko left the meeting in a furious and vengeful mood.

Khrushchev's own reaction was even more extreme: he threatened to re-form the Academy as a 'Committee on science'. In fact it was this declared willingness to tear down the science base to support Lysenko that soon brought about his downfall. Not long afterwards, he learned that the Timiryazev Agricultural Academy had shaken free of Lysenko's institutional control and was opposing Lysenko's activities, and with an unauthorised personal order (later quickly rescinded), he shut the academy down.

On 13 October Yakov Rapoport, the Jewish doctor who had been caught up in the Doctors' Plot, received a strange phone call. Normally he was being threatened with dismissal. Now the agricultural section of the Central Executive Committee was asking him to write a page-long article on the achievements of genetics.

Even as the bemused Rapoport prepared his article, the Central Committee of the Communist Party was insisting upon

Khrushchev's resignation. The embarrassment over having to import ever-growing quantities of wheat from Canada and the USA was if anything greater than the embarrassment over the Cuba crisis. 'Many examples of Khrushchev's activities deserving of extreme censure were discussed,' writes the geneticist Raissa Berg, 'including his unconditional support of Lysenko and, in particular, the episode involving the attempt to elect Nuzhdin and Remeslo to the Academy of Sciences, with Khrushchev's subsequent desire to invoke sanctions against the Academy.'[13] The reward for Khrushchev's graceful exit was a quiet life. He lived peaceably in Moscow, penning his memoirs, and died on 11 September 1971.

The day after Khrushchev's removal, Rapoport's article was passed for publication in the newspaper *Rural Life*. The following February Lysenko was dismissed as director of the Academy's Institute of Genetics, replaced by his old rival Nikolai Dubinin. New textbooks of biology for schools were prepared, and Lysenko vanished from school and university programmes.

With his institutional clout gone, and his patron Khrushchev removed in a bloodless coup, Lysenko still wasn't finished. He remained the nation's favourite barefoot scientist, pouring out simple and folksy farming advice. His barnstorming approach to agriculture took him next into dairy farming – and it was here that he came irretrievably unstuck. There would be no coming back from this one.

In May 1957 Khrushchev had set the country the challenge of increasing milk output to overtake the USA. Lysenko announced his solution just two months later. At his Lenin Hills farm, he said, he had found away of crossing Jersey bulls and ordinary farm cows that would increase milk yields, generation on generation.

This last point is important. Cross a purebred Jersey bull with a cow of an unexceptional breed, and you will sire a new generation of cows that produce more milk. The problem is that the higher milk yields vanish in the generation after that. There is no way to fix the Jersey bull's hereditary advantage.

Lysenko disagreed: once certain precautions were observed, his cross-bred cows gave and kept on giving. All you needed was especially big cows. Feed them well during gestation, Lysenko said, and the unborn calf would develop the butterfat capabilities of a pure-bred Jersey. Subsequent generations didn't even need special feeding; the bulls could be used freely without fear of decline in milk yield or quality.

Collective and state farms put in immediate orders for the bulls from Lysenko's farm. The Ministry of Agriculture itself recommended their purchase. Lysenko's Lenin Hills farm grew wealthy on the business. And by January 1965 the scale of the calamity – a nation's purebred stock lost in a mess of crossbreeding – was unignorable. An eight-man committee from the Academy of Sciences went to conduct a thorough investigation of Lysenko's farm. The committee spent over five weeks going over its records, examining its crops and its cattle. Balance sheets, crop yields, fertiliser consumption, milk and egg output, purchase and sale data on cattle: for the first time, Lysenko's claims were subjected to a serious scientific and statistical analysis.

On 2 September 1965 the committee presented its conclusions: the Lenin Hills farm gave high yields and turned a good profit. But so it should: it boasted 500 hectares of arable land, and operated up to fifteen tractors, eleven cars, two bulldozers, two excavators, and two combine harvesters. It was more or less exempt from having to supply grain to the government. It received more favourable terms, more investment, more *electricity* than its neighbours. Taking all this into account, the committee reported that the yields from the Lenin Hills farm were, at the very best, mediocre.

Even this Lysenko might have managed to flannel through. He had got through worse. But the cows did for him: in the past ten years, the committee reported, the average milk yield per cow had dropped from nearly 7,000 kilograms to less than 4,500. Milk yields were falling away with every generation. Lysenko's low-pedigree bulls had ruined herds of higher purity. Repairing the damage would require decades.

Lysenko's reputation never recovered. He got to keep his experimental farm but his journal, *Agrobiology*, was terminated. Two new journals of genetics, *Genetika* and *Ontogenez*, came into being. And the bust of Gregor Mendel, the father of classical genetics, which had been languishing in storage since 1949, was returned to the place the great Ivan Pavlov had chosen for it: right in front of his institute in Koltushi.

EPILOGUE : Spoil

Without science democracy has no future.
Maxim Gorky, April 1917

Careless of his personal reputation, Nikolai Fedorovich Fedorov never had any ambitions for his writing, and his works survive only because his acolytes diligently gathered together the scraps left behind on his death in 1903.[1]

A former schoolteacher, Fedorov worked at Moscow's Rumiantsev Museum. 'He is sixty,' Leo Tolstoy wrote, 'a pauper, gives away all he has, is always cheerful and meek.' He was also one of the most powerful cult figures in pre-revolutionary Russia, numbering among his followers the mystic Peter Ouspensky, the novelist Fyodor Dostoyevsky and the rocket theorist Konstantin Eduardovich Tsiolkovsky.

Unusually for a Russian, Fedorov took seriously the population theories of the English historian Thomas Malthus. Malthus's 'Essay on the Principle of Population' of 1798 had argued that human beings, like any other animal, would consume resources to the point where poverty, starvation, disease and war would

'These limits are real; they are not imaginary and they are not theoretical.' 2003: a ship rusts away in what was once the Aral Sea.

virtually exterminate them. Where utopians like William Godwin believed that God would provide food for all mouths, Malthus argued that God had already provided more than enough mouths for all the food.

Fedorov, too, worried that the earth would become over-crowded, and he came up with a novel solution to the problem: colonise outer space. (He also made one or two remarks about taking to the oceans.) In *The Philosophy of the Common Cause*, Fedorov explains that as humans evolve hand in hand with their technology, they will be able to get rid of their digestive and sexual parts, harvest cosmic energy for their food, achieve psychological perfection, and become immortal.

Fedorov's 'cosmism' succeeded, not because it was wild and apocalyptic, but because it gave the millennial outpourings, cults and beliefs flying around Russia in the nineteenth century a rational-seeming explanation.

In fact Fedorov rivalled Karl Marx as an intellectual driver of the 1917 revolution, and his influence over the Bolsheviks themselves was tremendous and long-lasting. You hardly need pan the outpourings of men like Lenin, Bogdanov, Gastev or Lunacharsky for more than a few minutes before you catch a glint of Fedorovian gold.

'Man will finally begin to really harmonise himself,' prophesied another Old Bolshevik, Leon Trotsky, in 1922:

He will put forward the task to introduce into the movement of his own organs – during work, walk, play – the highest precision, expediency, economy, and thus beauty. He will want to control semi-unconscious, and then unconscious processes in his own organism: breathing, blood circulation, digestion, and reproduction – and will subjugate them to the control of reason and will. Life, even purely physiological life, will become collectively-experimental. The humankind, frozen *Homo sapiens*, will again enter into a radical reconstruction and will become – under his own fingers – an object of most complicated methods of artificial selection and psycho-physical training ... Man will put forward a goal . . . to raise himself to a new level – to create a higher socio-biological type, an *Ubermensch*, if you will.[2]

Trotsky was a notorious blow-hard, of course – but then, who among that crowd was not? Here is the character Rybin in Maxim Gorky's *Mother* (1906): 'Let thousands of us die to resurrect millions of people all over the earth!' he declaims. 'That's what: dying's easy for the sake of resurrection! If only the people rise!'

In his two-volume *Religion and Socialism* (1908 and 1911) and other writings, Anatoly Lunacharsky called for a collectivist culture that would teach the masses to 'die for the common good' and to 'sacrifice to realise this conception' of socialism, which 'starts not with his "I" but with our "we"'. In a future socialist society, one of Alexander Bogdanov's followers wrote, 'people will be immortal' insofar as they 'develop their "I" beyond the limits of individualism into a tendency toward community'.

Perhaps they should have been more careful what they wished for; and a handful of the old guard did try to apply the brake.

Alexander Bogdanov simply jumped off the train.

When Anatoly Lunacharsky, his brother-in-law and for a long time his comrade-in-arms, was appointed education commissar in Lenin's government, one of his first acts was to offer Bogdanov a government post. Bogdanov's reply was terse: 'The bayonet is not a creative instrument, however widespread its application.'[3] Bogdanov, who had played no part in the 1917 coup, nursed little hope for Lenin's administration.

It means nothing to me that this soldiers' socialism is being implemented by a crude chess player like Lenin, or by a conceited actor like Trotsky. What saddens me is that someone like you should have become involved in this affair; firstly because your disenchantment will be much greater than theirs; and secondly because you could have achieved something different, something less conspicuous at the present moment but no less important and more lasting.

Bogdanov was referring to the men's long-nursed plans for encouraging a truly working-class culture. Bogdanov held that without a cultural revolution, a political revolution was pointless. As he argued elsewhere: 'if [proletarian culture] were beyond one's strength, the working class would have nothing to count on, except the transition from one enslavement to another; from under the yoke of capitalists to the yoke of engineers and the educated'.[4]

A lesser man – or a less indulgent one[5] – would have taken umbrage at Bogdanov's letter. Lunacharsky's reaction was quite different: he made room for his brother-in-law's experiment in his Commissariat of Education. The 'proletarian cultural and educational organisations' of Proletkult, held their first All-Russian Conference in September 1918. Incredibly, the project, meant to bootstrap an entire proletarian culture out of little more than a heap of good intentions, met with immediate success.[6]

Proletkult cells were established in every factory. Studios were set up around the country, where 80,000 workers learned and practised the arts and sciences.

Proletkult grew so fast, Bogdanov had no chance to give it a real ideological direction. It remained a rag-bag collection of studios, clubs, theatres and workshops. But this was not necessarily a bad thing: at its height, Bogdanov's cultural movement won the public allegiance of virtually every significant artist, musician and writer in Russia. Experimental art, music and theatre flourished. The Tula Proletkult music section had a symphony orchestra, a brass band, an orchestra for folk instruments, and special classes in solo and operatic singing. Even tiny Arctic Archangel, perched on the coast of the White Sea, boasted a choir.

Besides, ideological uniformity was hardly Bogdanov's style: workers' culture was supposed to be anti-authoritarian, and – most important of all – organised from the bottom.[7]

In failing to lead Proletkult in any meaningful way, Bogdanov managed to disseminate tektology, his Ernst Mach-flavoured 'science of organisation', to around half a million working people. When Lenin, attending a conference of Soviet educators in May 1919, discovered the great influence of Bogdanov's philosophy, he declared 'merciless hostility' toward the movement, and at Proletkult's Second Congress in 1920, he invoked the authority of the Party's Central Committee to dismantle it.[8] Lunacharsky's own commissariat was closed down soon after.

Alexander Bogdanov was one of those people who win every battle on the way to losing the war. At Gorky's villa on Capri, he had thrashed Lenin at chess but failed to heal the rift that had opened up between him and his friend – if indeed he really tried. Shortly after, his became by far the most dominant faction of the Party, but he lost interest in politics, even as control came within his grasp, because he wanted more time for his science-fiction writing. Those two novels of his, *Red Star* and *Engineer Menni*, contained the seeds of a fully worked-out philosophy of systems he called 'tektology'. But by the time he got around to publishing the full work he had become, as he himself put it, 'an official devil who had to be foresworn'.

Proletkult withered away over a couple of years but Bogdanov continued to publish. Bogdanov believed that, ultimately, everything was explainable in terms of everything else. This did not make him a scientist. If anything, it made him a magician. When Bogdanov left off his Martian prophesying and attempted to do real science, as head of a 'scientific research institute of blood transfusion', he resembled an hermetic philosopher of the sixteenth century. He relied on risky analogies, treated hypotheses as facts, and – most telling of all – never offered up his studies to the review of his peers. In fact he never wrote any actual science at all, just propaganda for the popular press. In this, he resembled no one so much as Trofim Lysenko. Indeed, it was his example made Trofim Lysenko politically possible.

Bogdanov was a follower of Lamarck. He believed that characteristics acquired during life were passed down to one's offspring. In his utopian science fiction, blood exchanges among his Martian protagonists levelled out their individual and sexual differences and extended their life span through the inheritance of acquired characteristics (an idea comprehensively dashed by experiments on rabbits conducted by Francis Galton forty years earlier). Bogdanov's scientific fantasies took an experimental turn in 1921 when Leonid Krasin, who like Bogdanov also been expelled from the Bolshevik Party in 1909, took him on a trade junket to London. There, Bogdanov happened across *Blood Transfusion*, a book by Geoffrey Keynes (younger brother of the economist).

Two years of private experiments followed.[9] When Krasin, now Russia's ambassador to Britain, became ill, Bogdanov treated him with a transfusion of blood. Krasin's condition improved markedly, and this miraculous cure won Bogdanov an appointment with the Communist Party's general secretary, Joseph Stalin. Up until this time, Russia lacked any kind of blood transfusion service, and Bogdanov was quickly installed as head of a new research institute, housed in the resplendent, Empire-style Igumnov mansion in Moscow. (It is now part of the French embassy.) The locks were changed overnight to keep out the factory workers who had been using the house as a cultural club.

Bogdanov promoted blood transfusion as a 'universal' technique for bodily rejuvenation. Blood, Bogdanov claimed, was a universal tissue that unified all other organs, tissues and cells. Blood's chemical composition reflected the activity of the whole organism. Transfusions offered the client better sleep, a fresher complexion, a change in eyeglass prescriptions, and greater resistance to fatigue, and although his institute had already managed to give a client syphilis, Bogdanov claimed that his transfusions were safe.

On 24 March 1928, Bogdanov conducted a typically Martian experiment, mutually transfusing blood with a 21-year-old male student. The men's blood groups matched, but Bogdanov suffered a massive transfusion reaction, probably because, after eleven previous experiments in this line, his blood was riddled with foreign antibodies. Despite vigorous therapy, Alexander Bogdanov died two weeks later at the age of fifty-four.

Other kinds of new Soviet person were imagined. For the poet and ergonomist Gastev, the revolution promised a new way of life and a new kind of culture. 'In building anew,' he once asked, 'do we need to take those "treasures" that have come down to us from the nobles' culture, with its "monuments" of art, habits, and way of life, or should we uncover the face of a new culture, born from technique and production?'[10]

The future beckoned: 'It is time to stop being sentimental about human nature,' Gastev wrote, in the pages of his institute's quarterly journal; 'It is time not just to study and analyse society, but recreate it. To introduce new social norms, make a new choreography of movement and new sets of social behaviour.' Gastev's industrial training institute was 'the battlefield where a new, technical culture went to battle against our historically formed humanitarian dreaminess' – and it won. The Stakhanovite movement fitted perfectly with Gastev's plans. As early as 1923 he had been putting forward the idea of a labour championship 'in which a finely performed labour operation will be honoured with a decoration before thousands of professional workers'.[11]

Bogdanov took Gastev to task for equating work habits and production behaviour with culture. The life of the working class, Bogdanov argued, wasn't just work. It made no sense 'to break off one piece, even if it is very important'. Bogdanov nursed a deeper worry, too – that, in following Gastev's 'sinister and fantastic' agenda, all the creative, original functions of the economy would fall in the laps of a few educated engineers.[12]

He was right to worry. Ultimately, the Soviet state never threw up lasting, ideologically correct versions of any of the basic rituals by which a community celebrates a human life. There was never any Soviet marriage ceremony worth the name. No burial service. No secular christening.[13]

Under Stalin's rule the 'new Soviet man' took its final shape, and the chief and most dreadful feature of this creature was that he was ageless.

By this, I do not mean so much that Stalin was interested in longevity research, though that was certainly one of his pet projects. A fascination with life extension was par for the period, and the Soviet version was really just an ideologically boosted version of a desire – universal to the rich and powerful (if they can't take it with them, then they're not bloody going) – to cling on forever.[14] Aged and worn men who religiously believed in science, Bolsheviks like Gorky and Lunacharsky, were suckers for the latest rejuvenation technique: implantations of monkey glands, vasectomies, Olga Lepeshinskaya's soda baths (let's not forget *them*) and gravidan, a curious substance extracted from pregnant women's urine. Soviet leaders, meanwhile, shunned retirement, left the country baffled as to succession, and weathered terrible illnesses and handicaps in their determination to go on, and on, and on, into the impossible future.

Never mind all that: the truly hideous thing about immortality is the way it erodes life at both ends. It banishes senescence and death, yes, but at the same stroke it erases birth, and renders youth so fleeting as to be irrelevant – a sort of pupa stage. There was, for many years, no youth culture in Soviet Russia. There were, on the other hand, many means of managing and regimenting and indoctrinating the young, not so much into a living ideology, as into an approved adult way of behaving. The Party's youth wing dragged you by the collar into adulthood and kept you there. Once past your babbling infant stage, the point of you was to be the responsible young man or woman on the poster.

You can see the process unfolding in the pages of Alexei Gastev's journal. The early editions are intellectually voracious, full of references to what ergonomists and industrial designers are up to in Germany and the USA. Then, as the First Five-Year Plan takes effect, Gastev's institute starts to focus less on people and more on machines, on production planning and economic output. In a way this is exciting. Never mind how to swing a hammer: now the institute is issuing manuals on how to start up entire factories, complete with everything from architectural plans to management schedules! But along with that vaunting ambition comes an ever-more removed view of what the worker's life is actually like. More worrying still, the pages begin to be filled with pictures of Stalin, with quotes from Stalin's speeches, with invocations to Stalin and what amount to hymns of praise.

The last issue of Gastev's journal came out in 1938, by which time it was clear that disarmament and declarations of loyalty were no defence against Stalin's pogrom of Party intellectuals. On the night of 7 September Gastev was taken away by the NKVD. His apartment, which was located in the same building as his Central Institute of Labour, was thoroughly searched and his papers were confiscated. The following spring he was sentenced to ten years' hard labour and died, shot by firing squad, in a Moscow suburb on 1 October 1941, by which time the Soviet Union was already beginning to resemble a ghastly, grim parody of his poem 'Express'.

Numbers, acronyms and fake names littered the Soviet landscape, as gulag camps and colonies sprang up to fuel the NKVD's slave economy. By the 1950s, the gulag's dehumanising nomenclature had passed into the everyday vocabulary of the Soviet defence industry, an enormous archipelago of institutes, factories and ministries that, by some accounts, made up 40 per cent of the Soviet economy. Hundred of thousands of workers played out limited lives in post-office boxes, special facilities, secret cities and fictitious towns.

The transition from the gulag economy to the military–industrial economy was an easy one. In many cases, prison science teams working under guard became 'free' enterprises overnight simply by allowing the scientists and engineers to go home at the end of the day. As Sergei Korolev once joked, the guards who protected him in his high position as the chief designer of the Soviet space programme were probably the same ones who watched over him in the *sharashka*. In both cases, the guards were doing the same job – maintaining the wall that separated scientists and the outside world.

Scientists themselves eventually breached that wall, led by Andrei Sakharov, for whom science and civics were virtually the same thing, and scientists the country's only remaining champions of civil rights. In 1968, in his essay 'Reflections on Progress, Peaceful Coexistence, and Intellectual Freedom', Sakharov, speaking from a place of Cold War privilege, called for the norms of scientific discussion to be transferred to public life. He wrote: 'We consider "scientific" that method which is based on a profound study of facts, theories and views, presupposing unprejudiced and open discussion, which is dispassionate in its conclusions.' For Sakharov, as for Gorky, as for Bogdanov and Lenin and Marx, science was the model for politics. All politicians had to do was learn how to be good scientists.[15]

But Sakharov's belief in the civic power of science would not triumph. His and others' dissident activities, which had been meant to reform the state, served in the end only to weaken it and speed its eventual collapse. Why should this have happened, in a nation dedicated to the ideal of scientific government, and whose very founding figures spoke – as Maxim Gorky spoke in 1917, in a lecture titled 'Science and Democracy' – of a society rooted in 'the soil of exact observation, directed by the iron logic of mathematics'?

Gorky had imagined, just as Sakharov imagined half a century later, that a society devoted to science could not help but develop civic institutions overnight. He imagined the revolution creating a Baconian 'city of science, a series of temples where each scholar is a priest who is free to serve his god'. And turning science into politics was, naturally, the particular hope of the Bolsheviks, who wanted to use science to leap-frog all those painful years of civic development that post-tsarist Russia so desperately needed if it was to catch up with the West.

The 1917 revolution failed to realise that dream. Lenin failed. Stalin failed (for it was his dream, too). Half a century on, Sakharov was dreaming that same dream. And still it would not be realised.

The repeated failure of that dream was only partly historical accident, only partly bad timing, only partly to do with the failings of individual men and women. It was (and is) much more to do with the failure of the sciences themselves to cohere into a single, coherent discipline that politics might wield.

And the longest-running failure of all – the failure of psychology and physiology to 'meet in the middle' (to paraphrase Trotsky) – was soon to fill the wards of the country's mental hospitals with Sakharov, his friends and colleagues, hopeless dreamers, dissidents, 'sluggish schizophrenics' and the 'philosophically intoxicated'.

It comes as no surprise that the Soviet government found it expedient to incarcerate its political opponents in mental hospitals. The Russian state had been doing that long before the Bolsheviks were even thought of. Spending time in an asylum was considered preferable to exile to Siberia – it was a way the tsarist state showed leniency.

What changed, as the Soviet Union moved beyond Stalinism, was the specious way in which political protest was pathologised, and the blame for this lay, not with the politicians, but with Soviet psychiatrists themselves. It was Soviet psychiatry that made waves of fresh incarcerations possible, by promulgating models of human mental functioning that were savagely reductive, doctrinaire, rigid and plain wrong.

In the late 1940s Andrei Vladimirovich Snezhnevsky, a talented bureaucrat and clinician, was appointed to a panel of scholars tasked with preparing a Soviet psychiatric manual. The first issue of *Neuropathology and Psychiatry* edited by Snezhnevsky was full of predictable Stalinist phrases: readers were exhorted to 'pay particular attention to the fight for the scientific purity of our discipline, to the fight with all enemy ideas and their conduits, and to the unmasking of the capitalist anti-national essence of neuropathology and psychiatry of the capitalist countries.'

With Stalin out the way all that vanished, and in its place came chapters reflecting Snezhnevsky's admiration for the German Emil Kraepelin, who wanted to classify mental diseases into a workable and intelligible system.

So far, so good, one would have thought – except that Soviet psychiatrists had next to no understanding of human psychology. They had not been taught anything about it. They were supposed to be *Pavlovian* – and since Pavlov had not strayed into psychoanalysis, social psychology, or analytical or individual psychology, then no one else was entitled to, either.[16] Pavlov had once instituted a playful system of fines for juniors who psychologised the behaviour of his experimental dogs. In a grotesque over-extension of that idea, Soviet psychiatrists were taught that psychological terms and expressions were subjective, idealistic and non-scientific – and they were forbidden to use them.

Under the guidance of Snezhnevsky – and he was a do-er, not a thinker[17] – these narrowly trained specialists now tried to classify clients and patients according to taxonomies of mental illness that were, as they remain today, vague and vulnerable to fads and fashions.

If all you have is a hammer, everything looks like a nail. Armed with Snezhnevsky's vague manual, and with no grasp of psychology, Soviet psychiatrists came to regard everyone as a potential schizophrenic. And while the West was de-institutionalising the mentally ill between 1950 and 1970, Soviet hospital beds for the mentally ill almost tripled, while the number of psychiatrists almost quadrupled.

Among the symptoms of schizophrenia listed in Soviet textbooks was something called 'ambivalence', so God help you if you admitted to disliking a relative, lacking sympathy for some people or steering clear of official events. Extending this kind of baggy thinking to politics was easy, if not inevitable. The term 'philosophical intoxication' was used to diagnose schizophrenia in people who did not agree with the authorities and used Marxist terminology to criticise them. In late May 1970 the biologist Zhores Medvedev was diagnosed as suffering from 'creeping schizophrenia' and was placed, by force, in a psychiatric hospital in the town of Kaluga, about 150 kilometres from Moscow, for publishing the book *The Rise and Fall of T. D. Lysenko* in the United States. As one of his more vigorous defenders, Andrei Sakharov, recalled later, 'Medvedev's work in two disparate fields – biology and political science – was regarded as evidence of a split personality, and his conduct allegedly exhibited symptoms of social maladjustment.'[18] While forcibly confined, Medvedev met a middle-aged man imprisoned for pasting up handbills complaining about the local Party Committee. Diagnosed with 'poor adaptation to the conditions of the social environment', the complainer was administered 'two powerful depressant drugs' that the doctors promised would 'change the basic structure of his psyche'. Another inmate, about twenty-four years old, incarcerated for calling the Communist Youth League a bunch of paper-pushers, suffered from 'reformist delusion', apparently; for three months he had been undergoing 'periodic insulin shock'.

A woman who complained to the prosecutor's office, a young man who announced his intention of quitting the Party, a Lithuanian engineer who refused to participate in building a monument to the Soviet war dead unless the state also put up a monument to the victims of Stalin's terror, young 'hippies', religious believers, draft dodgers: all served time in mental hospitals.

By the time of its unexpected collapse, the Soviet Union had become what its founders had always dreamt it might become: a scientific state.

In the 1980s the Soviet Union and its satellites boasted twice as many scientists for their population as the USA and Western Europe, and operated the largest and best-funded scientific establishment in the world. It was a wounded giant, at once the glory and the laughing stock of twentieth-century thought. It boasted the first and largest national health service in the world, yet fully a quarter of a medical student's curriculum was devoted to politics – more time than was allocated to surgery.[19] It ran arguably the world's most successful and certainly its most consistent programme of space exploration, then allowed institutional and personal rivalries to dismantle its strong bid to be first to the Moon. It boasted the largest and most sophisticated planned economy on earth, yet the paucity of finished goods in its shops was an international joke. Its collectivised agriculture had for decades promised to be a breadbasket for the world – yet, from Khrushchev's era until 1991, Russia had to import grain from the USA and Canada to feed its own people.

Soviet science was extraordinary, and ought to have delivered many more miracles than it did. The entire Bolshevik project was extraordinary, piecing back together the remnants of Russia's empire until it was once again the largest territorial state on earth. As time passes, however, what astonishes is not the achievements – the cities of over a million in deepest Siberia, the mines and railways and dams – but rather the sheer waste those achievements generated.

The despoliation of Russian land was not a uniquely Bolshevik habit. As early as 1911, as he prospected for radioactive minerals around Lake Ilmen, Vladimir Vernadsky, geologist and liberal

429

politician, witnessed the appalling condition of a land already exploited in a desultory fashion for 200 years:

The Urals produce a heavy impression by the terrible plundering of its riches ... forests, mines of precious stones, roads, forms of life – all reflect the disorder of this antediluvian government structure and the anarchy which reigns everywhere! You cannot imagine what barbarism is found in the famous Murzinka region and its vicinity! Yet, the wealth here is enormous. In two hundred years there is not yet one respectable road! The forests burn and are wasted. In order to obtain precious stones, almost half are destroyed and future work is made almost impossible.[20]

In 1920, responding to Vernadsky's campaign to save the area, Lenin signed into being the first nature preserve in Soviet Russia, and for a while it seemed that the revolutionary government would embrace the nascent science of ecology and protect the Russian environment for its people.

These good intentions did not survive the desperate struggle to industrialise the nation. The engineer Peter Palchinsky pitilessly recorded the government's practical and moral failures. Its signature 'hero project', the Dneprostroi dam, was an environmental as well as a human disaster. No one had thought to prepare proper maps, so no one knew just how large an area the 35-metre-high dam would flood. No one had included the loss of farmland in the estimates of costs. Palchinsky's criticisms of the Magnitostroi steel plant were just as trenchant. There were no towns nearby, so an entire city would have to be built for the construction workers. The plant was named after the nearby 'Magnetic Mountain' – but no one had bothered to get a good estimate of how much ore was really sitting in that fabled peak. Worst of all, there was no coal nearby: it would have to be brought, over a phenomenal distance, by train.

The government ignored Palchinsky (for all his trouble he was executed in 1929) and brought in American engineers from Gary, Indiana, to help plan the mills. Their pretty enclave, nicknamed

Amerikanka, boasted tennis courts, while the 200,000 who were brought in to build the works lived in tents and mud huts surrounded by open sewers and directly in the path of fumes from the blast furnaces.

Vladimir Vernadsky, ever the visionary, understood what lay behind these myopic government actions. On 7 April 1926, addressing the All-Union Congress of KEPS, he pointed out that no matter how advanced a social system got, it could not hope to change the laws of nature:

As [natural productive] forces are . . . not exhaustible, we know that they have limits and that these limits are real; they are not imaginary and they are not theoretical. They may be ascertained by the scientific study of nature and represent for us an insuperable natural limit to your productive capacity.[21]

The Stalin Plan for the Great Transformation of Nature was but the most grandiloquent monument to the Soviet state's refusal to admit physical limits to its future. Natural resources were not included in its price system. They could not be owned privately, but at the same time they had no value: they were 'free goods'. Is it any wonder, then, that resources, energy and materials were used without regard for waste or loss?

Projects were praised more for their daring and scale than for their utility. The economically worthless White Sea Canal, built in record time, but not lined and not deep enough, and which cost the lives of thousands of labour camp prisoners, became in the media and art a unique success of Soviet industrialisation. Slave labour without the appropriate tools was sold to the nation as creative work by a politically conscious proletariat.

Nikita Khrushchev, inheriting the Plan from Stalin and with it an unimaginably vast area of dead and dying forests, took one look at the looming repair costs and issued a decree eliminating the Main Shelterbelt Administration. All programmes related to the health of forests fell into steep decline. Some idea of the scale of the problem can be acquired when one remembers that Russian forests counted for more than a fifth of the world's total. In the 1960s a fifth of all Russian timber ended up as wood chips, scraps and sawdust. Up to 40 per cent of the harvest was used as firewood, was thrown away, or simply decayed. Millions of cubic metres of wood rotted away in warehouses on the lower reaches of rivers.[22]

Khrushchev and his successor, Leonid Brezhnev, did not so much scale back the 1948 plan as overlay it with their own wildly ambitious projects. By the mid-1950s, under Khrushchev, virtually all Russia's European rivers had been tamed to power the Soviet economic machine. They had also become dead zones of pollution. The Plan's expanded irrigation system has been left unlined and loses half the water it carries. The lost water has turned an area bigger than France into unplanned swamps.[23]

The Aral Sea, its feedwaters diverted to feed a disastrous cotton-growing project, lost more than half its size from 1960 to 1991 – three-quarters of its volume. Inevitably, the climate around the lake has changed. Around 43 million tonnes of dust and salt are blown by storms annually. The salts are toxic to plants and make people sick.

And once the European USSR had been tamed, Khrushchev turned his attention to Siberia. The large-scale construction trusts that had been established in the Stalin era grew ever larger. Some ended up with around 80,000 employees, none of whom could be laid off, because the law did not allow it. The moment the planners finished one project, they had to announce another to keep their workers employed. The trusts rampaged across the landscape in search of the next geo-engineering project. If you look on a map of the USSR you can follow them up the rivers, as they built dams and locks, year by year and decade by decade.

Another major consequence of all this construction was pollution. Between 1948 and 1951, 76 million cubic metres of high-level and intermediate-level radioactive waste were discharged from Chelyabinsk-40 into the local river system. The Techa River and its floodlands were put beyond use, and in 1951 10,000 people were evacuated while prison labourers built dams and reservoirs to isolate the river.

In 1957 another nuclear accident – a massive explosion at the nuclear waste dump in Kyshtym, close to Sungul – contaminated the area with radioactive waste. The plume rose about a kilometre into the air, exposed about 100,000 people to measurably harmful levels of radiation, and created a dead zone of several hundred square kilometres. The 'East Ural Radioactive Trace' became a unique testing area for radio-ecological studies but cleaning up the mess was a real headache, especially when, during the hot, dry summer of 1967, Lake Karachai evaporated, and winds blew radioactive dust more than fifty kilometres from its source, affecting 41,000 people.

In 1968, the Soviet government finally hit upon a sure-fire way of preventing people wandering into this contaminated zone. They turned it into a nature reserve.

* * *

If you are pure and selfless and focused – this was the Bolshevik promise – then the physical world will shape itself to your will. Believe in your dreams. Invest yourself in them, wholeheartedly. Concentrate completely on your work. Do not allow yourself to be distracted by obstacles. Above all, do not accommodate reality: it will only scatter your effort.

Focus on doing what must be done. Fulfil your crazy dream. Do not despair.

Well, since Stalin's day the world has done its work and fulfilled, in patchwork fashion, its crazy dream of clean water, sewage disposal, hospitals, roads, schools and civic institutions. And we are just now beginning to reap the unforeseen consequences of this uneven, gargantuan effort at global development. As I write this, human consumption is exceeding long-term supply by about a quarter. In other words, it takes the earth a year and three months to regrow what people consume in a year.

Natural resources accumulate over time. Oil is the example everyone reaches for first, but forests, soils and aquifers of fresh water also accumulate, and once these resources are gone, they're *gone*. Simple waiting won't bring them back, because the very processes that generated them in the first place have been disrupted.

By the time this book is published we'll be needing one and a half earths to meet our demands. Ten years after that, we're going to need two and a half earths. (Bogdanov was there before us: the highlight of his science-fiction novel *Red Star* is a debate between two Martians over whether they should exterminate us all to get access to more natural resources.)

No wonder we live in millennial times. Some believe there will be an ending, a collapse, a rapture. The rest of us think we have generated so much knowledge and technology that we'll soon be able to ignore and forget the last 10,000 years of human experience of living on a planet. We are all little Stalinists now, convinced of the efficacy of science to bail us out of any and every crisis, regardless of what science can actually do, impatient of anything scientists might actually say.

There was, I believe, something piteously unavoidable, something admirably human, about the way the Soviet Union faced a world of scarcity and poverty, and tried to light up its land with the fitful glow of science. For all the terrors, follies and crimes of that time, I believe this has also been a story of courage, imagination and even genius.

I fear we will not acquit ourselves nearly so well.

London, 2016

Acknowledgements

The reader is in luck: not being an academic, I am under no pressing obligation to thank every scholar, department and library whose hand-towels I darkened with my sweat. If they're in the bibliography and they're alive, you can be sure I begged favours off them. They know who they are, and they know how much I owe them. (The people at the Wellcome Library in London are powerful minor gods.)

Better that I mention the people who suffered through my monomania, who indulged it and, in some cases, even encouraged and facilitated it. Obviously any errors in the text are mine; however, the project as a whole is entirely their fault.

Will Hammond at Penguin Books sowed the first seed, with a notion that a biography of the pioneering neuropsychologist Alexander Luria might lead me down interesting rabbit-holes. Anna Davis probably had the worst of it, as she commented on my countless false starts, 'one-pagers' and important statements of intent. My agent, Peter Tallack ('The Science Factory' to his friends), drove me to write a book-length proposal document and then won me extension after extension as this project accreted and flaked away, accreted and flaked away, like a particularly tortured figurine by Alberto Giacometti.

I burned through one editor completely (Neil Belton – you were brilliant, and kind, and you deserved better from me) and ended up in the tender care of Julian Loose at Faber, who, to my amazement, proved even more of a Soviet nut than Neil.

Some books are so complicated, they have to be performed before they can be written. A chain of happy chance led me from Stephen Dalziel, via some splendid parties, to Julian Gallant and the intellectual oasis that is Pushkin House, and the opportunity, in lectures, to off-piste through Russian science long before I had earned my licence. Rhidian Davis of the British Film Institute, Doug Millard of London's Science Museum and Louis Savy of Sci-Fi-London all afforded me vital opportunities to straighten out my story in public.

Conversations with friends, had I only had the wit to transcribe them, would have made a book better (though no shorter) than this one. Stephen Jelley, Simon Spanton, Megnah Jayanth, Matthew Cobb and Oliver Morton: thank you. (And Daniel Brown, obviously: my *éminence grise*.) At my day job, Sumit Paul-Choudhury and Liz Else have kept a quiet, coolly amused eye on this project, and without their encouragement and patience I never would have finished.

Finally Lydia Nicholas, who thinks, speaks, feels and acts faster than anyone else living. Never mind her practical assistance (which was essential): if, as I hope, this book carries the spark of life, then it's hers.

Notes

Preface

1. The English translation of *Malen'kaia knizhka o bol'shoi pamiati* first appeared in 1960 through Basic Books.
2. On the outbreak of war in August 1914, the Russian capital's Germanic name, Saint Petersburg, was changed to the more Russian-sounding equivalent, Petrograd.
3. If you convert Old Style Julian dates to their Gregorian equivalents (as, in a fit of tidiness, I have throughout) then the 'October Revolution' took place in November. Russia adopted the Gregorian calendar in 1918.

Prologue

1. Cited in Kendall E. Bailes, *Science and Russian Culture in an Age of Revolutions: V. I. Vernadsky and His Scientific School, 1863–1945*, p. 99.
2. Richard Pipes, *Russia Under the Old Regime* (Scribner, 1975), p. 84.
3. Grain not exported went mostly to make vodka. See Patricia Herlihy, *The Alcoholic Empire: Vodka and Politics in Late Imperial Russia* (Oxford University Press, 2002).
4. Soviet historians often dubbed this policy an 'Arakcheev regime' after Count Alexei Andreevich Arakcheev (1769–1834), a general famous for establishing colonies which combined agricultural hard labour with army-style discipline. See Richard Stites, *Revolutionary Dreams: Utopian Vision and Experimental Life in the Russian Revolution*.
5. Michael Haynes and Rumy Husan, *A Century of State Murder?: Death and Policy in Twentieth-Century Russia*, p. 28.
6. Jeffrey Burds, *Peasant Dreams and Market Politics: Labor Migration and the Russian Village, 1861–1905*, p. 34.
7. Ibid., p. 74
8. Edvard Radzinskii, *Alexander II: The Last Great Tsar*, p. 413.

PART ONE: CONTROL

1 Scholars

1. Quoted in David Holloway, *Stalin and the Bomb: The Soviet Union and Atomic Energy, 1939–1956*, p. 112.
2. Quoted in Bailes, *Science and Russian Culture*, p. 9.
3. Ibid., p. 17.
4. Ibid., p. 26.
5. The restrictions placed on women, and their efforts to get around them, generated some bizarre headlines, as when some Jewish women who had

registered as prostitutes in order to gain residence rights in St Petersburg were discovered to have enrolled as auditors in medical institutes and private higher courses for women. The story (which was perfectly true) combined high-mindedness and salaciousness in such exquisite proportions that it inspired four feature films by the end of the First World War. See Benjamin Nathans, *Beyond the Pale: The Jewish Encounter with Late Imperial Russia* (University of California Press, 2004). Also worth a look are Ann Koblitz, *Science, Women and Revolution in Russia*, and Samuel Kassow, *Students, Professors, and the State in Tsarist Russia*.

6. V. I. Vernadsky, *Geochemistry and the Biosphere*.
7. Vernadsky's visionary late works are discussed in Arsenii B. Roginsky, Felix F. Perchenok, and Vadim M. Borisov, 'Community as the Source of Vernadsky's Concept of Noosphere', *Configurations*, 1 (1993), p. 415; and Akop P. Nazaretyan, 'Big (Universal) History Paradigm: Versions and Approaches', *Social Evolution and History*, 4 (2005), pp. 61–86.
8. Bailes, *Science and Russian Culture*, p. 65.
9. David Moon, 'The Environmental History of the Russian Steppes: Vasilii Dokuchaev and the Harvest Failure of 1891', *Transactions of the Royal Historical Society* (Sixth Series) 15 (2005), p. 158.
10. The question was not particularly fair. Given the scale of the disaster, the government's measures proved effective. By 1893 the Russian economy was functioning as if the famine had never happened. But by then nobody was in a mood to hear good news. See James Simms, 'The Economic Impact of the Russian Famine of 1891–92', *Slavonic and East European Review* 60, no. 1 (1982), p. 70.
11. Quoted in Kendall E. Bailes, *Science and Russian Culture*, p. 85.
12. Kassow, *Students, Professors, and the State in Tsarist Russia*, p. 217.
13. Ibid., p. 227.
14. Ibid., p. 253.
15. Ibid., p. 266.
16. Ibid., p. 269.
17. Kurt Johansson and A. K. Gastev, *Aleksej Gastev, Proletarian Bard of the Machine Age*, p. 29.
18. Kassow, *Students, Professors, and the State*, p. 356.
19. Bailes, *Science and Russian Culture*, p. 140.
20. R. Fando, 'The Unknown About a Well-known Biologist', *Herald of the Russian Academy of Sciences* 78 (2), (1 April 2008), pp. 165–6.

2 Revolutionaries

1. Johansson & Gastev, *Aleksej Gastev*, p. 65–6.
2. In *Opticks* (1704), his famous work on the colours that make white light, Isaac Newton categorically states that his mechanics will completely and conspicuously fail to describe the colours his prism has thrown upon his wall; all his 'mechanics' can do is measure the width of the rainbow.
3. Marx was not alone. Across the globe, the First World War inspired ideas of 'rational government'. Governments in war had taken control of their

economies – why not in peace? In the USA, Thorstein Veblen wrote *The Engineers and the Price System* (1921), which called for committees (called soviets) to run a 'rational' economy.

4. Quoted in David Joravsky, *Soviet Marxism and Natural Science 1917–1932*, pp. 3–4.
5. Georgii D. Gloveli, '"Socialism of Science" versus "Socialism of Feelings": Bogdanov and Lunacharsky', *Studies in Soviet Thought*, 42, no. 1 (1 July 1991), p. 44.
6. Anthony Mansueto, 'From Dialectic to Organization: Bogdanov's Contribution to Social Theory', *Studies in East European Thought*, 48 (1996), pp. 37–61; and George E. Gorelik, 'Bogdanov's Tektology: Its Nature, Development and Influence', *Studies in Soviet Thought*, 26 (1983), pp. 39–57.
7. For more than this swift caricature, see Arran Gare, 'Aleksandr Bogdanov's History, Sociology and Philosophy of Science', *Studies in History and Philosophy of Science Part A*, 31 (2000), pp. 231–48.
8. 'Lenin' was a Party alias, acquired around 1900.
9. Robert C. Williams, 'Collective Immortality: The Syndicalist Origins of Proletarian Culture, 1905–1910', *Slavic Review* 39, no. 3 (1980), p. 391.
10. Henry Adams, *The Education of Henry Adams: An Autobiography*, p. 347.
11. The Nobel Prize-winning physiologist Ivan Pavlov and his colleagues had a terrible time getting consistent results from their salivating dogs. One young assistant felt he had to resign because the very sight of him caused the dogs to salivate copiously whether they were hungry or not. He went off to Switzerland and became a psychiatrist.
12. Katherine Arens, 'Mach's "Psychology of Investigation"', *Journal of the History of the Behavioral Sciences*, 21 (1985), pp. 151–68. Mach's main biographer is John T. Blackmore: see his *Ernst Mach; His Work, Life, and Influence*.
13. V. I. Lenin, *Materialism and Empirio-Criticism: Critical Comments on a Reactionary Philosophy*, chapter 5, part 2.
14. Maxim W. Gorky, 'V. I. Lenin', trans. David Walters.
15. Ibid.
16. The physicist Yakov Frenkel (1894–1952) said that it 'amounts to little more than the assertion of elementary truths that it's not worth breaking a lance over' and subsequent commentators have been no more kind. Writing in 1991, the historian Paul Josephson delivers a suitable capstone: '*Materialism and Empiriocriticism* is characterised by vulgar materialist generalisations, lengthy diatribes, and excessively long quotations of opponents followed by brief, haughty ridicule. Had the Bolsheviks not succeeded in 1917, it is doubtful that Lenin's book would be read today.' See Paul R. Josephson, *Physics and Politics in Revolutionary Russia*, p. 250.
17. Quoted in Diane Koenker, William Rosenberg and Ronald Suny, *Party, State and Society in the Russian Civil War: Explorations in Social History*, p. 286.
18. Quotation adapted from Bailes, *Science and Russian Culture*, p. 186.
19. Quoted in Bailes, *Technology and Society under Lenin and Stalin*, p. 55.
20. Ibid., p. 56.
21. Bailes, *Science and Russian Culture*, p. 151.

22. Matthew Rendle, 'Revolutionary Tribunals and the Origins of Terror in Early Soviet Russia', *Historical Research*, 84, no. 226 (1 November 2011), pp. 693–721.
23. Haynes & Husan, *A Century of State Murder?*, p. 53.
24. Solomon Volkov, *St Petersburg: A Cultural History*, p. 211.
25. Lennard D. Gerson, *The Secret Police in Lenin's Russia*, p. 183.
26. Douglas R. Weiner, more usually an environmental historian, neatly captures Dzerzhinsky's almost-moral transformation in 'Dzherzhinskii and the Gerd Case: The Politics of Intercession and the Evolution of "Iron Felix" in NEP Russia', *Kritika*, 7, no. 4 (22 September 2006), p. 759.
27. Vaclav Smil, *The Earth's Biosphere: Evolution, Dynamics, and Change*, p. 5.

3 Entrepreneurs

1. David Joravsky, *The Lysenko Affair*, p. 27.
2. No sooner had James P. Goodrich retired as Republican Governor of Indiana than President Herbert Hoover appointed him to run the Russian operations of the American Relief Administration. For more on his observations of Russia, and his lobbying for the Soviet Union, see Dane Starbuck's *The Goodriches: An American Family*.
3. Daniel P. Todes, 'Pavlov and the Bolsheviks', *History and Philosophy of the Life Sciences*, 17, no. 3 (1995), p. 384.
4. A microtome is a tool used to cut extremely thin slices of material, useful in preparing slides for microscopy. (A macrotome is a precision saw used to prepare bigger anatomical sections.)
5. Josephson, *Physics and Politics in Revolutionary Russia*, p. 77.
6. Nikolai L. Krementsov, *Stalinist Science*, p. 14.
7. Quoted in Igor G. Loskutov, *Vavilov and His Institute: A History of the World Collection of Plant Genetic Resources in Russia*, p. 18.
8. Quoted in Peter Pringle, *The Murder of Nikolai Vavilov: The Story of Stalin's Persecution of One of the Great Scientists of the Twentieth Century*, p. 92.
9. Quoted in Loskutov, *Vavilov and his Institute*, p. 23.
10. Leonid Rodin in the introduction to N. I. Vavilov, *Five Continents*, p. xxi.
11. Barbara Walker, 'Kruzhok Culture: The Meaning of Patronage in the Early Soviet Literary World', *Contemporary European History*, 11 (2002), pp. 107–23.
12. Viacheslav Ivanov, 'Why Did Stalin Kill Gorky?', *Russian Social Science Review*, 35, no. 6 (1 November 1994), pp. 52–3.
13. Pringle, *The Murder of Nikolai Vavilov*, p. 85.
14. Speech in opening the Eleventh Congress of the Party, 27 March 1922. In V. I. Lenin, *Collected Works*, Volume 33, pp. 237–42.
15. Loren R. Graham, *Science in Russia and the Soviet Union: A Short History*, p. 89.
16. Ivanov, 'Why Did Stalin Kill Gorky?', p. 55.
17. Lenin to Gorky, 15 September 1919; available from the US Library of Congress at http://1.usa.gov/1NBxNon.
18. Stuart Finkel, 'Purging the Public Intellectual: The 1922 Expulsions from Soviet Russia', *Russian Review*, 62 (2003), p. 589.

19. A. E. Fersman, 'From a speech delivered at the Annual Meeting of the Geographical Institute', *Problems of scholarly organisation in the work of Soviet scientists, 1917–1930* [in Russian], p. 201.

4 Workers

1. 'Narodnaia vypravka' (1922), from Alexei Gastev, *Poeziia rabochego udara* [*Poetry of the Factory Floor*] (Moscow, 1971), p. 258.
2. Kendall E. Bailes, 'Alexei Gastev and the Soviet Controversy over Taylorism, 1918–24', *Soviet Studies* 29, no. 3 (1 July 1977), pp. 373–94.
3. 'From a Tram-Driver's Diary' (1910), in Johansson & Gastev, *Aleksej Gastev*, p. 26.
4. Williams, 'Collective Immortality: The Syndicalist Origins of Proletarian Culture, 1905–1910', p. 395.
5. Stites, *Revolutionary Dreams*, pp. 150–1.
6. Patricia Carden, 'Utopia and Anti-Utopia: Aleksei Gastev and Evgeny Zamyatin', *Russian Review* 46, no. 1 (1987), p. 5.
7. Ibid. p. 4.
8. The Bolsheviks sent 'agitprop' trains, trucks and ships into Russia's remotest corners. Public speakers, writers, books and pamphlets, even printing presses and movie projectors, were used in an effort to spread information about the new regime.
9. Quotation adapted from Johansson & Gastev, *Aleksej Gastev*, p. 68.
10. Evgenij I. Zamyatin, *We*, trans. Bernard Guilbert Guerney and Michael Glenny, p. 181. It is likely that Zamyatin came upon *We*'s other big literary influence here – 'The New Utopia', a remarkably creepy short story by Jerome K. Jerome, better known as the author of *Three Men in a Boat*.
11. Alexei Gastev, *Poetry of the Factory Floor*, p. 258.
12. Johansson & Gastev, *Aleksej Gastev*, p. 105.
13. Ibid. p. 110. Oblomov, the eponymous hero of a nineteenth-century novel by Ivan Goncharov, is a young but feckless nobleman who takes fifty pages to get from his bed to his chair. He is not ill. He is simply incapable of making the smallest decision.
14. V. I. Lenin, 'The Taylor System – Man's Enslavement by the Machine', *Put Pravdy* No. 35 (13 March 1914). In *Lenin: Collected Works* Vol. 20, pp. 152–154; available from Marxists Internet Archive at http://bit.ly/1OZnxrU.
15. N. Krupskaya, 'Sistema Teilora i organizatsiya raboty sovetskikh uchrezhdenii' ['The Taylor System and Soviet Labour Practices'], *Krasnaya Nov'*, Vol. 1 (1921), pp. 140–6.
16. Quoted in R. S. Schultz, 'Industrial Psychology in the Soviet Union', *Journal of Applied Psychology* 19, no. 3 (1935), p. 269.
17. Quoted in Vera L. Talis, 'New Pages in the Biography of Nikolai Alexandrovich Bernstein', *Anticipation: Learning from the Past* (Hanse-Wissenschaftskolleg Institute for Advanced Study, 2014), p. 2.
18. Gastev's son talked about his father's Social Engineering Machine in 'The Engineers' Plot', the opening episode of Adam Curtis's unmissable, so-concise-as-to-be-paranoid TV documentary *Pandora's Box* (BBC, 1992).

19. Bailes, 'Alexei Gastev and the Soviet Controversy over Taylorism, 1918–24', pp. 373–94; also Daniel A. Wren, and Arthur G. Bedeian, 'The Taylorization of Lenin: Rhetoric or Reality?', *International Journal of Social Economics* 31, no. 3 (1 March 2004), pp. 287–99. For a while, the Taylorist discipline of 'psychotechnics' promised to put the workplace under rational, enlightened management. For a contemporary account see Schultz, 'Industrial Psychology in the Soviet Union', pp. 265–308. Zenovia A. Sochor provides a more recent perspective in 'Soviet Taylorism Revisited', *Soviet Studies* 33, no. 2 (1981), pp. 246–64.

20. Quoted in Jean-Gaël Barbara and Jean-Claude Dupont, *History of the Neurosciences in France and Russia: From Charcot and Sechenov to IBRO* (Hermann, 2011), p. 188.

21. Julia Kursell, 'Piano Mécanique and Piano Biologique: Nikolai Bernstein's Neurophysiological Study of Piano Touch', *Configurations* 14, no. 3 (2006), pp. 245–73.

22. Irina E. Sirotkina and Elena V Biryukova, 'Futurism in Physiology: Nikolai Bernstein, Anticipation, and Kinaesthetic Imagination', *Anticipation: Learning from the Past: The Russian/Soviet Contributions to the Science of Anticipation*, ed. Mihai Nadin.

23. Onno G. Meijer and Sjoerd M. Bruijn, 'The Loyal Dissident: N. A. Bernstein and the Double-Edged Sword of Stalinism', *Journal of the History of the Neurosciences* 16, no. 1–2 (2007), p. 209.

24. N. A. Bernstein, 'Biomekhanika i fiziologiya dvizheniy' ['Biomechanics and Physiology of Movement'], *Izbrannye Psikhologicheskie Trudy* [*Selected Psychological Works*], p. 462.

25. Alex Kozulin, *Psychology in Utopia: Toward a Social History of Soviet Psychology* (MIT Press, 1984), p. 67.

27. Ibid., p. 65.

5 Exploring the mind

1. Lev Vygotsky, 'Soznanie kak problema psikhologii povedeniia' ['Consciousness as a Problem of Behavioural Psychology'; 1925], *Sobranie sochinenii*, vol. 1 (Pedagogika, 1982), pp. 78–98.

2. Daniel P. Todes, *Pavlov's Physiology Factory: Experiment, Interpretation, Laboratory Enterprise*, p. 57.

3. David Joravsky, *Russian Psychology: A Critical History*, p. 80.

4. S. V. Anichkov, 'How I Became a Pharmacologist', *Annual Review of Pharmacology* 15 (1975), pp. 1–10.

5. William James, *The Principles of Psychology*, Vol. 1 (Holt, 1890), p. 192.

6. William James, *Psychology: The Briefer Course* (Courier Corporation, 2012), p. 335.

7. *Trotsky's Notebooks, 1933–1935, Writings on Lenin, Dialectics and Evolutionism*, trans. and intro. Philip Pomper, p. 49.

8. Leon Trotsky, 'Culture and Socialism' [1927], *Problems of Everyday Life: And Other Writings on Culture and Science*, p. 298.

9. Galina Kichigina, *The Imperial Laboratory: Experimental Physiology and Clinical Medicine in Post-Crimean Russia.*

10. Bekhterev's energetic, scattergun approach to his subject is neatly caught in M. A. Akimenko, 'Vladimir Mikhailovich Bekhterev', *Journal of the History of the Neurosciences*, 16 (2007). Bekhterev's incredible capacity for work was quite the equal of Pavlov's. Between his lectures he ran hypnosis sessions in the auditorium next door. His evening consultations ran into the small hours of the night, and could involve seeing forty-odd patients.

11. Ivan P. Pavlov, *Polnoe Sobranie Sochinenni* [*Complete Works*], vol. 3, part 1, p. 23.

12. B. F. Lomov, V. A. Koltsova and E. I. Stepanova, 'Ocherk Zhizni i nauchnoi deyatelnosti Vladimira Mikhailovicha Bekhtereva' ['Essay of Life and Works of V. M. Bekhterev'], *Ob'ektivnaya Psykhologia* [*Objective Psychology*], ed. V. A. Koltsova (Nauka, 1991), pp. 424–44. For more about the men's rivalry, see Robert Boakes, *From Darwin to Behaviourism: Psychology and the Minds of Animals*; and Slava Gerovitch, 'Love–Hate for Man–Machine Metaphors in Soviet Physiology: From Pavlov to "Physiological Cybernetics"', *Science in Context*, 15 (2002), pp. 339–74.

13. Todes, 'Pavlov and the Bolsheviks', pp. 386–7.

14. I have slightly adapted this quotation from Abbott Gleason and Richard Stites, *Bolshevik Culture: Experiment and Order in the Russian Revolution*, p. 102.

15. Todes, 'Pavlov and the Bolsheviks', p. 392.

16. After his dismissal, Chelpanov found work at the State Academy of Aesthetic Sciences, lecturing and publishing on aesthetic perception, 'primitive' creativity, children's drawings and related themes. The academy was closed in 1930, leaving him out of a job once again. The misfortunes of his last years left Chelpanov embittered. One of his daughters died, another emigrated, and his son was arrested and shot in the Great Purge. Chelpanov died in poverty in 1936.

17. Irina Sirotkina, 'When Did "Scientific Psychology" Begin in Russia?', *Physis; Rivista Internazionale Di Storia Della Scienza* (2006), 43, pp. 1–2.

18. The Bolsheviks' increasingly fraught relationship with psychoanalysis is the subject of Alexander M. Etkind's *Eros of the Impossible: The History of Psychoanalysis in Russia*. See also Martin A. Miller, 'Freudian Theory under Bolshevik Rule: The Theoretical Controversy during the 1920s', *Slavic Review*, 44 (1985), pp. 625–46; and Alberto Angelini, 'History of the Unconscious in Soviet Russia: From Its Origins to the Fall of the Soviet Union', *International Journal of Psychoanalysis*, 89 (2008), pp. 369–88.

19. A. R. Luria, *The Autobiography of Alexander Luria: A Dialogue with The Making of Mind*, p. 22.

20. Ibid., p. 5.

21. Ibid., pp. 38–9.

22. Early collaborations between Lev Vygotsky and Alexander Luria are explored in T. V. Akhutina, 'L. S. Vygotsky and A. R. Luria: Foundations of Neuropsychology', *Journal of Russian and East European Psychology*, 41 (2010), pp. 159–90.

23. Vygotsky's student thesis formed the basis of one of his better-known books, *The Psychology of Art* (1925). Published in English translation by MIT Press in 1971, it is available online from Marxists Internet Archive at http://bit.ly/1QzCg1f.
24. Kozulin, *Psychology in Utopia*, p. 82.
25. Ibid., p. 108
26. Gita L. Vygodskaya, '[Lev Vygotsky] His Life', *School Psychology International*, vol 16, no. 2 (1 May 1995), pp. 105–16.
27. There has been a lot of recent work published about Sabina Spielrein, and even a movie, *A Dangerous Method*, directed by David Cronenberg. Few champions are as spirited as Jerry Aldridge: see 'Another Woman Gets Robbed? What Jung, Freud, Piaget, and Vygotsky Took from Sabina Spielrein', *Childhood Education*, 85 (2009), p. 318.
28. Frank Brenner, 'Intrepid Thought: Psychoanalysis in the Soviet Union'; available at http://bit.ly/20qQK6i.
29. Vygotsky to Lev Sakharov, 15 February 1926, in L. S. Vygotsky, *The Vygotsky Reader*.
30. Quoted in Joravsky, *Russian Psychology*, p. 229.
31. Miller, 'Freudian Theory under Bolshevik Rule: The Theoretical Controversy during the 1920s', p. 644
32. Quoted in Akhutina, 'L. S. Vygotsky and A. R. Luria', p. 169.
33. A. R. Luria and L. S. Vygotsky, *Ape, Primitive Man, and Child: Essays in the History of Behavior*, p. 15.
34. Ibid. pp. 77–8.
35. Luria's best account of his Uzbek expedition is in *Cognitive Development*, based on the Russian version published in 1974.
36. V. Nell, 'Luria in Uzbekistan: The Vicissitudes of Cross-Cultural Neuropsychology', *Neuropsychology Review* 9, no. 1 (1999), p. 51.
37. L. S. Vygotsky, 'Letters to Students and Colleagues', *Journal of Russian and East European Psychology*, vol. 45, no. 2 (2007), pp. 11–60; quoted in Jennifer Fraser and Anton Yasnitsky, 'Deconstructing Vygotsky's Victimization Narrative: A Re-Examination of the "Stalinist Suppression" of Vygotskian Theory', n. 43.

6 Understanding evolution

1. In Conway Zirkle (ed.), *Death of a Science in Russia: The Fate of Genetics as Described in Pravda and Elsewhere*, p. 51
2. This objection to natural selection – that useful novelties would vanish like a drop in the ocean long before natural selection had a chance to make them spread – was first voiced in a review of *The Origin of Species* by Fleeming Jenkin, a professor of engineering at Edinburgh University, whose remarks so spooked Darwin that he inserted a new chapter into the sixth edition of *The Origin*, resuscitating Lamarck's ideas about the inheritance of acquired characteristics. See Arthur Koestler, *The Case of the Midwife Toad*, p. 43.
3. The effort was maintained throughout the Soviet era. In 1947 a visiting academic, Eric Ashby, reckoned that 'not even in America, is there such

a widespread interest in science among the common people as there is in Russia. Science is kept before the people through newspapers, books, lectures, films, exhibitions in parks and museums, and through frequent public festivals in honour of scientists and their discoveries. There is even an annual 'olympiad' of physics for Moscow schoolchildren in which they solve problems in a competition of the lines of a "knock-out" tennis tournament . . .' (Eric Ashby, *Scientist in Russia*, p. 186).

4. Édouard Herriot, *Eastward from Paris*, cited in Ashby, *Scientist in Russia*, p. 186.

5. S. P. Fyodorov, 'Surgery at the Crossroads', *Journal of Surgery*, ed. I. I. Grekov [in Russian] (1996), vol. 155, no. 6, p. 114.

6. Billed as the 'trial of the century,' the trial of John Thomas Scopes in Dayton, Tennessee, received international coverage. Scopes was convicted of teaching evolution in a state-funded school, in contravention of Tennessee's Butler Act of 1925, and fined $100. The verdict was overturned on appeal, and is remembered as a victory for proponents of evolution. In practical terms, it was anything but: the Butler statute was upheld as constitutional, and several other states followed suit and banned the teaching of evolution in the years that followed.

7. This nice turn of phrase is not mine. It belongs to a soldier in the First World War, responding to one of William Bateson's lectures to the troops on Darwinism. Bateson, a pioneering geneticist, dubbed it 'a flash of illiterate inspiration'. See Koestler, *The Case of the Midwife Toad*, p. 20.

8. Quotation adapted from A. E. Gaissinovitch, 'The Origins of Soviet Genetics and the Struggle with Lamarckism, 1922–1929', trans. Mark B. Adams, *Journal of the History of Biology* 13, no. 1 (1980), p. 17.

9. Ibid. p. 48.

10. Koestler, *The Case of the Midwife Toad*, p. 12.

11. Ibid.

12. Ibid., p. 1.

131. Quoted in David Joravsky, 'Soviet Marxism and Biology before Lysenko', *Journal of the History of Ideas* 20, no. 1 (1959), p. 93.

14. Serebrovsky was born into the family of a leftist architect acquainted with Alexander Bogdanov and the future commissar of education Anatoly Lunacharsky.

15. I. P. Pavlov, 'New Research on Conditioned Reflexes,' *Science*, 58, no. 1506 (1923), pp. 359–361. Cited in N. F. Suvorov and V. N. Andreeva, 'Problems of the Inheritance of Conditioned Reflexes in Pavlov's School', *Neuroscience and Behavioral Physiology* 21, no. 1 (1 January 1991), p. 9.

16. V. V. Babkov, 'The Theoretical–Biological Concept of Nikolai K. Kol'tsov', *Russian Journal of Developmental Biology*, 33 (2002), pp. 255–62. See also Sergei Fokin, 'Russian Zoologists in Naples', *Science in Russia* (2010), pp. 71–8.

17. D. A. Granin, *The Bison: A Novel About the Scientist Who Defied Stalin*, pp. 45–6.

18. Robert E. Kohler captures both the modest resources and global importance of that room in *Lords of the Fly: Drosophila Genetics and the Experimental Life*.

19. S. S. Chetverikov, 'Volny Zhizni' ['Waves of Life'], *Dnevnik zootdeleniia*, vol. 3, no. 6 (1905). Quoted in Mark B. Adams, 'Towards a Synthesis: Population Concepts in Russian Evolutionary Thought, 1925–1935', *Journal of the History of Biology*, 3, no. 1 (1970), pp. 107–29. For more on Chetverikov's work, see Adams's paper 'The Founding of Population Genetics: Contributions of the Chetverikov School 1924–1934', *Journal of the History of Biology* 1, no. 1 (1968), pp. 23–39.

20. For Muller, genes were the be-all and end-all of all life processes; he conjectured that they 'arose coincidentally with growth and "life" itself'. Thomas Hunt Morgan, on the contrary, absolutely refused to speculate about the nature of the gene. In his Nobel Prize lecture, delivered in 1931, he said: 'At the level at which the genetic experiments lie, it does not make the slightest difference whether the gene is a hypothetical unit, or whether the gene is a material particle.' This constant evasion drove his student Muller to distraction. See 'Thomas H. Morgan – Nobel Lecture: The Relation of Genetics to Physiology and Medicine', Nobelprize.org. Nobel Media AB 2014; available from Nobelprize.org at http://bit.ly/1SlUfs1.

21. The 'Sovmestnoe Oranie' shortened itself to the acronym 'So-or', added 'Droz' (for 'drosophilists') as a prefix, and is often referred to as the 'drozsoor'.

22. S. S. Chetverikov, 'On Certain Aspects of the Evolutionary Process from the Standpoint of Modern Genetics,' *Proceedings of the American Philosophical Society*, 105, no. 2 (21 April 1961), p. 171.

23. A tip of the hat to Charles Darwin. There are fifteen species of finches in the Galapagos Islands. Darwin came across them in 1835 and had his servant shoot samples for HMS *Beagle*'s collection. Their significance for his evolutionary theory dawned only later. The most important differences between the finches on the Galapagos are in the size and shape of their beaks; each is adapted to a different food source.

24. Marina Bentivoglio, 'Cortical Structure and Mental Skills: Oskar Vogt and the Legacy of Lenin's Brain', *Brain Research Bulletin* 47, no. 4 (1998), pp. 291–6.

25. History has not been particularly interested in Cécile, a pioneering investigator of Huntington's disease; her work tends to get folded in under her husband's name. Contemporaries were more canny: fellow neurologist Igor Klutzo remembered her as 'extremely intelligent, probably the most intelligent person I have met in my life'. (See G. W. Kreutzberg, 'If You Had Met Him, You Would Know', *Brain Pathology* 2, no. 4 [1 October 1992], pp. 365–9.) Her husband was only too happy to concede the point: Cécile is cited as the main author on many of the couple's joint papers.

26. Jochen Richter, 'Pantheon of Brains: The Moscow Brain Research Institute 1925–1936', *Journal of the History of the Neurosciences*, 16 (2007), pp. 138–49

27. Bentivoglio, 'Cortical Structure and Mental Skills', p. 293.

28. The second part of Nikolai's name, 'Ressovsky', indicated the place where he was born. The elder sons of the nobility often sported such double names, though few persisted with the affectation after 1917.

29. N. V. Timofeev-Ressovsky, *Vospominaniia: Istorii, rasskazannye im samim,*

s pis'mami, fotografiiami i dokumentami [*The Stories Told by Himself with Letters, Photos and Documents*], p. 106.

30. Granin, *Bison*, p. 56.

7 Shaping humanity

1. Quoted in Mark B. Adams, *The Wellborn Science: Eugenics in Germany, France, Brazil, and Russia*, p. 162.
2. Quoted in Karl Pearson, *The Life, Letters and Labours of Francis Galton*, p. 78.
3. Edward J. Larson, 'Biology and the Emergence of the Anglo-American Eugenics Movement', in Denis Alexander and Ronald L. Numbers, *Biology and Ideology from Descartes to Dawkins*, p. 173.
4. Nikolai Krementsov, 'From "Beastly Philosophy" to Medical Genetics: Eugenics in Russia and the Soviet Union', *Annals of Science* 68, no. 1 (January 2011), p. 66.
5. Speech at the annual meeting of the Russian Eugenics Society on 20 October 1921. Published in 1922 in *Russkiy evgenicheskiy zhurnal* [*Russian Eugenics Journal*] 1(1), pp. 3–27. English translation in V. V. Babkov, *The Dawn of Human Genetics*, p. 71.
6. B. M. Zavadovsky in *Pod znamenem marksizma* [*Under the Banner of Marxism*], vol. 10 no. 11 (1925), pp. 100, 106. English translation in Babkov, *The Dawn of Human Genetics*, p. 475.
7. Karl Günter Zimmer, Max Delbrück and N. V. Timofeev-Resovskii, *Creating a Physical Biology: The Three-Man Paper and Early Molecular Biology*, p. 20.
8. H. J. Muller, 'The Measurement of Gene Mutation Rate in *Drosophila*, its High Variability, and its Dependence upon Temperature', *Genetics* vol. 1, no. 3 (1928), pp. 279–357.
9. Quoted in Gaissinovitch, *The Origins of Soviet Genetics*, p. 49.
10. In *Mediko-biologicheskiy zhurnal* [*Medical–Biological Journal*], no. 4–5 (1929). English translation in Babkov, *The Dawn of Human Genetics*, pp. 552–65.
11. Ibid., pp. 505–16.
12. By 1959 there were pop songs. 'Hold on, geologist', ran one popular number, 'hold out, geologist, you are brother of the wind and sun!' See Alla Bolotova, 'Colonization of Nature in the Soviet Union. State Ideology, Public Discourse, and the Experience of Geologists', *Historische Sozialforschung* [*Historical Social Research*] 29, part 3 (2004), pp. 104–23.
13. Bedny's story is fascinating in itself: see Robert Horvath, 'The Poet of Terror: Dem'ian Bednyi and Stalinist Culture', *Russian Review* 65, no. 1 (1 January 2006), pp. 53–71.
14. One obscure regional official got all hot under the collar at the possibilities, declaring, 'Every girl above the age of eighteen is hereby declared to be state property. Every unmarried girl who has reached the age of eighteen is obliged, on pain of severe penalty, to register with the Free Love Office of the Welfare Commissariat. A woman registered with the Free Love Office has the right to choose from men who have reached the age of eighteen.

Interested persons may choose a husband or wife once a month . . . The offspring of such collaboration become the property of the Republic.' See Kozulin, *Psychology in Utopia*, p. 84.

15. Kirill Rossiianov, 'Beyond Species: Il'ya Ivanov and His Experiments on Cross-Breeding Humans with Anthropoid Apes', *Science in Context* 15, no. 2 (2002), p. 286.

16. One, 'G' from Leningrad, wrote to Ivanov on 16 March 1928: 'Dear Professor . . . With my private life in ruins, I don't see any sense in my further existence . . . But when I think that I could do a service for science, I feel enough courage to contact you. I beg you, don't refuse me . . . I ask you to accept me for the experiment.' The attempt had to be postponed in June 1929: Ivanov cabled to G, 'The orang has died, we are looking for a replacement.' But nothing came of it. See Rossiianov, 'Beyond Species', pp. 306–7.

17. Gaissinovitch, *The Origins of Soviet Genetics*, p. 17.

18. In *Konferentsiya po meditsinskoy genetike* [*The Conference on Medical Genetics*], 1934, p. 16. English translation in Babkov, *The Dawn of Human Genetics*, p. 546.

PART TWO : POWER

8 'Storming the fortress of science'

1. Josif Stalin, 'Rech' Na VIII S'ezde VLKSM' ['Speech at the Eighth Congress of the Communist Youth], *Sobranie Sochinenii* [*Collected Works*], vol. 11, p. 77; quoted in Krementsov, *Stalinist Science*, pp. 29–30.

2. Finkel, 'Purging the Public Intellectual', p. 604.

3. Kendall E. Bailes, 'The Politics of Technology: Stalin and Technocratic Thinking among Soviet Engineers', *American Historical Review* 79, no. 2 (1 April 1974), p. 462.

4. Hiroaki Kuromiya, 'The Crisis of Proletarian Identity in the Soviet Factory, 1928–1929', *Slavic Review* 44, no. 2 (1985), pp. 286–7.

5. Bailes, *Technology and Society under Lenin and Stalin*, p. 170.

6. Weiner, 'Dzherzhinskii and the Gerd Case', n. 125.

7. Quoted in Bailes, *Technology and Society under Lenin and Stalin*, p. 81.

8. Kendall E. Bailes puts the many grotesque absurdities of that trial in political context in 'The Politics of Technology: Stalin and Technocratic Thinking among Soviet Engineers'.

9. Eugene Lyons, *Assignment in Utopia*, p. 127.

10. Sheila Fitzpatrick, *Cultural Revolution in Russia, 1928–1931*, p. 10.

11. Quoted in Bailes, *Technology and Society under Lenin and Stalin* , p. 88.

12. Sheila Fitzpatrick, *The Cultural Front: Power and Culture in Revolutionary Russia*, p. 54.

13. A. V. Lunacharsky, 'Intelligentsiia i ee mesto v sotsialisticheskom stroitel'stve' ['The Intelligentsia and its Role in Soviet Construction'], *Revoliutsiia i kultura*, no. 1 (15 November 1927), p. 32–3.

14. James T. Andrews, *Science for the Masses: The Bolshevik State, Public Science, and the Popular Imagination in Soviet Russia, 1917–1934*, p. 138.

15. Following his exile, in 1935 Chetverikov reestablished a biological career in the city of Gorki, and devoted himself to research on silkworms. In the winter of 1937 he was approached by the scientific secretary of the Ministry of Agriculture with the proposal that he breed silkworms for Russian conditions: silk was needed for parachutes, and Japan was no longer regarded as a reliable supplier. In 1944 a monograph summarising six years of research was sent to the Lenin Academy for publication. The Supreme Soviet awarded him a medal of excellence. It did not save him: in 1948 he was fired from his job at Gorki University and at the time of this death in 1959, he was penniless, blind and virtually forgotten.

16. Andrews, *Science for the Masses*, p. 139.

17. Michael David-Fox, 'Symbiosis to Synthesis: The Communist Academy and the Bolshevization of the Russian Academy of Sciences, 1918–1929', *Symbiosis*, 219 (1998), p. 43.

18. Aleksey E. Levin, 'Expedient Catastrophe: a Reconsideration of the 1929 Crisis at the Soviet Academy of Science', *Slavic Review* 47, no. 2 (1988), p. 265.

19. Todes, 'Pavlov and the Bolsheviks', pp. 399–400.

20. Daniel P. Todes, *Ivan Pavlov: A Russian Life in Science*, p. 577.

9 Eccentrics

1. Karl Marx, 'Private Property and Communism', *Economic and Philosophic Manuscripts of 1844*, pp. 135–43.

2. Kees Boterbloem, *The Life and Times of Andrei Zhdanov, 1896–1948*, p. 65.

3. Quoted in Fitzpatrick, *The Cultural Front*, p. 112.

4. Bailes, *Technology and Society under Lenin and Stalin*, p. 363.

5. Joravsky, *The Lysenko Affair*, p. 42.

6. Ibid., p. 49.

7. Ibid., p. 80.

8. Ashby, *Scientist in Russia*, p. 33.

9. At the meetings Stalin instigated at the Society of Materialist Biologists Boris Tokin, an embryologist and the society's newly appointed chairman, asserted that 'the extreme Morganists have become prisoners of modern bourgeois genetics'. He accused them of accepting the 'precedence of the gene' and of having a noncritical 'view of Weismann's teaching on the continuity of germplasm'. During the late 1800s and early 1900s, the German cytologist August Weismann had argued that while most of the body ('the soma') undergoes changes while alive, the hereditary material – which he labelled the 'germ plasm' – remains unchanged more or less indefinitely.

10. Vadim J. Birstein, *The Perversion of Knowledge: The True Story of Soviet Science*, pp. 257–8.

11. Douglas R. Weiner, 'Community Ecology in Stalin's Russia: "Socialist" and "Bourgeois" Science', *Isis* 75, no. 4 (1984), p. 694.

12. Nils Roll-Hansen, *The Lysenko Effect: The Politics of Science*, p. 87.

13. Dialectical materialism recognises no absolute or fixed categories. The world is not a ready-made thing: it is a complex interplay of processes, all of which are ebbing and flowing, coming into being and passing away. (Its author,

Friedrich Engels, had great fun with the duckbilled platypus, asking where *that* fitted into any rigid scheme of things.) There is no 'essence' hiding behind a cloud of steam, a puddle of water or a block of ice. There are only structures, succeeding each other in response to changes in the local conditions. In the same way, there is no 'essence' to thought. No soul. No recording angel. Just a dance of interactions among processes that spread out in a web through an individual's whole nervous system, across society, and into the physical world. Changes stick, and time moves inexorably forward. To put it another way (and this is what makes dialectical materialism such strong medicine for the sciences) time is real and everything has a history. As a scientist at the turn of the twentieth century, and depending on your field, this claim would have struck you as (for biologists) self-evident, (for chemists) disturbing, or (for the physicists of those days) plain mad. Ernst Mayr's short, sympathetic account, 'Roots of Dialectical Materialism', appears in *Sovetskaia Biologiaa v 20–30kh Godakh*, ed. Edouard I. Kolchinsky (1997), pp. 12–17; also available online at http://bit.ly/1mhUW8Y.

14. Nikolai Krementsov, 'Darwinism, Marxism, and Genetics in the Soviet Union', in Alexander & Numbers, *Biology and Ideology from Descartes to Dawkins*.
15. Joravsky, *The Lysenko Affair*, p. 238.
16. Ted Benton, *The Greening of Marxism*, p. 124.
17. Douglas R. Weiner, *Models of Nature: Ecology, Conservation, and Cultural Revolution in Soviet Russia*, p. 182.
18. Boris Konstantinovich Fortunatov, 'O general'nom plane rekonstruktsii promyslovoi fauny evropeiskoi chasti SSSR i Ukrainy', *Priroda i sotsialisticheskoe khoziaistvo* (1933), no. 6, pp. 90–109.
19. Weiner, 'Community Ecology in Stalin's Russia', pp. 684–96.

10 The primacy of practice

1. Kendall E. Bailes, 'Soviet Science in the Stalin Period: The Case of V. I. Vernadskii and His Scientific School, 1928–1945', *Slavic Review* 45, no. 1 (1986), pp. 33–4.
2. Margarita Fofanova, *O Vladimire Ilyiche Lenine: Vospominaniya* [*About Vladimir Ilyich Lenin. Reminiscences*], pp. 175–6.
3. Nils Roll-Hansen, 'Wishful Science: The Persistence of T. D. Lysenko's Agrobiology in the Politics of Science', *Intelligentsia Science: The Russian Century, 1860–1960*, p. 171.
4. Maurice Hindus, 'Henry Ford Conquers Russia', *The Outlook* (29 June 1927), p. 282. See also Kendall E. Bailes, 'The American Connection: Ideology and the Transfer of American Technology to the Soviet Union, 1917–1941', *Comparative Studies in Society and History*, 23 (1981), pp. 421–48.
5. L. C. Dunn, 'Soviet Biology', *Science*, New Series, 99, no. 2561 (28 January 1944), pp. 65–7.
6. Pringle, *The Murder of Nikolai Vavilov*, p. 36.
7. Joravsky, *The Lysenko Affair*, p. 31.
8. F. Kh. Bakhteyev, 'Reminiscences of N. I. Vavilov (1887–1943) on the

Eightieth Anniversary of His Birthday', *Theoretical and Applied Genetics: International Journal of Plant Breeding Research* 38, no. 3 (1968), p. 81.

9. Joravsky, *The Lysenko Affair*, p. 25.
10. Roll-Hansen, *The Lysenko Effect*, p. 90.
11. Pringle, *The Murder of Nikolai Vavilov*, p. 153.
12. 'Russia: Collective Congress', *Time*, 25 February 1935; available at http://www.time.com/time/magazine/article/0,9171,754543,00.html.
13. Quoted in Zhores A. Medvedev, *The Rise and Fall of T. D. Lysenko*, p. 11.
14. Semen Reznik, *Nikolai Vavilov*, p. 267.
15. The richness of that literature – soon to be subsumed, championed (and, later, caricatured) as 'Michurinism' – is made evident in Douglas R. Weiner's 'The Roots of "Michurinism": Transformist Biology and Acclimatization as Currents in the Russian Life Sciences', *Annals of Science*, 42 (1985), pp. 243–60.
16. Joravsky, *The Lysenko Affair*, p. 190.
17. Roll-Hansen, *The Lysenko Effect*, p. 118.
18. Vavilov, *Five Continents*, p. 58.
19. Roll-Hansen, *The Lysenko Effect*, p. 135.

11 Kooperatorka

1. Quoted in Valery N. Soyfer, 'New Light on the Lysenko Era', *Nature* 339, no. 6224 (1989), p. 417.
2. V. Safonov, *Land in Bloom*.
3. Audra J. Wolfe, 'What Does It Mean to Go Public? The American Response to Lysenkoism, Reconsidered', *Historical Studies in the Natural Sciences* 40, no. 1 (2010), p. 53.
4. William deJong-Lambert, 'Hermann J. Muller, Theodosius Dobzhansky, Leslie Clarence Dunn, and the Reaction to Lysenkoism in the United States', *Journal of Cold War Studies* 15, no. 1 (2013), p. 87.
5. Barry Mendel Cohen, 'Nikolai Ivanovich Vavilov: The Explorer and Plant Collector', *Economic Botany* 45 (1969), pp. 38–46.
6. *Proceedings of the International Congress of Genetics* (1932), p. 150.
7. Gary Paul Nabhan, *Where Our Food Comes from: Retracing Nikolay Vavilov's Quest to End Famine*, p. 129.
8. Roll-Hansen, *The Lysenko Effect*, p. 134.
9. Quoted in Loskutov, *Vavilov and His Institute*, p. 97.
10. N. L. Krementsov, *International Science between the World Wars: The Case of Genetics*, p. 41.
11. Nabhan, *Where Our Food Comes From*, p. 164.
12. James F. Crow, 'Sixty Years Ago: The 1932 International Congress of Genetics', *Genetics* 131, no. 4 (1992), pp. 761–8.
13. Leslie C. Dunn, *The Reminiscences of Leslie Clarence Dunn*.
14. William deJong-Lambert, *The Cold War Politics of Genetic Research: An Introduction to The Lysenko Affair*, p. 32.
15. Burds, *Peasant Dreams and Market Politics*, p. 91.
16. Maurice G. Hindus, *Red Bread: Collectivization in a Russian Village*, p. 53.

17. Fitzpatrick, *Cultural Revolution in Russia, 1928–1931*, p. 2.
18. As this book went to press, Loren Graham's *Lysenko's Ghost* (2016) appeared, offering a fascinating re-evaluation of Soviet "Lamarckism".
19. Joravsky, *Soviet Marxism and Natural Science*, p. 245.
20. Pringle, *The Murder of Nikolai Vavilov*, p. 177.
21. Soyfer, 'New Light on the Lysenko Era', p. 417.
22. Roll-Hansen, *The Lysenko Effect*, p. 162.
23. Quoted in Loskutov, *Vavilov and His Institute*, p. 97.
24. Roll-Hansen, *The Lysenko Effect*, p. 167.
25. O. M. Targulian, *Controversial questions on genetics and selection: transactions of the 4th session of the Academy, 19–27 December 1936*; quoted in Medvedev, *The Rise and Fall of T. D. Lysenko*, p. 29.
26. Joravsky, *The Lysenko Affair*, p. 90.
27. Roll-Hansen, *The Lysenko Effect*, p. 175.
28. Ibid. p. 181.
29. Quoted in Medvedev, *The Rise and Fall of T. D. Lysenko*, pp. 16–17.
30. Valery N. Soyfer, 'Tragic History of the VII International Congress of Genetics', *Genetics* 165, no. 1 (September 2003), p. 4.

12 The great patron

1. In Joseph Stalin, *Problems of Leninism*, p. 213.
2. Ivanov, 'Why Did Stalin Kill Gorky?', p. 61.
3. Maxim Gorky, *The White Sea Canal: Being an Account of the Construction of the New Canal between the White Sea and the Baltic Sea*.
4. Maxim Gorky, *Sobranie sochinenij* [*Collected Works*], vol. 27, p. 43. Bolotova, 'Colonization of Nature in the Soviet Union', p. 110.
5. Schultz, 'Industrial Psychology in the Soviet Union', p. 265.
6. Helen Rappaport, *Joseph Stalin: a Biographical Companion*, p. 258.
7. V. Andrle discusses the political uses of Stakhanovism in 'How Backward Workers Became Soviet: Industrialization of Labour and the Politics of Efficiency under the Second Five-Year Plan, 1933–1937', *Social History*, 10 (1985), pp. 147–69. Aside from anything else, the ferment took workers' minds off their money troubles: over the course of the First Five-Year Plan, the average industrial worker's wage plummeted.
8. One intriguing attempt to settle Siberia with volunteers was the creation of Birobidzhan, a region on the Amur River in the Far East, which was established as a homeland for the Jews of the USSR in the 1920s. Many sincere and elaborate inducements were offered; still, Birobidzhan failed.
9. Edwin Bacon, *The Gulag at War: Stalin's Forced Labour System in the Light of the Archives*, p. 48.
10. Ivanov, 'Why Did Stalin Kill Gorky?', p. 64.
11. By spinning a tale of murder and conspiracy two years after Gorky's death, Stalin also managed to get rid of one of the doctors who had refused to put his name to the post mortem issued on the death of his wife, Nadezhda Alliluyeva. Pletnev had been reluctant to accept that Nadezhda had died from appendicitis. She had in fact shot herself in the head.

12. Simon Sebag Montefiore, *Stalin: The Court of the Red Tsar*, p. 553.

13. Holloway, *Stalin and the Bomb*, p. 27.

14. See Robert A. McCutcheon, 'The 1936–1937 Purge of Soviet Astronomers', *Slavic Review*, 50 (1991), pp. 100–17; and A. I. Eremeeva, 'Political Repression and Personality: The History of Political Repression Against Soviet Astronomers', *Journal for the History of Astronomy* 26 (1995), p. 297.

15. Jeroen Van Dongen, 'On Einstein's Opponents, and Other Crackpots', *Studies in History and Philosophy of Science Part B: Studies in History and Philosophy of Modern Physics*, 41 (2010), p. 78.

16. Milena Wazeck, *Einstein's Opponents: The Public Controversy about the Theory of Relativity in the 1920s*, pp. 59, 250.

17. Josephson, *Physics and Politics in Revolutionary Russia*, pp. 229–31.

18. Loren R. Graham, 'The Socio-Political Roots of Boris Hessen: Soviet Marxism and the History of Science', *Social Studies of Science* 15, no. 4 (1985), p. 711.

19. N. I. Vavilov, 'The Problem of the Origin of the World's Agriculture in the Light of the Latest Investigations', *Science at the Crossroads*; available at https://www.marxists.org/subject/science/essays/vavilov.htm. This and the other papers were edited by the leader of the delegation, Nikolai Bukharin, in the space of just a fortnight, for immediate multilingual publication.

20. Roll-Hansen, 'Wishful Science', p. 175.

21. In works that sold in their tens of thousands and were frequently reprinted, Eddington and Jeans had each contrived to find comfort in the way relativity and quantum theory scotched hopes that science might provide mutually coherent answers to everything. The quantum mechanics revolution, according to Sir Arthur Eddington, revealed that 'all was vanity, that unreason lay at the very basis of reality – in the quanta of action and the behaviour of electrons'. Sir James Jeans asserted that 'the universe begins to took more like a great thought than like a great machine'. Mind was no longer 'an accidental intruder', but 'the creator and governor of the realm of matter'. The genre they were writing into was already well-developed; in 1920 the popular science writer John Sullivan skewered it neatly as 'works [which] seem to result from a close collaboration between, say, a professor of physics and a Bond Street Crystal Gazer'.

22. Newton's laws depend on formal logic. This is the kind of logic that says that one plus one equals two, and never three. Formal logic sees us through most everyday situations, so it seems like common sense. But it breaks down when it tries to deal with change. One egg plus another egg equals two eggs, unless one of them has hatched into a chicken. And if there are two eggs, and you have one, and your brother-in-law has the other, and he's late for supper, what are the words 'plus' and 'equals' worth? These are facetious examples of a serious problem for formal logic and for the Newtonian model: nothing ever stays still. 'The fundamental flaw in vulgar thought', wrote Trotsky, 'lies in the fact that it wishes to content itself with motionless imprints of reality which consists of eternal motion.'

23. Ioffe's grand style was already a matter of unhelpful legend. In 1928 the Russian Association of Physicists organised its sixth national congress

to examine the experimental results in support of quantum mechanics. After four days of discussions in Leningrad's Scholars' Club, its members boarded a train to Nizhny Novgorod and then sailed down the Volga on the steamship *Alexei Rykov* to Stalingrad, stopping in Nizhny, Kazan and Saratov for popular lectures and discussions with local audiences. The participants were delighted. 'How many new acquaintances were tied together on the Volga!' the physicist and historian Torichan Kravets enthused. 'How many interesting general gatherings were conducted in the salons of the ship! How many private conversations which were rich in content individual scholars held, strolling along the decks and slowly feasting their eyes on the broad, melancholy vistas flowing by!' But to an increasingly class-conscious Party, the boat ride down the Volga represented these physicists' elitism and aloofness, and the Russian Association of Physicists would convene only one more time – to arrange for its own dissolution.

24. Karl Hall, 'The Schooling of Lev Landau: The European Context of Postrevolutionary Soviet Theoretical Physics', *Osiris* 23, no. 1 (1 January 2008), pp. 230–59.

25. Josephson, *Physics and Politics in Revolutionary Russia*, p. 317.

26. With fellow physics students Gamow, Ivanenko and Bronstein, Lev Landau had made life hell for the tenured professors of Leningrad University. Their 'Physikalische Dummheiten' ('Physical Gibberish') posters, devoted to the mistaken calculations and misguided assertions of their betters, were pasted up on the walls to greet senior physicists as they arrived to lecture. The satire was often not very well focused: they sent a scoffing telegram to Boris Hessen even as he was defending Einstein against a barrage of reactionary criticism. It was this embarrassing and self-defeating gaffe which led to a public reprimand from Ioffe and Landau's decision to leave Leningrad.

27. Peter Leonidovich Kapitsa, *Letters to Mother: The Early Cambridge Period*, p. 46.

28. Kapitsa was immensely proud of his machinery. He once demonstrated the power of his magnetic fields by shooting a glass rod immersed in liquid oxygen (glass is strongly magnetic in that state) up to the ceiling, where it shattered. His visitor, the American physicist Robert Wood, refused to be outdone. He went up to the coil, took out the vessel of liquid oxygen, and calmly drank Kapitsa's health. (Or appeared to: he spat the liquid out again after a few seconds – a neat conjuring trick, not to be tried at home.)

29. Peter Kapitsa may have brought his 'detention' upon himself. He omitted on this occasion to collect his usual reassurance of return from the Russian consulate. And, according to his sponsor and Cambridge employer Ernest Rutherford, Kapitsa, 'in one of his expansive moods in Russia, told the Soviet engineers that he himself would be able to alter the whole face of electrical engineeering in his lifetime' – a tantalising claim to make in a country assured by Lenin that electrical power alone could power the success of socialism.

30. Lawrence Badash, *Kapitza, Rutherford, and the Kremlin*, p. 54.

31. A. B. Kojevnikov, *Stalin's Great Science: The Times and Adventures of Soviet Physicists* (Imperial College Press, 2004), p. 100.

32. Stalin's team attempted to maintain the practice of 'criticism and self-criticism', even at the very highest echelons of government. See J. A. Getty, 'Samokritika Rituals in the Stalinist Central Committee, 1933–38', *Russian Review*, 58 (1999), pp. 49–70.
33. Letter to Valery Mezhlauk, 5 July 1935, in Petr L. Kapitsa et al., *Kapitza in Cambridge and Moscow: Life and Letters of a Russian Physicist*, p. 328.
34. Krementsov, *Stalinist Science*, pp. 40–1.
35. Joravsky, *Soviet Marxism and Natural Science*, p. 258.

13 'Fascist links'

1. In Zimmer, Delbrück and Timofeev-Resovskii, *Creating a Physical Biology*, p. 61.
2. Muller's early trials and tribulations are very well handled by his biographer Elof A. Carlson in his brief memoir *Hermann Joseph Muller 1890–1967* (National Academy of Sciences, 2009); available at http://bit.ly/1L3wiPY.
3. H. J. Muller, 'The Dominance of Economics over Eugenics', *Scientific Monthly* 37 (1933), pp. 40–7.
4. In Donald F. Jones, *Proceedings of the Sixth International Congress of Genetics, Ithaca, New York, 1932*; available at http://www.esp.org/books/6th-congress/facsimile/contents/6th-cong-p213-muller.pdf.
5. Yakov G. Rokityanskij, 'N. V. Timofeeff-Ressovsky in Germany (July 1925–September 1945)', *Journal of Biosciences* 30, no. 5 (December 2005), pp. 573–80.
6. 'On the nature of gene mutation and gene structure' was given a 'funeral first class' – as Delbrück later remarked – published in the reports of the biology section of the journal of the Göttingen Academy of Sciences, which folded after three issues. It was rescued from obscurity by the physicist Erwin Schrödinger who sang its praises and summarised its contents (none too accurately) in his book *What Is Life?*

    Years later, Delbrück wrote to Timofeev-Ressovsky to say that re-reading the paper – bound in that peculiar, desaturated green card (it goes by the name 'Green Pamphlet' sometimes; more often as 'The Three-Man Paper') 'brings back to memory the idyllic and enthusiastic sessions at your house and at our house where we delighted in our first adventures at bringing genetics and physics together' (Max Delbrück to Nikolai Timofeev-Ressovsky, 1 October 1970; in Zimmer et al., *Creating a Physical Biology*, p. 61).
7. Susan Gross Solomon, *Doing Medicine Together: Germany and Russia between the Wars*, p. 354, n. 10.
8. H. J. Muller, *Out of the Night: A Biologist's View of the Future*, pp. 113 and 122.
9. Quoted in Babkov, *The Dawn of Human Genetics*, p. 643. See also Larson, 'Biology and the Emergence of the Anglo-American Eugenics Movement'. See also Diane B. Paul, '"Our Load of Mutations" Revisited', *Journal of the History of Biology* 20, no. 3 (1 September 1987), pp. 321–35.
10. One report has it that the secretary who translated the manuscript into Russian was later arrested and shot.

11. See Krementsov, *International Science between the World Wars*.
12. Leon Trotsky, *My Life: An Attempt at an Autobiography*, p. 315.
13. Joravsky, *The Lysenko Affair*, p. 218.
14. Loren R. Graham, *Science and Philosophy in the Soviet Union*, n. 58.
15. Lysenko's linguistic tactics are unpicked by Dmitri Stanchevici in *Stalinist Genetics: The Constitutional Rhetoric of T. D. Lysenko*. See also Paul M. Dombrowski, 'Plastic Language for Plastic Science: The Rhetoric of Comrade Lysenko', *Journal of Technical Writing and Communication* 31, no. 3 (2001), pp. 293–333.
16. Quoted in Joravsky, *The Lysenko Affair*, p. 104.
17. Roll-Hansen, *The Lysenko Effect*, p. 203.
18. Pringle, *The Murder of Nikolai Vavilov*, p. 211.
19. H. J. Muller to Julian Huxley, 9 and 11 March 1937, Lilly Library. See Mark B. Adams, 'The Politics of Human Heredity in the USSR, 1920–1940', *Genome* 31, no. 2 (1989), pp. 879–84.
20. The message was repeated a year later, following an official instruction to return home. Nikolai Koltsov managed to smuggle a letter to Timofeev-Ressovsky through the Swedish embassy: 'Of all the methods of suicide, you have chosen the most agonising and difficult. And this not only for yourself, but also for your family . . . If you do decide to come back, though, then book your ticket straight through to Siberia!'
21. Muller stayed in Spain through the siege of Madrid. He joined the International Brigade and served with the Canadian physician Norman Bethune, testing a new form of blood transfusion involving extracting blood from cadavers. With the Republican Army on the verge of defeat, Muller's friend Julian Huxley heard of his difficulties and contacted Francis Crew, the director of the Institute for Animal Genetics at the University of Edinburgh. Crew was not at all taken with the idea of having Muller in Edinburgh: as he wrote to a colleague, 'Huxley now writes to me to find Muller a job, but I doubt very much that I can, and I am not quite sure that I particularly want to.' But the move proved successful. Muller met and married his second wife, Thea Kantorowicz, and for the first time, and quite unexpectedly, seems to have experienced real happiness.

14 Office politics

1. Loren R. Graham, *Moscow Stories*, p. 124.
2. Roll-Hansen, *The Lysenko Effect*, p. 212.
3. Babkov, *The Dawn of Human Genetics*, p. 679.
4. Mark Popovsky, *The Vavilov Affair*, p. 152.
5. John Hawkes to Igor Loskutov, 1995; quoted in Loskutov, *Vavilov and His Institute*. For Lysenko's involvement in attacks on Nikolai Vavilov, see Bakhteyev, 'Reminiscences of N. I. Vavilov', pp. 79–84.
6. Calvin O. Qualset, 'Jack R. Harlan (1917–1998), Plant Explorer, Archaeobotanist, Geneticist, and Plant Breeder', *The Origins of Agriculture and Crop Domestication*, p. 1. See also Jack Harlan's *The Living Fields: Our Agricultural Heritage*.

7. In N. A. Grigorian, 'N. K. Koltsov i experimental'naya genetika vysshei nervoi deyatel'nosti' ['N. K. Koltsov and the Genetics of Higher Nervous Activity'], *Priroda* 6 (1992), pp. 93–7.
8. Babkov, *The Dawn of Human Genetics*, pp. 686–8.

15 'We shall go to the pyre'

1. Quoted in Pringle, *The Murder of Nikolai Vavilov*, p. 238.
2. Joravsky, *The Lysenko Affair*, p. 110.
3. Roll-Hansen, *The Lysenko Effect*, pp. 69–70.
4. Krementsov, *Stalinist Science*, p. 76.
5. A word of caution: there is some doubt over Yefrem Yakushevsky's account of these days. If Vavilov's meeting with Stalin actually occurred, Stalin never made a note of it, which is highly unusual. See Loskutov, *Vavilov and His Institute*.
6. Krementsov, *International Science between the World Wars*, pp. 66–8.
7. Gennady M. Andreev-Khomiakov and Ann Healy, *Bitter Waters: Life And Work In Stalin's Russia*, 1998.
8. Bakhteyev, 'Reminiscences of N. I. Vavilov', pp. 79–84.
9. Sergei Vavilov, Secretary of the Academy of Sciences from 1945 to his death in 1951, managed to perform his job without committing any major political mistakes. Sergei's position was ambiguous – dangerously so – but not uncommon. Even Politburo members like Molotov and Kalinin had convicts among their closest relatives. Sergei was a solitary intellectual whose hobby was searching for rare books among the heaps in second-hand bookstores. He translated Isaac Newton's *Lectiones Opticae* into Russian. He also (quietly and unofficially) countermanded decisions to which he himself had put his official stamp. The physicist and industrialist Dmitry Rozhdestvensky once noted: 'He carries in his pocket a letter of resignation, already signed but without a date on it. When the moment comes and he is forced to agree with something totally unacceptable, he is prepared to take this letter out of his pocket. But he can never decide whether this moment has already come.' Alexei Kojevnikov, 'President of Stalin's Academy: The Mask and Responsibility of Sergei Vavilov', *Isis* 87, no. 1 (1996), pp. 18–50.
10. Quoted in Pringle, *The Murder of Nikolai Vavilov*, p. 250.
11. A detailed and distressing account of Vavilov's questioning is provided by Birstein in *The Perversion of Knowledge*, pp. 223–9.

PART THREE : DOMINION

16 'Lucky stiffs'

1. Quotation adapted from Robert Conquest, *Stalin: Breaker of Nations* (Weidenfeld and Nicolson, 1991). See also Vladimir Hachinski, 'Stalin's Last Years: Delusions or Dementia?', *European Journal of Neurology* 6, no. 2 (1 March 1999).
2. Andreev-Khomiakov & Healy, *Bitter Waters*, pp. 163, 166.

3. The American journalist Harrison Salisbury remembered Zhdanov as a 'dark-haired man with brown eyes and, in his early years, considerable physical attraction. But as with many Soviet functionaries the ceaseless hours of work (often at night because of Stalin's habit of keeping late evening hours), the lack of physical exercise, the multitude of ceremonial banquets took their toll. By the eve of the war Zhdanov was overweight, pasty-faced and prey to severe asthmatic attacks. He was a chain smoker, lighting one Belomor after the other until the pepelnitsa [ashtray] on his desk was cluttered with stubs.' Quoted in Boterbloem, *The Life and Times of Andrei Zhdanov*, p. 219.

4. Boterbloem, *The Life and Times of Andrei Zhdanov*, p. 235.

5. Directive to the Chief of Staff of German Navy on the destruction of Leningrad, 22 September 1941. Available at bit.ly/1nTEqNi.

6. In February 1945, Brücher had been ordered to destroy the eighteen research facilities under his control to avoid their capture by advancing Soviet forces. He refused. After the war he took part of the Russian material with him during his move to South America. On 17 December 1991, at the age of seventy-five, Brücher was murdered at his vineyard in the Mendoza district of Argentina. Burglary seems the most likely motive, though the police report hinted that cocaine traffickers were responsible. Just before his death, Brücher had discussed in public the possibility of using a viral disease – *Estella*, or coca wilt – to destroy illicit Andean coca crops without damaging other plants. See Carl-Gustaf Thornstrom and Uwe Hossfeld, 'Instant Appropriation: Heinz Brücher and the SS Botanical Collecting Commando to Russia 1943', *Plant Genetic Resources Newsletter*, no. 129 (March 2002), pp. 54–57; and Daniel W. Gade, 'Converging Ethnobiology and Ethnobiography: Cultivated Plants, Heinz Brücher, and Nazi Ideology', *Journal of Ethnobiology* 26, no. 1 (1 March 2006), pp. 82–106.

7. Lev Kopelev, *Ease My Sorrows: A Memoir*, pp. 4–5.

8. This, at any rate, is the persuasive argument of Oleg V. Khlevnyuk in 'The Objectives of the Great Terror, 1937–1938', in *Stalinism*, ed. David L. Hoffmann, pp. 82–104.

9. In 1946, all Soviet commissariats were renamed 'ministries'.

10. Benjamin Zajicek, 'Scientific Psychiatry in Stalin's Soviet Union: The Politics of Modern Medicine and the Struggle to Define "Pavlovian" Psychiatry, 1939–1953', p. 211.

17 'Can I go to the reactor?'

1. Richard L. Garwin, 'Enrico Fermi and Ethical Problems in Scientific Research', presented at *Enrico Fermi and Modern Physics*, Pisa, Italy (19 October 2001).

2. Rokityanskij, *N. V. Timofeeff-Ressovsky in Germany*, p. 576.

3. Interview of Nikolaus Riehl by Mark Walker. Transcript, 17 October 1985. American Institute of Physics. https://www.aip.org/history-programs/niels-bohr-library/oral-histories/4844–2.

4. Georgy S. Levit and Uwe Hossfeld, 'From Molecules to the Biosphere:

Nikolai V. Timoféeff-Ressovsky's (1900–1981) Research Program within a Totalitarian Landscape', *Theory in Biosciences* 128, no. 4 (16 October 2009), pp. 237–48.

5. Nikolaus Riehl, *Stalin's Captive: Nikolaus Riehl and the Soviet Race for the Bomb.*

6. Quoted in Rokityanskij, *N. V. Timofeeff-Ressovsky in Germany*, p. 578.

7. See for example A. Kuzmin, 'K kakomu khramu ischchem mi dorogu?' ['Toward What Sort of Cathedral Are We Searching for a Road?'], *Nash Sovremennik* 1988, Vol. 3, pp. 154–64.

8. Timofeev-Ressovsky, *The Stories Told by Himself with Letters, Photos and Documents*, p. 360.

9. Vadim A. Ratner, 'Nikolay Vladimirovich Timofeeff-Ressovsky (1900–1981): Twin of the Century of Genetics', in *Genetics* 158, no. 3 (July 1, 2001), p. 936.

10. In the run up to the Trinity test, while they were waiting for suitable weather, Enrico Fermi broke the tension among the American bomb development team by offering a wager on whether or not their bomb would ignite the atmosphere, and if so, whether it would merely destroy New Mexico or destroy the world.

11. Holloway, *Stalin and the Bomb*, pp. 60–9.

12. Josephson, *Physics and Politics in Revolutionary Russia*, p. 182.

13. Kojevnikov, *Stalin's Great Science: The Times and Adventures of Soviet Physicists*, p. 135.

14. Klaus Fuchs returned to Britain in June 1946 and became head of the theoretical physics division of the Atomic Energy Establishment at Harwell where he helped the British to build their own bomb. Fuchs's reputation as a spy has overshadowed a much more striking fact about him: that he contributed significantly to three bomb projects – the American, the British, and the Russian. See Matin Zuberi, 'Stalin and the Bomb', *Strategic Analysis* 23, no. 7 (1999), pp. 1133–53.

15. On paper, Zavenyagin was a high-placed gulag bureaucrat, answering to the Chief Directorate of Camps for Mining and Metallurgy Enterprises. As head of the Ninth Chief Directorate of the NKVD, however, he had special responsibilities. Among them was the supervision of Laboratory No. 2. For more on Zavenyagin's activities in Berlin, see Pavel V. Oleynikov, 'German Scientists in the Soviet Atomic Project', *Nonproliferation Review* 7, no. 2 (2000), pp. 1–30.

16. Riehl, *Stalin's Captive*, p. 164.

17. Quoted in Oleynikov, 'German Scientists in the Soviet Atomic Project', p. 10.

18. Interview of Nikolaus Riehl by Mark Walker. Transcript, 17 October 1985. American Institute of Physics. https://www.aip.org/history-programs/niels-bohr-library/oral-histories/4844-2.

19. Once when the foreign intelligence expert Leonid Kvasnikov was reporting to Beria on the latest data from overseas, Beria threatened him: 'If this is disinformation, I'll put you all in the cellar.' See Holloway, *Stalin and the Bomb*, p. 211.

20. Kurchatov's laboratory equipment churns out 24 kilowatts of electricity to

this day, making it the world's oldest operating reactor.

21. Montefiore, *Stalin*, p. 513.
22. Alexei B. Kojevnikov, 'Piotr Kapitza and Stalin's Government: A Study in Moral Choice', *Historical Studies in the Physical and Biological Sciences*, 22 (1991), pp. 131–64. See also David Holloway, 'The Scientist and the Tyrant', *New York Review of Books* (1 March 1990).
23. Holloway, *Stalin and the Bomb*, p. 139.

18 'How did anyone dare insult Comrade Lysenko?'

1. Quoted in Pringle, *The Murder of Nikolai Vavilov*, p. 290, from N. S. Khrushchev, *Khrushchev Remembers* (André Deutsch, 1971), pp. 200–15.
2. Ashby, *Scientist in Russia*, p. 129.
3. Raissa Berg, *On the History of Genetics in the Soviet Union: Science and Politics: The Insight of a Witness: Final Report*, pp. 28–9.
4. Robert C. Cook, 'Lysenko's Marxist Genetics: Science or Religion?', *Journal of Heredity* 40, no. 7 (1949).
5. Nikolai Krementsov, 'A "Second Front" in Soviet Genetics: The International Dimension of the Lysenko Controversy, 1944–1947', *Journal of the History of Biology* 29, no. 2 (1 June 1996), p. 239.
6. The most succinct summation of *Heredity and Its Variability*, and the funniest, was Eric Ashby's: 'When it is stated on the dust jacket of the book that "a careful study of these papers will amply reward every serious student of biology" we can only utter a startled assent.'
7. Quoted in Krementsov, 'A "Second Front" in Soviet Genetics', p. 241.
8. See, for example, H. J. Muller, 'It Still Isn't a Science. A Reply to George Bernard Shaw', *Saturday Review of Literature*, 16 April 1949, 11–12, p. 61.
9. Shostakovich's reputation was not helped by persistent rumours surrounding *Orango*, a satirical opera, begun in 1932 and hurriedly aborted, about the nightclub adventures of an experimental ape-man. See Gerard McBurney's optimistically titled 'Some Frequently Asked Questions about Shostakovich's "Orango"', *Tempo*, 64 (2010), pp. 38–40.
10. Montefiore, *Stalin*, p. 555.
11. Krementsov, *Stalinist Science*, p. 137.
12. At the outbreak of war Lysenko's younger brother Pavel was an industrial chemist working at the Institute of Coal Chemistry in Kharkov. His research into converting coal into coke went well, conjured envy in his colleagues, and led to his denunciation and arrest. Pavel seems to have shared his brother's difficulty in making friends: when war came his colleagues even tried to get him drafted (though, as a Scientific Worker, he was exempt). He defected to the Germans when they overran the city and impressed them so much that they made him mayor. After their defeat the Germans took Pavel with them as they retreated, and he finally ended up in the American Zone in Munich. US authorities found him running a factory that turned horse chestnuts into artificial honey.
13. Krementsov, *Stalinist Science*, p. 112.
14. Ethan Pollock, *Stalin and the Soviet Science Wars*, pp. 47–8.

15. Ibid.
16. Douglas R. Weiner, *A Little Corner of Freedom: Russian Nature Protection from Stalin to Gorbachev*, p. 458 n. 36.
17. Ibid., p. 74.
18. Ibid.
19. Ibid.
20. Pollock, *Stalin and the Soviet Science Wars*, p. 50.
21. Krementsov, *Stalinist Science*, p. 165.
22. Zhores A. Medvedev, *The Unknown Stalin*, p. 183.
23. Two papers by Kirill O. Rossianov reveal Stalin's editorial style: 'Editing Nature: Joseph Stalin and the "New" Soviet Biology', *Isis*, 84 (1993), pp. 728–45; and 'Stalin as Lysenko's Editor: Reshaping Political Discourse in Soviet Science', *Configurations*, 1 (1993), pp. 439–56.
24. Mark A. Popovsky, *Manipulated Science: The Crisis of Science and Scientists in the Soviet Union Today*, p. 149.
25. Krementsov, *Stalinist Science*, p. 172.
26. Quoted in Cook, 'Lysenko's Marxist Genetics', p. 188.
27. Quoted in Pollock, *Stalin and the Soviet Science Wars*, p.65.
28. Stephen Jay Gould, *Hen's Teeth and Horse's Toes* (Norton, 1983), p. 135.

19 Higher nervous activity

1. Quoted in John E. Bowlt and Olga Matich, *Laboratory of Dreams: The Russian Avant-Garde and Cultural Experiment*, p. 92, from an article in *Samarskaia Gazeta*.
2. One particularly egregious example was a manual written by Nikolai Turbin, dean of biology at Leningrad University. *Genetics and Selection*, published in 1950 and in official circulation until the mid-1960s, included such topics as 'The struggle of the progressive Michurin theory in genetics with the reactionary genetics of Mendel and Morgan'; 'Reactionary distortions in bourgeois genetics originating from the class ideology of the imperialistic bourgeois'; 'Complete bankruptcy of modern Morganism in theory and practice'; 'The Golden Age of Michurin's genetics and selection in the USSR', and so on. This exercise in cant sickened even the author. From the moment of his book's publication Turbin began, bit by bit, claim by claim, to defect from Lysenko's cause. He eventually left his post for the Belorussian Academy of Sciences in Minsk, and went on, in the mid-fifties, to build up a perfectly respectable centre of maize genetics there. See S. M. Gershenson, 'The Grim Heritage of Lysenkoism: Four Personal Accounts. IV. Difficult Years in Soviet Genetics', *Quarterly Review of Biology* 65, no. 4 (1990), pp. 447–56.
3. Berg, *On the History of Genetics in the Soviet Union*.
4. Hermann Muller's return to the USA was a happy one. His boss Fernandus Payne, a *Drosophila* geneticist who had worked in the Morgan lab, had the measure of his new employee. To a doubtful friend he explained that he already had several prima donnas on his staff and one more wouldn't matter. See J. F. Crow and S. Abrahamson. 'Seventy Years Ago: Mutation Becomes Experimental', *Genetics* vol. 147, no. 4 (1997), pp. 1491–6.

5. Wolfe, 'What Does It Mean to Go Public?', pp. 64–5.
6. Lina Shtern's career is summarised in Alla A. Vein, 'Science and Fate: Lina Stern (1878–1968), A Neurophysiologist and Biochemist', *Journal of the History of the Neurosciences* 17, no. 2 (2008), 195–206.
7. W. B. Gratzer, *The Undergrowth of Science*, p. 193.
8. Quoted in David Joravsky, 'The Mechanical Spirit: The Stalinist Marriage of Pavlov to Marx', *Theory and Society* 4, no. 4 (1977), p. 473.
9. Pollock, *Stalin and the Soviet Science Wars*, p. 144.
10. Quoted in Krementsov, *Stalinist Science*, p. 266.
11. Suvorov and Andreeva describe the ideological struggle over Pavlov's legacy in 'Problems of the Inheritance of Conditioned Reflexes in Pavlov's School', pp. 8–16.
12. Pollock, *Stalin and the Soviet Science Wars*, p. 146.
13. Ibid.
14. Ibid., p. 141.
15. Ibid., p. 161.
16. See George Windholz, 'The 1950 Joint Scientific Session: Pavlovians as the Accusers and the Accused', *Journal of the History of the Behavioral Sciences*, 33 (1997), pp. 61–81. The consequences of the session are the subject of a later paper: 'Soviet Psychiatrists under Stalinist Duress: The Design for a "New Soviet Psychiatry" and its Demise', *History of Psychiatry*, 10 (1999), pp. 329–47.
17. Joravsky, *Russian Psychology: A Critical History*, p. 410.
18. Vance Kepley, 'The Scientist as Magician: Dovzhenko's "Michurin" and the Lysenko Cult', *Journal of Popular Film and Television* 8, no. 2 (1 January 1980), p. 19.
19. Giordano Bruno was an Italian philosopher burned at the stake by the Inquisition in 1600 for claiming that stars were distant suns and that the universe, being infinite, lacked a centre.
20. Trofim Lysenko, 'Stalin i michurinskaya biologiya', quoted in Medvedev, *The Rise and Fall of T. D. Lysenko*, p. 134.

Chapter 20 'The death agony was horrible'

1. Cited in Mira Liehm and Antonín J. Liehm, *The Most Important Art: Eastern European Film after 1945*, p. 55.
2. *Michurin* is the name the film went by domestically; under the title *Life in Bloom* it was given extensive overseas distribution, and the *New York Times* ran a positive review.
3. *Memoirs of Nikita Khrushchev*, Volume 1: *Commissar, 1918–1945*, p. 522.
4. Stephen Brain, 'The Great Stalin Plan for the Transformation of Nature', *Environmental History* 15, no. 4 (1 October 2010), p. 684.
5. Quoted in Weiner, *A Little Corner of Freedom*, pp. 89–90.
6. Marcel Penant, a French scientist and communist, writing in 1957 in *La Pensée* remembers he was so startled by Lysenko's claim, he assumed it was a mistranslation, 'until Lysenko repeated it to me word for word in the course of a discussion with which he favoured me in 1950 . . . I allowed myself to put

a question to him: "I admit that young trees should be planted in a cluster; they may thus be better protected at first; but is it not necessary to remove some of them after a few years?" "No," replied Lysenko, explaining: "They will sacrifice themselves for one." "Do you mean," I replied, "'that one will turn out to be stronger and the others will weaken or perish?" "No," he repeated, "they will sacrifice themselves for the good of the species," and he entered into a long and very hazy discourse, completely overwhelming me with a "materialistic" explanation which would have been acceptable to Bernardin de Saint Pierre, and which was very close to a belief in divine Providence.'

7. Peter Kneen, 'Physics, Genetics and the Zhdanovshchina', *Europe–Asia Studies*, 50 (1998), pp. 1183–1202

8. Zuberi, 'Stalin and the Bomb', p. 1142.

9. Ibid., p. 1147

10. Holloway, *Stalin and the Bomb*, p. 273.

11. Caesar P. Korolenko and Dennis V. Kensin, 'Reflections on the Past and Present State of Russian Psychiatry', *Anthropology and Medicine* 9, no. 1 (2002), p. 54.

12. For a full account see Benjamin Tromly, 'The Leningrad Affair and Soviet Patronage Politics, 1949–1950', *Europe–Asia Studies* 56, no. 5 (July 2004), pp. 707–29.

13. Joshua Rubenstein and V. P. Naumov (eds), *Stalin's Secret Pogrom: The Postwar Inquisition of the Jewish Anti-Fascist Committee*.

14. Stalin's death saved Lina Shtern from the full rigour of the judgment. She spent only ten months in Dzhambul. Released from exile, she refused to speak about her ordeal.

15. In 1948 Prezent had become chair of the Darwinian Departments at Moscow and Leningrad. Students and professors alike were obliged to attend Prezent's lectures in Michurinist biology. The exams at the end of this course of 're-education' acquired a bad reputation. Prezent had a habit of inviting female students to his apartment for the exam. The wise ones turned up with strong male chaperones.

16. In Birstein, *The Perversion of Knowledge*, p. 260.

17. Yakov L. Rapoport, *The Doctors' Plot*, pp. 74, 79.

18. S. Allilueva, *Twenty Letters to a Friend*, p. 18.

21 Succession

1. Richard Levins and Richard C. Lewontin, *The Dialectical Biologist*, p. 65.

2. David Joravsky, 'The Stalinist Mentality and the Higher Learning', *Slavic Review* 42, no. 4 (1983), p. 580.

3. G. G. Polikarpov, 'Sketches to the Portrait of Nikolai Timofeeff-Ressovsky'; available at http://bit.ly/1PEsLgl .

4. V. I. Korogodin, G. G. Polikarpov and V. V. Velkov, 'The Blazing Life of N. V. Timofeeff-Ressovsky', *Journal of Biosciences* 25, no. 2 (June 2000), pp. 125–31.

5. V. I. Ivanov, 'A True Scientist and a Most Amiable Person: To the Centenary of the Birthday of H. A. Timofeeff-Ressovsky', trans. Irina V. Kronshtadtova

(12 January 1999); available at http://bit.ly/1Pn7uEo.

6. See Slava Gerovitch, *From Newspeak to Cyberspeak: A History of Soviet Cybernetics.*

7. Nikolai and Elena Timofeev-Ressovsky to Liapunov, 5 October 1957, in Timofeev-Ressovsky, *The Stories Told by Himself with Letters, Photos and Documents*, p. 489.

8. Sergei L. Sobolev and Aleksei A. Liapunov, 'Kibernetika i estestvoznanie' ['Cybernetics and Science'], in *Filosofskie problemy sovremennogo estestvoznaniia: Trudy Vsesoiuznogo soveshchaniia po filosofskim voprosam estestvoznaniia* [*Philosophical Problems of Modern Science: Proceedings of the All-Union Conference on the Philosophical Problems of the Natural Sciences*], ed. P. Fedoseev (Izd-vo Akademii nauk SSSR, 1959), pp. 251–2.

9. Arnost Kolman, 'A Life-Time in Soviet Science Reconsidered: The Adventure of Cybernetics in the Soviet Union', *Minerva* 16, no. 3 (1 September 1978), p. 418.

10. Ibid., p. 422}

11. Graham, *Science and Philosophy in the Soviet Union*, p. 241.

12. Andrei D. Sakharov, *Memoirs* (Knopf, 1990), p. 236.

13. Berg, *On the History of Genetics in the Soviet Union.*

Epilogue: Spoil

1. N. F. Fedorov, *What Was Man Created for? The Philosophy of the Common Task: Selected Works.*

2. Quoted in Krementsov, 'From "Beastly Philosophy" to Medical Genetics', p. 90.

3. Quoted in Gloveli, '"Socialism of Science" versus "Socialism of Feelings"', pp. 43, 49.

4. Arran Gare, 'Aleksandr Bogdanov and Systems Theory', *Democracy and Nature* 6, no. 3 (2000), p. 355.

5. According to Leon Trotsky, 'For Lunacharsky each and every new philosophical and political toy had its attraction and he could not resist playing with them.' Gloveli, '"Socialism of Science" versus "Socialism of Feelings"', p. 33.

6. John Biggart, *Alexander Bogdanov, left-Bolshevism and the Proletkult 1904–1932.*

7. James D. White, 'Alexander Bogdanov's Conception of Proletarian Culture', *Revolutionary Russia* 26, no. 1 (2013), pp. 52–70.

8. John Biggart, 'Bukharin and the Origins of the "Proletarian Culture" Debate', *Soviet Studies*, 39 (1987), pp. 229–46.

9. Bogdanov's own account of this work has been translated into English by Douglas W. Huestis as *The Struggle for Viability: Collectivism through Blood Exchange.*

10. A. Gastev, *Ustanovka rabochei sily*, 1929, no. 1–2, p. 11. Irina Lunacharskaia, 'Why Did Commissar of Enlightenment A. V. Lunacharskii Resign?', trans. Kurt S. Schultz, *Russian Review* 51, no. 3 (1992), p. 326.

11. Johansson and Gastev, *Aleksej Gastev*, p. 129, n. 30.

12. A. Bogdanov, 'O tendentsiyakh proletarskoi kul'tury (ovtet A. Gastevu)' ['Regarding Trends in Proletarian Culture (the Role of A. Gastev)'], *Proletarskaya kul'tura*, nos. 9–10. 1919, pp. 46–52. Cited in Sochor, 'Soviet Taylorism Revisited', p. 249.

13. Alexander M. Etkind, 'Psychological Culture: Ambivalence and Resistance to Social Change', *First Nevada Conference on Russian Culture*, p. 15.

14. Another Fedorovian nugget: not content with his assertion that death is neither necessary nor inevitable, cosmism's founder insisted that it was our moral duty to ensure that the living live forever, He even expected us bring our dead ancestors back to life.

15. Holloway, *Stalin and the Bomb*, p. 367.

16. Korolenko and Kensin, 'Reflections on the Past and Present State of Russian Psychiatry'.

17. There is a striking and none-too-flattering portrait of Snezhevsky to be had in 'The World of Soviet Psychiatry', Walter Reich's anecdotal article in the *New York Times* (30 January 1983).

18. Sakharov, *Memoirs*, pp. 309–12.

19. Sidney Bloch, 'Psychiatry as Ideology in the USSR', *Journal of Medical Ethics* 4, no. 3 (1978), p. 128.

20. Quoted in Bailes, *Science and Russian Culture in an Age of Revolutions*, p. 125.

21. Weiner, *Models of Nature*, p. 44.

22. Paul R. Josephson, *Industrialized Nature: Brute Force Technology and the Transformation of the Natural World*.

23. B. S. Richter, 'Nature Mastered by Man: Ideology and Water in the Soviet Union', *Environment and History* 3, no. 1 (1997) pp. 69–96.

Bibliography

Abraham, Ralph, 'A Review of "Geochemistry and the Biosphere: Essays by Vladimir I. Vernadsky"', *World Futures*, 65 (2009), 436–41

Adams, Henry, *The Education of Henry Adams: An Autobiography* (Modern Library, 1931)

Adams, Mark B., 'Biology after Stalin: A Case History', *Survey*, 1977, 53–80

——, 'Genetics and Molecular Biology in Khrushchev's Russia' (Harvard, 1973)

——, 'The Founding of Population Genetics: Contributions of the Chetverikov School 1924–1934', *Journal of the History of Biology*, 1 (1968), 23–39

——, 'The Politics of Human Heredity in the USSR, 1920–1940', *Genome*, 31 (1989), 879–84

——, *The Wellborn Science: Eugenics in Germany, France, Brazil, and Russia* (Oxford University Press, 1990)

——, 'Towards a Synthesis: Population Concepts in Russian Evolutionary Thought, 1925–1935', *Journal of the History of Biology*, 3 (1970), 107–29

Adams, Mark B., Garland E. Allen and Sheila Faith Weiss, 'Human Heredity and Politics', *Osiris*, 2nd Series, 20 (2005), 232–62

Akhutina, T. V., 'L. S. Vygotsky and A. R. Luria: Foundations of Neuropsychology', *Journal of Russian and East European Psychology*, 41 (2010), 159–90

Akimenko, M. A., 'Vladimir Mikhailovich Bekhterev', *Journal of the History of the Neurosciences*, 16 (2007), 1–2

Aldridge, Jerry, 'Another Woman Gets Robbed? What Jung, Freud, Piaget, and Vygotsky Took from Sabina Spielrein', *Childhood Education*, 85 (2009), 318

Alexander, Denis, and Ronald L. Numbers, *Biology and Ideology from Descartes to Dawkins* (University of Chicago Press, 2010)

Allilueva, Svetlana, *Twenty Letters to a Friend*, trans. Priscilla Johnson McMillan (Heinemann, 1968)

All Music Guide, 'Dmitri Shostakovich: Michurin ("Life in Bloom"), Op. 78', *All Music Guide*, 2008

Andreev-Khomiakov, Gennady M., and Ann Healy, *Bitter Waters: Life And Work In Stalin's Russia* (Westview Press, 1998)

Andrews, James T., *Science for the Masses: The Bolshevik State, Public Science, and the Popular Imagination in Soviet Russia, 1917–1934* (Texas A&M University Press, 2003)

466

Bibliography

Andrle, V., 'How Backward Workers Became Soviet: Industrialization of Labour and the Politics of Efficiency under the Second Five-Year Plan, 1933–1937', *Social History*, 10 (1985), 147–69

Angelini, Alberto, 'History of the Unconscious in Soviet Russia: From Its Origins to the Fall of the Soviet Union', *The International Journal of Psychoanalysis*, 89 (2008), 369–88

Anichkov S. V., 'How I Became a Pharmacologist', *Annual Review of Pharmacology*, 15 (1975), 1–10

Arens, Katherine, 'Mach's "Psychology of Investigation"', *Journal of the History of the Behavioral Sciences*, 21 (1985), 151–68

Ashby, Eric, *Scientist in Russia* (Penguin Books, 1947)

A. W., 'Michurin (1948)', *New York Times*, 9 May 1949

Babkov, V. V., *The Dawn of Human Genetics*, ed. by James Schwartz, trans. by Victor Fet (Cold Spring Harbor Laboratory Press, 2013)

——, 'The Theoretical–Biological Concept of Nikolai K. Kol'tsov', *Russian Journal of Developmental Biology*, 33 (2002), 255–62

Bacon, Edwin, *The Gulag at War: Stalin's Forced Labour System in the Light of the Archives* (Palgrave Macmillan, 1996)

Badash, Lawrence, *Kapitza, Rutherford, and the Kremlin* (Yale University Press, 1985)

Bailes, Kendall E., 'Alexei Gastev and the Soviet Controversy over Taylorism, 1918-24', *Soviet Studies*, 29 (1977), 373–94

——, *Science and Russian Culture in an Age of Revolutions: V. I. Vernadsky and His Scientific School, 1863–1945* (John Wiley & Sons, 1990)

——, 'Soviet Science in the Stalin Period: The Case of V. I. Vernadskii and His Scientific School, 1928–1945', *Slavic Review*, 45 (1986), 20–37

——, *Technology and Society under Lenin and Stalin: Origins of the Soviet Technical Intelligentsia, 1917–1941* (Princeton University Press, 1978)

——, 'The American Connection: Ideology and the Transfer of American Technology to the Soviet Union, 1917–1941', *Comparative Studies in Society and History*, 23 (1981), 421–48

——, 'The Politics of Technology: Stalin and Technocratic Thinking among Soviet Engineers', *The American Historical Review*, 79 (1974), 445–69

Baker, David B., *The Oxford Handbook of the History of Psychology: Global Perspectives* (Oxford University Press, 2012)

Bakhteyev, F. Kh., 'Reminiscences of N. I. Vavilov (1887–1943) on the Eightieth Anniversary of His Birthday', *Theoretical and Applied Genetics: International Journal of Plant Breeding Research*, 38 (1968), 79–84

Barbara, Jean-Gaël and Jean-Claude Dupont, *History of the Neurosciences in France and Russia: From Charcot and Sechenov to IBRO* (Hermann, 2011)

Bedeian, Arthur G., and Carl R. Phillips, 'Scientific Management and Stakhanovism in the Soviet Union: A Historical Perspective', *International Journal of Social Economics*, 17 (1990), 28–35

Bibliography

Bekhterev, Vladimir Mikhailovich, *Collective Reflexology: The Complete Edition*, ed. Lloyd H. Strickland, trans. Eugenia and Alisa Lockwood (Transaction Publishers, 2001)

Bengtsson, Bengt Olle, and Anna Tunlid, *The 1948 International Congress of Genetics in Sweden: People and Politics* (Genetics Society of America, 1948)

Bentivoglio, Marina, 'Cortical Structure and Mental Skills: Oskar Vogt and the Legacy of Lenin's Brain', *Brain Research Bulletin*, 47 (1998), 291–6

Benton, Ted, *The Greening of Marxism* (Routledge, 1996)

Berg, Raissa L., *Acquired Traits: Memoirs of a Geneticist from the Soviet Union* (Viking, 1988)

——, 'In Defense of Timoféeff-Ressovsky', *The Quarterly Review of Biology*, 65 (1990), 457–79

——, *On the History of Genetics in the Soviet Union: Science and Politics: The Insight of a Witness: Final Report to National Council for Soviet and East European Research* (National Council for Soviet and East European Research, 1983)

Bernal, J. D., 'The Biological Controversy in the Soviet Union and Its Implications', *Modern Quarterly*, 4 (1949), 204

Bernstein, N. A., 'Biomekhanika i fiziologiya dvizheniy' ['Biomechanics and Physiology of Movement'], *Izbrannye Psikhologicheskie Trudy* [*Selected Psychological Works*] (MPSI, 2008)

Biggart, John, *Alexander Bogdanov, left-Bolshevism and the Proletkult 1904–1932* (University of East Anglia, 1989)

——, 'Bukharin and the Origins of the "Proletarian Culture" Debate', *Soviet Studies*, 39 (1987), 229–46

Birstein, Vadim J., *The Perversion of Knowledge: The True Story of Soviet Science* (Westview Press, 2001)

Blackmore, John T., *Ernst Mach: His Work, Life, and Influence* (University of California Press, 1972)

Bloch, Sidney, 'Psychiatry as Ideology in the USSR', *Journal of Medical Ethics*, 4 (1978)

Boakes, Robert, *From Darwin to Behaviourism: Psychology and the Minds of Animals* (Cambridge University Press, 1984)

Bogdanov, A. A., *The Struggle for Viability: Collectivism through Blood Exchange*, trans. Douglas W. Huestis (Xlibris, 2001)

——, *Bogdanovs' Tektology*, ed. Peter Dudley (Centre for Systems Study, University of Hull, 1996)

Bolotova, Alla, 'Colonization of Nature in the Soviet Union. State Ideology, Public Discourse, and the Experience of Geologists', *Historical Social Research / Historische Sozialforschung*, 29 (2004), 104–23

Boterbloem, Kees, *The Life and Times of Andrei Zhdanov, 1896–1948* (McGill-Queen's University Press, 2004)

Bowlt, John E., and Olga Matich, *Laboratory of Dreams: The Russian Avant-Garde and Cultural Experiment* (Stanford University Press, 1996)

Bibliography

Brackman, Roman, *The Secret File of Joseph Stalin: A Hidden Life* (Routledge, 2003)

Brain, Stephen, *Song of the Forest: Russian Forestry and Stalinist Environmentalism, 1905–1953* (University of Pittsburgh Press, 2011)

———, 'Stalin's Environmentalism', *Russian Review*, 69 (2010), 93–118

———, 'The Great Stalin Plan for the Transformation of Nature', *Environmental History*, 15 (2010), 670–700

Brandenberger, David, 'Stalin, the Leningrad Affair, and the Limits of Postwar Soviet Russocentrism', *Russian Review*, 63 (2004), 241–55

Brenner, Frank, 'Intrepid Thought: Psychoanalysis in the Soviet Union' (World Socialist Web Site, 1999)

Brown, Julie V., 'Heroes and Non-Heroes: Recurring Themes in the Historiography of Russian-Soviet Psychiatry', *Discovering the History of Psychiatry*, ed. Mark S. Micale and Roy Porter (Oxford University Press, 1994), 297–307

Bryant, Louise, *Six Red Months in Russia* (Arno Press, 1970)

Burds, Jeffrey, *Peasant Dreams and Market Politics: Labor Migration and the Russian Village, 1861–1905* (University of Pittsburgh Press, 1998)

Buzin, V. N., 'Psychoanalysis in the Soviet Union: On the History of a Defeat', *Russian Social Science Review*, 36 (1995), 65–73

Carden, Patricia, 'Utopia and Anti-Utopia: Aleksei Gastev and Evgeny Zamyatin', *Russian Review*, 46 (1987), 1–18

Carlson, Elof A., *Genes, Radiation, and Society: The Life and Work of H. J. Muller* (Cornell University Press, 1981)

———, *Hermann Joseph Muller 1890–1967* (National Academy of Sciences, 2009)

———, *The Gene: A Critical History* (Saunders, 1966)

Carter, Huntly, 'Moscow's House of Science', *The Sociological Review*, 19, 2 (1927), 168–71

Casimir, H. B. G., *Haphazard Reality: Half a Century of Science* (Amsterdam University Press, 1983)

Chamberlain, Lesley, *The Philosophy Steamer: Lenin and the Exile of the Intelligentsia* (Atlantic Books, 2006)

Charnley, Berris, 'Experiments in Empire-Building: Mendelian Genetics as a National, Imperial, and Global Agricultural Enterprise', *Studies in History and Philosophy of Science Part A*, 44 (2013), 292–300

Chernyshevsky, Nikolai, *What Is to Be Done?*, trans. Michael R. Katz (Cornell University Press, 2014)

Chetverikov, S. S., 'On Certain Aspects of the Evolutionary Process from the Standpoint of Modern Genetics', ed. I. Michael Lerner, trans. Malina Barker, *Proceedings of the American Philosophical Society*, 105 (1961), 167–95

———, 'Volny Zhizni' ['Waves of Life'], *Dnevnik zootdeleniia*, vol. 3, no. 6 (1905)

Cifali, Mireille, 'Sabina Spielrein, a Woman Psychoanalyst: Another Picture', *Journal of Analytical Psychology*, 46 (2001), 129–38

Cigna, Arrigo A., and Marco Durante, *Radiation Risk Estimates in Normal and Emergency Situations* (Springer, 2006)

Clarke, Ruscoe, and Leonard Crome, 'A Visit to Koltushi: The Study of Conditioned Reflexes', *The Lancet*; originally published as volume 1, issue 6866, 265 (1955), 712–16

Clark, Katerina, *Petersburg, Crucible of Cultural Revolution* (Harvard University Press, 1995)

Cohen, Barry Mendel, 'Nikolai Ivanovich Vavilov: The Explorer and Plant Collector', *Economic Botany: A Publication of the Society for Economic Botany*, 45 (1969), 38–46

Conquest, Robert, *Harvest of Sorrow: Soviet Collectivisation and the Terror-Famine* (Pimlico, 2002)

——, *The Great Terror: A Reassessment* (Oxford University Press, 2007)

——, *Stalin: Breaker of Nations* (Weidenfeld and Nicolson, 1991).

Cook, Robert C., 'Lysenko's Marxist Genetics: Science or Religion?', *Journal of Heredity*, 40 (1949)

Covington, Coline, and Barbara Wharton, *Sabina Spielrein: Forgotten Pioneer of Psychoanalysis* (Brunner-Routledge, 2003)

Crow, James F., 'A Diamond Anniversary: The First Chromosomal Map', *Genetics*, 118 (1988), 1–3

——, 'N. I. Vavilov, Martyr to Genetic Truth', *Genetics*, 134 (1993), 1–4

——, 'Plant Breeding Giants: Burbank the Artist; Vavilov, the Scientist', *Genetics*, 158 (2001), 1391–5

——, 'Sixty Years Ago: The 1932 International Congress of Genetics', *Genetics*, 131 (1992), 761–8

Crow, James F., and S. Abrahamson, 'Seventy Years Ago: Mutation Becomes Experimental', *Genetics*, 147 (1997), 1491–6

Curtis, Adam, 'Pandora's Box', *The Engineers' Plot* (BBC, 1992)

Daniels, Harry, ed., *Introduction to Vygotsky, Second Edition* (Routledge, 2005)

Daniels, Harry, Michael Cole, and James V. Wertsch, *The Cambridge Companion to Vygotsky* (Cambridge University Press, 2007)

Darlington, Cyril D., 'Psychology, Genetics and the Process of History', *British Journal of Psychology*, 54 (1963), 293–98

——, *The Conflict of Science and Society* (Watts, 1948)

Darlington, Cyril D., and Kenneth Mather, *Genes, Plants and People: Essays on Genetics* (Allen & Unwin, 1950)

David-Fox, Michael, *Revolution of the Mind: Higher Learning among the Bolsheviks, 1918–1929* (Cornell University Press, 1997)

——, 'Symbiosis to Synthesis: The Communist Academy and the Bolshevization of the Russian Academy of Sciences, 1918–1929', *Symbiosis*, 219 (1998), 43

Davies, Sarah, and James R. Harris, *Stalin: A New History* (Cambridge University Press, 2005)

deJong-Lambert, William, 'Hermann J. Muller, Theodosius Dobzhansky, Leslie Clarence Dunn, and the Reaction to Lysenkoism in the United States', *Journal of Cold War Studies*, 15 (2013), 78–118

——, *The Cold War Politics of Genetic Research: An Introduction to the Lysenko Affair* (Springer, 2012)

Delbrück, Max, and N. W. Timofeev-Ressovsky, 'Cosmic Rays and the Origin of Species', *Nature*, 137 (1936), 358–9

Dobzhansky, Th., 'Animal Breeding under Lysenko', *American Naturalist*, 88 (1954), 165–7

Dombrowski, Paul M., 'Plastic Language for Plastic Science: The Rhetoric of Comrade Lysenko', *Journal of Technical Writing and Communication*, 31 (2001), 293–333

Dorozynski, Alexandre, *The Man They Wouldn't Let Die* (Macmillan, 1965)

Dronin, Nikolai M., and Edward G. Bellinger, *Climate Dependence and Food Problems in Russia: 1900–1990: The Interaction of Climate and Agricultural Policy and Their Effect on Food Problems* (Central European University Press, 2005)

Dunn, L. C., 'Soviet Biology', *Science*, New Series, 99 (1944), 65–7

——, *The Reminiscences of Leslie Clarence Dunn*, Columbia University Oral History Collection, Part 4, No. 5 (Columbia University Press, 1975)

Easter, Gerald M., *Reconstructing the State: Personal Networks and Elite Identity in Soviet Russia* (Cambridge University Press, 2007)

Eaton, Katherine Bliss, *Daily Life in the Soviet Union* (Greenwood, 2004)

Eilam, Gavriela, 'The Philosophical Foundations of Aleksandr R. Luria's Neuropsychology', *Science in Context*, 16 (2003), 551–77

Engels, Friedrich, *Anti-Dühring: Herr Dühring's Revolution in Science* (Foreign Languages Publishing House, 1959)

——, *The Part Played by Labour in the Transition from Ape to Man* (Foreign Languages Publishing House, 1953)

Engels, Friedrich, C. P. Dutt and J. B. S. Haldane, *Dialectics of Nature* (International Publishers, 1940)

Eremeeva, A. I., 'Political Repression and Personality: The History of Political Repression Against Soviet Astronomers', *Journal for the History of Astronomy*, 26 (1995), 297

Etkind, Alexander M., 'Beyond Eugenics: The Forgotten Scandal of Hybridizing Humans and Apes', *Studies in History and Philosophy of Biological and Biomedical Sciences*, 39 (2008), 205–10

——, *Eros of the Impossible: The History of Psychoanalysis in Russia* (Westview Press, 1997)

——, 'Psychological Culture: Ambivalence and Resistance to Social Change', in *First Nevada Conference on Russian Culture* (UNLV Center for Democratic Culture, 1992)

——, 'Trotsky and Psychoanalysis', *Partisan Review*, 61 (1994), 303–8

Fando, R., 'The Unknown About a Well-Known Biologist', *Herald of the Russian Academy of Sciences* 78 (2), (1 April 2008), pp. 165–6.

Fando, R. A., and I. A. Zakharov, 'An Unknown Page in the History of Russian Genetics: S. I. Alikhanyan's Letter to I. V. Stalin', *Russian Journal of Genetics*, 42 (2006), 1329–40

Fedorov, N. F., *What Was Man Created for? The Philosophy of the Common Task: Selected Works*, ed. Marilyn Minto, trans. Elisabeth Koutaissoff (Honeyglen, 1990)

Fersman, A. E., 'Iz rechi, proiznesennoi na godichnom akte Geograficheskogo Instituta', in *Problemy organizatsy nauki v trudakh sovetskikh uchenykh, 1917–1930* (Nauka, 1990)

Feuer, Lewis S., 'Dialectical Materialism and Soviet Science', *Philosophy of Science*, 16 (1949), 105–24

Field, Mark G., *Soviet Socialized Medicine: An Introduction* (Free Press, 1967)

Figes, Orlando, *The Whisperers: Private Life in Stalin's Russia* (Allen Lane, 2007)

Filner, Robert E., 'Science and Marxism in England, 1930–1945', *Science and Nature*, 3 (1980), 60–9

Finkel, Stuart, *On the Ideological Front: The Russian Intelligentsia and the Making of the Soviet Public Sphere* (Yale University Press, 2007)

——, 'Purging the Public Intellectual: The 1922 Expulsions from Soviet Russia', *The Russian Review*, 62 (2003), 589–613

——, '*The Brains of the Nation*': *The Expulsion of Intellectuals and the Politics of Culture in Soviet Russia, 1920–1924* (Stanford University, 2001)

Fisher, R. A., 'What Sort of Man Is Lysenko?', *Occasional Pamphlet of the Society for Freedom in Science*, 1948, 6–9

Fitzpatrick, Sheila, *Cultural Revolution in Russia, 1928–1931* (Indiana University Press, 1978)

——, *Education and Social Mobility in the Soviet Union, 1921–1934* (Cambridge University Press, 1979)

——, *Everyday Stalinism: Ordinary Life in Extraordinary Times: Soviet Russia in the 1930s* (Oxford University Press, 2000)

——, *Stalinism: New Directions* (Routledge, 1999)

——, *Stalin's Peasants: Resistance and Survival in the Russian Village after Collectivization* (Oxford University Press, 1994)

——, *The Commissariat of Enlightenment: Soviet Organization of Education and the Arts under Lunacharsky, October 1917–1921* (Cambridge University Press, 2002)

——, *The Cultural Front: Power and Culture in Revolutionary Russia* (Cornell University Press, 1992)

Bibliography

Fitzpatrick, Sheila, Alexander Rabinowitch and Richard Stites, *Russia in the Era of NEP: Explorations in Soviet Society and Culture* (Indiana University Press, 1991)

Fitzpatrick, Sheila, and Katherine Verdery, 'Tear Off the Masks! Identity and Imposture in Twentieth-Century Russia', *The Journal of Modern History*, 79 (2007), 959

Fofanova, Margarita, *O Vladimire Hyiche Lenine: Vospominaniya* [*About Vladimir Llyich Lenin: Reminiscences*] (Gospolitizdat, 1963)

Fokin, Sergei, 'Russian Zoologists in Naples', *Science in Russia* (2010), 71–8

Fortescue, Stephen, *The Communist Party and Soviet Science* (Palgrave Macmillan, 1986)

Fortunatov, Boris Konstantinovich, 'O General'nom Plane Rekonstruktsii Promyslovoi Fauny Evropeiskoi Chasti SSSR I Ukrainy', *Priroda i Sotsialisticheskoe Khoziaistvo* (1933), 90–109

Fraser, Jennifer, and Anton Yasnitsky, 'Deconstructing Vygotsky's Victimization Narrative: A Re-Examination of the "Stalinist Suppression" of Vygotskian Theory' (University of Toronto, 2014)

Fridman, E. P., and D. M. Bowden, 'The Russian Primate Research Center – A Survivor', *Laboratory Primate Newsletter*, 48 (2009)

Fueloep-Miller, René, *The Mind and Face of Bolshevism* (Harper & Row, 1965)

Fyodorov, S. P., 'Surgery at the Crossroads', *Journal of Surgery*, ed. I. I. Grekov [in Russian] (1996), vol. 155, no. 6, p. 114.

Gade, Daniel W., 'Converging Ethnobiology and Ethnobiography: Cultivated Plants, Heinz Brücher, and Nazi Ideology', *Journal of Ethnobiology*, 26 (2006), 82–106

Gaissinovitch A. E., 'The Origins of Soviet Genetics and the Struggle with Lamarckism, 1922–1929', trans. Mark B Adams, *Journal of the History of Biology*, 13 (1980), 1–51

Gajewski, W., 'The Grim Heritage of Lysenkoism: Four Personal Accounts. II. Lysenkoism in Poland', *Quarterly Review of Biology*, 65 (1990), 423–34

Gall, Yasha, 'The Botanist V. N. Sukachev and the Development of Darwin's Ideas in Russia', *Ludus Vitalis*, 17 (2009), 25–32

Ganson, Nicholas, *The Soviet Famine of 1946–47 in Global and Historical Perspective* (Palgrave Macmillan, 2009)

Gantt, W. Horsley, 'Reminiscences of Pavlov', *Journal of the Experimental Analysis of Behavior*, 20 (1973), 131–6

Gare, Arran, 'Aleksandr Bogdanov and Systems Theory', *Democracy and Nature*, 6 (2000), 341–59

——, 'Aleksandr Bogdanov: Proletkult and Conservation', *Capitalism Nature Socialism*, 5 (1994), 65–94

——, 'Aleksandr Bogdanov's History, Sociology and Philosophy of Science', *Studies in History and Philosophy of Science Part A*, 31 (2000), 231–48

Gastev, A. K., *Poeziia rabochego udara* [*Poetry of the Factory Floor*] (Khudozh. lit., 1971)

——, *Ustanovka rabochei sily*, 1929, no. 1–2, p. 11.

Gates, R. Ruggles, 'International Congress of Genetics', *Nature*, 1927, 495–6

Geldern, James von, *Bolshevik Festivals, 1917-1920* (University of California Press, 1993)

Gerovitch, Slava, *From Newspeak to Cyberspeak: A History of Soviet Cybernetics* (MIT, 2004)

——, 'Love–Hate for Man–Machine Metaphors in Soviet Physiology: From Pavlov to "Physiological Cybernetics"', *Science in Context*, 15 (2002), 339–74

——, '"Russian Scandals": Soviet Readings of American Cybernetics in the Early Years of the Cold War', *The Russian Review*, 60 (2001), 545–68

Gershenson, S. M., 'The Grim Heritage of Lysenkoism: Four Personal Accounts. IV. Difficult Years in Soviet Genetics', *Quarterly Review of Biology* 65, no. 4 (1990), 447–56

Gerson, Lennard D., *The Secret Police in Lenins's Russia* (Temple University Press, 1976)

Getty, John Archibald, *Origins of the Great Purges: The Soviet Communist Party Reconsidered, 1933–1938* (Cambridge University Press, 1987)

——, 'Samokritika Rituals in the Stalinist Central Committee, 1933–38', *The Russian Review*, 58 (1999), 49–70

Gill, Graeme J., *The Origins of the Stalinist Political System* (Cambridge University Press, 2002)

Glad, John, 'Hermann J. Muller's 1936 Letter to Stalin', *Mankind Quarterly*, 43 (2003)

Glants, Musya, and Joyce Stetson Toomre, *Food in Russian History and Culture* (Indiana University Press, 1997)

Gleason, Abbott, and Richard Stites, *Bolshevik Culture: Experiment and Order in the Russian Revolution* (Indiana University Press, 1989)

Gloveli, Georgii D., '"Socialism of Science" versus "Socialism of Feelings": Bogdanov and Lunacharsky', *Studies in Soviet Thought*, 42 (1991), 29–55

Golubev, G. N., *Nikolai Vavilov: The Great Sower: Pages from the Life of the Scientist* (Mir, 1987)

Gordin, Michael D., Karl Hall and A. B. Kojevnikov (eds), *Intelligentsia Science: The Russian Century, 1860–1960* (University of Chicago Press, 2008)

Gorelik, George E., 'Bogdanov's Tektology: Its Nature, Development and Influence', *Studies in Soviet Thought*, 26 (1983), 39–57

Gorelik, George E., and Antonina W. Bouis, *The World of Andrei Sakharov: A Russian Physicist's Path to Freedom* (Oxford University Press, 2005)

Gorky, Maxim W., *The White Sea Canal: Being an Account of the Construction of the New Canal between the White Sea and the Baltic Sea,* ed. Amabel Williams-Ellis, trans. L Averbakh and S. G. Firin (John Lane, 1935)

———, 'V. I. Lenin', trans. David Walters, 1924, Lenin Museum and Maxim Gorky Internet Archive (https://www.marxists.org/archive/gorky-maxim/1924/01/x01.htm)

———, *Sobranie sochinenij [Collected Works]*, vol. 27, p. 43.

Gormley, Melinda, 'Geneticist L. C. Dunn: Politics, Activism, and Community' (Oregon State University, 2006)

Graham, Loren R., *Moscow Stories* (Indiana University Press, 2006)

———, *Science and Philosophy in the Soviet Union* (Knopf, 1972)

———, *Science and the Soviet Social Order* (Harvard University Press, 1990)

———, 'Science and Values: The Eugenics Movement in Germany and Russia in the 1920s', *American Historical Review*, 82 (1977), 1133–64

———, *Science in Russia and the Soviet Union: A Short History* (Cambridge University Press, 1994)

———, *Science, Philosophy, and Human Behavior in the Soviet Union* (Columbia University Press, 1987)

———, 'The Formation of Soviet Research Institutes: A Combination of Revolutionary Innovation and International Borrowing', *Social Studies of Science*, 5 (1975), 303–29

———, *The Ghost of the Executed Engineer: Technology and the Fall of the Soviet Union* (Harvard University Press, 1993)

———, 'The Socio-Political Roots of Boris Hessen: Soviet Marxism and the History of Science', *Social Studies of Science*, 15 (1985)

———, *What Have We Learned about Science and Technology from the Russian Experience?* (Stanford University Press, 1998)

———, *Lysenko's Ghost: Epigenetics and Russia* (Harvard University Press, 2016)

Granin, Daniil Aleksandrovich, *The Bison: A Novel About the Scientist Who Defied Stalin* (Doubleday, 1990)

Gratzer, W. B., *The Undergrowth of Science: Delusion, Self-Deception, and Human Frailty* (Oxford University Press, 2000)

Green, Judy, and Jeanne LaDuke, *Pioneering Women in American Mathematics: The Pre-1940 PhD's* (American Mathematical Society, 2009)

Gregory, Paul R., *Lenin's Brain and Other Tales from the Secret Soviet Archives* (Hoover Institution Press, 2008)

Grigorenko, Elena L., Patricia Ruzgis and Robert J. Sternberg, *Psychology of Russia: Past, Present, Future* (Nova Science Publishers, 1997)

Grigorian, N. A., 'L. A. Orbeli – Outstanding Physiologist and Science Leader of the Twentieth Century', *Journal of the History of the Neurosciences*, 16 (2007), 181–93

———, 'N. K. Koltsov I Experimental'naya Genetika Vysshei Nervoi Deyatel'nosti' ['N. K. Koltsov and the Genetics of Higher Nervous Activity'], *Priroda* (1992), 93–7

Grigoriev, A. I., and N. A. Grigorian, 'The Difficult Years of the Leader of Physiology', *Herald of the Russian Academy of Sciences*, 77 (2007), 254–61

Hachinski, Vladimir, 'Stalin's Last Years: Delusions or Dementia?', *European Journal of Neurology*, 6 (1999), 129–32

Hagemeister, Michael, 'Die Eroberung des Raums und die Beherrschung der Zeit: utopische, apokalyptische und magisch-okkulte Elemente in den Zukunftsentwürfen der Sowjetzeit', *Die Musen der Macht*, 2003, 257–84

Haldane, J. B. S, 'In Defense of Genetics', *Modern Quarterly*, 1949, 14–20

——, 'Lysenko and Genetics', *Science and Society*, 4 (1940)

Hall, Karl, 'The Schooling of Lev Landau: The European Context of Postrevolutionary Soviet Theoretical Physics', *Osiris*, 23 (2008), 230–59

Hamilton, David, *The Monkey Gland Affair* (Chatto & Windus, 1986)

Hardcastle, John, 'Of Dogs and Martyrs. Sherrington, Richards, Pavlov and Vygotsky', *Changing English*, 12 (2005), 31–42

Harlan, Jack R., *The Living Fields: Our Agricultural Heritage* (Cambridge University Press, 1995)

Harland, S. C., 'Nicolai Ivanovitch Vavilov, 1885–1942', *Obituary Notices of Fellows of the Royal Society*, 9 (1954), 259–64

Harré, Rom, *Pavlov's Dogs and Schrödinger's Cat: Tales from the Living Laboratory* (Oxford University Press, 2008)

Harrison, Mark, *The Soviet Economy in the 1920's and 1930's: A Survey of New Research in Britain and the U.S.A (1966–1976)* (Department of Economics, University of Warwick, 1977)

Harwood, William Sumner, *The New Earth: A Recital of the Triumphs of Modern Agriculture in America* (Macmillan, 1906)

Haymaker, W., 'Cécile and Oskar Vogt, on the Occasion of Her 75th and His 80th Birthday', *Neurology*, 1 (1951)

Haynal, André, *Psychoanalysis and the Sciences: Epistemology–History* (University of California Press, 1993)

Haynes, Michael, and Rumy Husan, *A Century of State Murder?: Death and Policy in Twentieth-Century Russia* (Pluto Press, 2003)

Healey, Dan, 'Russian and Soviet Forensic Psychiatry: Troubled and Troubling', *International Journal of Law and Psychiatry*, 37 (2014), 71–81

Hellebust, Rolf, *Flesh to Metal: Soviet Literature and the Alchemy of Revolution* (Cornell University Press, 2003)

Herlihy, Patricia, *The Alcoholic Empire: Vodka and Politics in Late Imperial Russia* (Oxford University Press, 2002).

Herriot, Édouard, *Eastward from Paris,* trans. Phyllis Marks Mégroz (V. Gollancz, 1934)

Hessen, Boris, Henryk Grossmann, Gideon Freudenthal and Peter McLaughlin, *The Social and Economic Roots of the Scientific Revolution: Texts by Boris Hessen and Henryk Grossmann* (Springer, 2009)

Hill, Alexander, 'Soviet Planning for War, 1928–June 1941', in *A Companion to World War II*, ed. Thomas W. Zeiler and Daniel M. DuBois (Blackwell, 2012), 91–101

Hill, Fiona, and Clifford G. Gaddy, *The Siberian Curse: How Communist Planners Left Russia out in the Cold* (Brookings Institution Press, 2003)

Hindus, Maurice, *Broken Earth* (Read Books, 2006)

——, 'Henry Ford Conquers Russia', *The Outlook*, 1927, 280–2

——, *Red Bread: Collectivization in a Russian Village* (J. Cape & H. Smith, 1931)

Hoffmann, David L., *Cultivating the Masses: Modern State Practices and Soviet Socialism, 1914–1939* (Cornell University Press, 2011)

Holloway, David, 'Scientific Truth and Political Authority in the Soviet Union', *Government and Opposition*, 5 (1970), 345–67

——, *Stalin and the Bomb: The Soviet Union and Atomic Energy, 1939–1956* (Yale University Press, 1994)

——, 'The Scientist and the Tyrant', *New York Review of Books* (1 March 1990)

Horvath, Robert, 'The Poet of Terror: Dem'ian Bednyi and Stalinist Culture', *The Russian Review*, 65 (2006), 53–71

Huestis, Douglas W., 'Alexander Bogdanov: The Forgotten Pioneer of Blood Transfusion', *Transfusion Medicine Reviews*, 21 (2007), 337–40

Husband, William B., '"Correcting Nature's Mistakes": Transforming the Environment and Soviet Children's Literature, 1928–1941', *Environmental History*, 11 (2006), 300–18

Huschtscha, Lily, '"I Am Not a Supporter of Simplistic Explanations . . .", An Interview with Zhores Medvedev', *Biogerontology*, 5 (2004), 129–36

Huxley, Julian, *Soviet Genetics and World Science: Lysenko and the Meaning of Heredity* (Chatto & Windus, 1949)

'"I Lived a Happy Life" – In Honor of the 90th Anniversary of the Birth of Timofeev-Resovskij', *Priroda* (1990), 68–104

International Congress of Genetics, and Reginald Crundall Punnett, eds., *Proceedings of the Seventh International Genetical Congress: Edinburgh, Scotland, 23–30 August 1939* (Cambridge University Press, 1941)

Ivanov, Valery I., 'A True Scientist and a Most Amiable Person: To the Centenary of the Birthday of H. A. Timofeeff-Ressovsky', trans. Irina V. Kronshtadtova, 1999; available at http://bit.ly/1Pn7uEo

Ivanov, Valery I., and N. A. Liapunova, 'Nikolay W. Timofeeff-Ressovsky (1900–1981): An Essay on His Life and Scientific Achievements', *Advances in Mutagenesis Research*, ed. Günter Obe, Vol. 4 (Springer, 1993)

Ivanov, Viacheslav, 'Why Did Stalin Kill Gorky?', *Russian Social Science Review*, 35 (1994), 49–92

James, William, *Psychology: The Briefer Course* (Dover Publications, 2003)

——, *The Principles of Psychology: Volume 1* (Dover Publications, 2000)

Jensen, Kenneth Martin, *Beyond Marx and Mach: Aleksandr Bogdanov's Philosophy of Living Experience* (Springer Science & Business, 1978)

——, 'Red Star: Bogdanov Builds a Utopia', *Studies in East European Thought*, 23 (1982), 1–34

Johansson, Kurt, and A. K. Gastev, *Aleksej Gastev, Proletarian Bard of the Machine Age* (Almqvist & Wiksell International, 1983)

Jones, Donald F., *Proceedings of the Sixth International Congress of Genetics: Ithaca, New York, 1932*, 1 1 (Brooklyn Botanic Garden, 1932)

Jones, Gareth Stedman, 'Engels and the Genesis of Marxism', *New Left Review*, 1 (1977)

Joravsky, David, *Cultural Revolution and the Fortress Mentality* (Wilson Center, Kennan Institute for Advanced Russian Studies, 1981)

——, *Russian Psychology: A Critical History* (Blackwell, 1989)

——, 'Soviet Marxism and Biology before Lysenko', *Journal of the History of Ideas*, 20 (1959), 85–104

——, *Soviet Marxism and Natural Science 1917–1932* (Columbia University Press, 1961)

——, 'Soviet Views on the History of Science', *Isis*, 46 (1955), 3–13

——, 'The Impossible Project of Ivan Pavlov (and William James and Sigmund Freud)', *Science in Context*, 5 (1992), 265–80

——, *The Lysenko Affair* (Harvard University Press, 1970)

——, 'The Mechanical Spirit: The Stalinist Marriage of Pavlov to Marx', *Theory and Society*, 4 (1977)

——, 'The Stalinist Mentality and the Higher Learning', *Slavic Review*, 42 (1983), 575–600

Josephson, Paul R., *Industrialized Nature: Brute Force Technology and the Transformation of the Natural World* (Island Press/Shearwater Books, 2002)

——, *Physics and Politics in Revolutionary Russia* (University of California Press, 1991)

——, *Resources under Regimes: Technology, Environment, and the State* (Harvard University Press, 2006)

——, 'Science Policy in the Soviet Union, 1917–1927', *Minerva*, 26 (1988), 342–69

——, *Would Trotsky Wear a Bluetooth?: Technological Utopianism under Socialism, 1917–1989* (Johns Hopkins University Press, 2010)

Kammerer, Paul, *The Inheritance of Acquired Characteristics* (Boni & Liveright, 1924)

Kapitsa, Petr Leonidovich, *Letters to Mother: The Early Cambridge Period*, ed. David J. Lockwood (National Research Council Canada, 1989)

Kapitsa, Petr Leonidovich, J. W. Boag, P. E. Rubinin and D. Shoenberg, *Kapitza in Cambridge and Moscow: Life and Letters of a Russian Physicist* (Elsevier, 1990)

Kassow, Samuel D., *Students, Professors, and the State in Tsarist Russia* (University of California Press, 1989)

Kats, Yefim, 'Bogdanov, Marx, and the Limits to Growth Debate', *The European Legacy*, 9 (2004), 305–16

Kedrov, F. B., *Kapitza – Life and discoveries*, trans. Mark Fradkin, ed. John Crowfoot (Mir, 1984)

Kepley, Vance, 'The Scientist as Magician: Dovzhenko's "Michurin" and the Lysenko Cult', *Journal of Popular Film and Television*, 8 (1980), 19

Kevles, Daniel J., *In the Name of Eugenics: Genetics and the Uses of Human Heredity* (Harvard University Press, 2001)

Khlevnyuk, Oleg V., 'The Objectives of the Great Terror, 1937–1938', in *Stalinism*, ed. David L. Hoffmann (Blackwell, 2003), 82–104

Khlevnyuk, Oleg V., and Nora Seligman Favorov, *Stalin: New Biography of a dictator*, 2015

Khomskaia, E. D., and David E. Tupper, *Alexander Romanovich Luria: A Scientific Biography* (Kluwer Academic/Plenum Publishers, 2001)

Khrushchev, N. S., *Khrushchev Remembers* (André Deutsch, 1971), pp. 200–15.

——, *Memoirs of Nikita Khrushchev*, Volume 1: *Commissar, 1918–1945*, ed. Sergei Khrushchev (Penn State University Press, 2004)

Kichigina, Galina, *The Imperial Laboratory: Experimental Physiology and Clinical Medicine in Post-Crimean Russia* (Rodopi, 2009)

Kirschenmann, P., 'On the Kinship of Cybernetics to Dialectical Materialism', *Studies in Soviet Thought*, 6 (1966), 37–41

Kitson, H. D., 'Report of the Seventh International Congress on Psychotechnics', *Journal of Applied Psychology*, 15 (1931), 593

Kneen, Peter, 'Physics, Genetics and the Zhdanovshchina', *Europe–Asia Studies*, 50 (1998), 1183–1202

Koblitz, Ann, *Science, Women and Revolution in Russia* (Routledge, 2014)

Koenker, Diane, William Rosenberg and Ronald Suny, *Party, State and Society in the Russian Civil War: Explorations in Social History* (Indiana University Press, 1989)

Koestler, Arthur, *The Case of the Midwife Toad* (Picador, 1975)

Kohler, Robert E., *Lords of the Fly: Drosophila Genetics and the Experimental Life* (University of Chicago Press, 1994)

Kojevnikov, Alexei B., 'Piotr Kapitza and Stalin's Government: A Study in Moral Choice', *Historical Studies in the Physical and Biological Sciences*, 22 (1991), 131–64

——, 'President of Stalin's Academy: The Mask and Responsibility of Sergei Vavilov', *Isis*, 87 (1996), 18–50

——, 'Rituals of Stalinist Culture at Work: Science and the Games of Intraparty Democracy circa 1948', *Russian Review*, 57 (1998), 25–52

——, *Stalin's Great Science: The Times and Adventures of Soviet Physicists* (Imperial College Press, 2004)

——, 'The Great War, the Russian Civil War, and the Invention of Big Science', *Science in Context*, 15 (2002), 239–75

Bibliography

Kolchinsky, Eduard I., 'Nikolai Vavilov in the Years of Stalin's "Revolution from Above" (1929–1932)', *Centaurus*, Vol. 56, Issue 4 (2014)

Kolman, Arnost, 'A Life-Time in Soviet Science Reconsidered: The Adventure of Cybernetics in the Soviet Union', *Minerva*, 16 (1978), 416–24

——, *Zhizn' I Tekhnika Budushchego* (Moscow, 1928)

Kopelev, Lev, *Ease My Sorrows: A Memoir* (Random House, 1983)

Kornilov, K. N., *Psikhologiia i marksizm: sbornik stateĭ sotrudnikov moskovskogo gosudarstvennogo instituta ėksperimental'nogo psikhologii* (Gos. izd-vo, 1925)

Korochkin, L. I., 'Behavioral Genetics and Neurogenetics and Their Development in Russia', *Russian Journal of Genetics*, 40 (2004), 585–9

Korogodin, V. I., 'Master: Recollection about N.V. Timofeeff-Ressovsky' (http://wwwinfo.jinr.ru/~drrr/Timofeeff/auto/korogodin_e.html)

Korogodin, V. I., G. G. Polikarpov and V. V. Velkov, 'The Blazing Life of N. V. Timofeeff-Ressovsky', *Journal of Biosciences*, 25 (2000), 125–31

Korolenko, Caesar P., and Dennis V. Kensin, 'Reflections on the Past and Present State of Russian Psychiatry', *Anthropology and Medicine*, 9 (2002), 51–64

Koshtoyants, Kh. S., and Donald B. Lindsey, *Essays on the History of Physiology in Russia* (American Institute of Biological Sciences, 1964)

Kotkin, Stephen, *Magnetic Mountain: Stalinism as a Civilization* (University of California Press, 1995)

——, *Steeltown, USSR: Soviet Society in the Gorbachev Era* (University of California Press, 1991)

Kozulin, Alex, *Psychology in Utopia: Toward a Social History of Soviet Psychology* (MIT Press, 1984)

Krementsov, Nikolai L., *A Martian Stranded on Earth: Alexander Bogdanov, Blood Transfusions, and Proletarian Science* (University of Chicago Press, 2011)

——, 'A "Second Front" in Soviet Genetics: The International Dimension of the Lysenko Controversy, 1944–1947', *Journal of the History of Biology*, 29 (1996), 229–50

——, 'Big Revolution, Little Revolution: Science and Politics in Bolshevik Russia', *Social Research*, 73 (2006), 1173–1204

——, 'Darwinism, Marxism, and Genetics in the Soviet Union', in Denis Alexander and Ronald L. Numbers, *Biology and Ideology from Descartes to Dawkins* (Chicago University Press, 2010)

——, 'From "Beastly Philosophy" to Medical Genetics: Eugenics in Russia and the Soviet Union', *Annals of Science*, 68 (2011), 61–92

——, 'Hormones and the Bolsheviks: From Organotherapy to Experimental Endocrinology, 1918–1929', *Isis*, 99 (2008), 486–518

——, *International Science between the World Wars: The Case of Genetics* (Routledge, 2005)

——, *Stalinist Science* (Princeton University Press, 1997)

——, *The Cure: A Story of Cancer and Politics from the Annals of the Cold War* (University of Chicago Press, 2004)

Bibliography

Kreutzberg, G. W., 'If You Had Met Him, You Would Know', *Brain Pathology* 2, no. 4 [1 October 1992], pp. 365–9.

Kreutzberg, Georg W., Igor Klatzo and Paul Kleihues, 'Oskar and Cécile Vogt, Lenin's Brain and the Bumble-Bees of the Black Forest', *Brain Pathology*, 2 (1992), 363–9

Kropotkin, Petr Alekseyevich, *Mutual Aid: A Factor of Evolution* (Heinemann, 1904)

Krupskaya, N., 'Sistema Teilora I Organizatsiya Raboty Sovetskikh Uchrezhdenii', *Krasnaya Nov'*, 1 (1921)

Kupzow, Alexander J., 'Vavilov's Law of Homologous Series at the Fiftieth Anniversary of Its Formulation', *Economic Botany*, 29 (1975), 372–9

Kuromiya, Hiroaki, *Stalin's Industrial Revolution: Politics and Workers, 1928–1932* (Cambridge University Press, 1988)

——, 'The Crisis of Proletarian Identity in the Soviet Factory, 1928–1929', *Slavic Review* 44, no. 2 (1985), pp. 286–7.

Kursell, Julia, 'Piano Mécanique and Piano Biologique: Nikolai Bernstein's Neurophysiological Study of Piano Touch', *Configurations*, 14 (2006), 245–73

Kuzmin, A., 'K kakomu khramu ischchem mi dorogu?' ['Toward What Sort of Cathedral are We Searching for a Road?'], *Nash Sovremennik* 1988, Vol. 3, pp. 154–64.

Langdon-Davies, John, *Russia Puts the Clock Back; a Study of Soviet Science and Some British Scientists* (Gollancz, 1949)

Lanska, D. J., 'Vogt, Cécile and Oskar', in *Encyclopedia of the Neurological Sciences (Second Edition)*, ed. Michael J. Aminoff and Robert B. Daroff (Academic Press, 2014), 722–5

Lapo, Andrei V., 'Why to Paris?', *Noosphere*, 14 (2002)

Larson, Edward J., 'Biology and the Emergence of the Anglo-American Eugenics Movement', in Denis Alexander and Ronald L Numbers, *Biology and Ideology from Descartes to Dawkins* (University of Chicago Press, 2010)

Latash, Mark L., 'Bernstein's "Desired Future" and Physics of Human Movement', in Mihai Nadin, *Anticipation: Learning from the Past* (Springer, 2015)

Lavretsky, H., 'The Russian Concept of Schizophrenia: A Review of the Literature', *Schizophrenia Bulletin*, 24 (1998), 537–57

Lazcano, Antonio, 'Historical Development of Origins Research', *Cold Spring Harbor Perspectives in Biology*, 2 (2010)

Lecourt, Dominique, *Proletarian Science?: The Case of Lysenko* (NLB Humanities Press, 1977)

Lenin Academy of Agricultural Sciences of the U.S.S.R., *The situation in biological science* (Foreign Languages Publishing House, 1949)

Lenin, V. I., *Collected Works*, Volume 14: *1908* (Lawrence and Wishart, 1962)

——, *Collected Works*, Volume 20: *December 1913–August 1914* (Lawrence & Wishart, 1972)

Bibliography

——, *Collected Works*, Volume 33: *August 1921–March 1923* (Progress, 1966)

——, 'Letter to Aleksei Maksimovich Gorky', 15 September 1919, US Library of Congress, http://www.loc.gov/exhibits/archives/g2aleks.html

——, *Materialism and Empirio-Criticism: Critical Comments on a Reactionary Philosophy* (Progress, 1970)

Lepeshinskaya, Olga Borisovna, *The Origin of Cells from Living Substance* (Foreign Languages Publishing House, 1954)

Lerner, Vladimir, Jacob Margolin and Eliezer Witztum, 'Vladimir Bekhterev: His Life, His Work and the Mystery of His Death', *History of Psychiatry*, 16 (2005), 217–27

——, 'Vladimir Mikhailovich Bekhterev (1857–1927)', *Journal of Neurology*, 253 (2006), 1518–19

Levin, Aleksey E., 'Expedient Catastrophe: A Reconsideration of the 1929 Crisis at the Soviet Academy of Science', *Slavic Review*, 47 (1988), 261–79

Levina, Elena S., Vladimir D. Yesakov and Lev L. Kisselev, 'Nikolai Vavilov: Life in the Cause of Science or Science at a Cost of Life', *Comprehensive Biochemistry*, 44 (2005), 345–410

Levins, Richard, and Richard C. Lewontin, *The Dialectical Biologist* (Harvard University Press, 1985)

Levit, Georgy S., *Biogeochemistry – Biosphere – Noosphere: The Growth of the Theoretical System of Vladimir Ivanovich Vernadsky* (Verlag für Wissenschaft und Bildung, 2001)

Levit, Georgy S., and Uwe Hossfeld, 'From Molecules to the Biosphere: Nikolai V. Timoféeff-Ressovsky's (1900–1981) Research Program within a Totalitarian Landscape', *Theory in Biosciences*, 128 (2009), 237–48

Levitin, Karl, and Vasiliĭ Vasil'evich Davydov, *One Is Not Born a Personality: Profiles of Soviet Education Psychologists* (Progress Publishers, 1982)

Leyda, Jay, *Kino: A History of the Russian and Soviet Film* (Princeton University Press, 1983)

Liehm, Mira, and Antonín J. Liehm, *The Most Important Art: Eastern European Film after 1945* (University of California Press, 1977)

Lomov, B. F., V. A. Koltsova and E. I. Stepanova, 'Ocherk Zhizni i nauchnoi deyatelnosti Vladimira Mikhailovicha Bekhtereva' ['Essay of life and works of V. M. Bekhterev'], *Ob'ektivnaya Psykhologia* [*Objective Psychology*], ed. V. A. Koltsova (Nauka, 1991), pp. 424–44.

Loskutov, Igor G., *Vavilov and His Institute: A History of the World Collection of Plant Genetic Resources in Russia* (International Plant Genetic Resources Institute, 1999)

Lubrano, Linda L., 'The Hidden Structure of Soviet Science', *Science, Technology, and Human Values*, 18 (1993), 147–75

Lubrano, Linda L., and Susan Gross Solomon, *The Social Context of Soviet Science* (Westview Press, 1980)

Lunacharskaia, Irina, 'Why Did Commissar of Enlightenment A. V. Lunacharskii Resign?', trans. Kurt S. Schultz, *Russian Review*, 51 (1992), 319–42

Lunacharsky, A. V., 'Intelligentsiia i ee mesto v sotsialisticheskom stroitel'stve', *Revoliutsiia i kul'tura*, 1927, 32–3

——, 'Kak Voznik Stsenarii "Salamandra"', *Sovetskiiekran*, 1 (1929)

Luria, A. R., *Cognitive Development, Its Cultural and Social Foundations* (Harvard University Press, 1976)

——, 'L. S. Vygotsky', *Journal of Personality*, 3 (1935), 238–40

——, 'Neuropsychological Studies in the USSR. A Review (Part I)', *Proceedings of the National Academy of Sciences of the United States of America*, 70 (1973), 959–64

——, 'Neuropsychological Studies in the USSR. A Review (Part II)', *Proceedings of the National Academy of Sciences of the United States of America*, 70 (1973), 1278–83

——, *The Autobiography of Alexander Luria: A Dialogue with The Making of Mind*, ed. Michael Cole, trans. Karl E. Levitin (Erlbaum, 2006)

——, *The Mind of a Mnemonist*, ed. Jerome S. Bruner, trans. Lynn Solotaroff (Basic Books, 1960)

Luria, A. R., and L. S. Vygotsky, *Ape, Primitive Man, and Child: Essays in the History of Behavior* (Harvester/Wheatsheaf, 1992)

Lyons, Eugene, *Assignment in Utopia* (Transaction Publishers, 1991)

Lysenko, Trofim Denisovich, *Agrobiology: Essays on Problems of Genetics, Plant Breeding and Seed Growing* (Foreign Languages Publishing House, 1954)

——, *Heredity and Its Variability*, trans. Theodosius Dobzhansky (King's Crown Press, 1946)

——, *New Developments in the Science of Biological Species* (Foreign Languages Publishing House, 1951)

Maier, Charles S., *Between Taylorism and Technocracy: European Ideologies and the Vision of Industrial Productivity in the 1920s* (Weidenfeld & Nicolson, 1970)

Mansueto, Anthony, 'From Dialectic to Organization: Bogdanov's Contribution to Social Theory', *Studies in East European Thought*, 48 (1996), 37–61

Marx, Karl, 'Private Property and Communism', in *Economic and Philosophic Manuscripts of 1844* (International Publishers, 1964), 135–43

Mayr, E., 'Roots of Dialectical Materialism', ed. Eduard I. Kolchinsky, *Na Perelome: Sovetskaia Biologiaa v 20–30kh Godakh*, 20 (1997), 12–17

McBurney, Gerard, 'Some Frequently Asked Questions about Shostakovich's "Orango"', *TEMPO*, 64 (2010), 38–40

McCannon, John, *Red Arctic: Polar Exploration and the Myth of the North in the Soviet Union, 1932–1939* (Oxford University Press, 1998)

——, 'The Commissariat of Ice: The Main Administration of the Northern Sea Route (GUSMP) and Stalinist Exploitation of the Arctic, 1932–1939', *Journal of Slavic Military Studies*, 20 (2007), 393–419

McCutcheon, Robert A., 'The 1936–1937 Purge of Soviet Astronomers', *Slavic Review*, 50 (1991), 100–17

Bibliography

Medvedev, Zhores A., 'Nikolai Wladimirovich Timoféeff-Ressovsky (1900–1981)', *Genetics*, 100 (1982), 1–5

——, *The Medvedev Papers*, trans. Vera Rich (St Martin's Press, 1971)

——, *The Rise and Fall of T. D. Lysenko*, trans. Michael I. Lerner (Columbia University Press, 1969)

——, *The Unknown Stalin* (I. B. Tauris, 2003)

Meijer, Onno G., and Sjoerd M. Bruijn, 'The Loyal Dissident: N. A. Bernstein and the Double-Edged Sword of Stalinism', *Journal of the History of the Neurosciences*, 16 (2007), 206–24

Miller, Martin A., 'Freudian Theory under Bolshevik Rule: The Theoretical Controversy during the 1920s', *Slavic Review*, 44 (1985), 625–46

Montefiore, Simon Sebag, *Stalin: The Court of the Red Tsar* (Weidenfeld & Nicolson, 2003)

Moon, David, 'The Environmental History of the Russian Steppes: Vasilii Dokuchaev and the Harvest Failure of 1891', *Transactions of the Royal Historical Society* (Sixth Series), 15 (2005), 149–74

Morgan, Thomas Hunt, *Evolution and Genetics* (Princeton University Press, 1925)

——, *The Mechanism of Mendelian Heredity* (Holt, 1915)

——, 'The Relation of Genetics to Physiology and Medicine', *Prix Nobel En 1933*, 1935

——, *The Theory of the Gene* (Hafner, 1964)

Muller, H. J., 'Artificial Transmutation of the Gene', *Science*, New Series, 66 (1927), 84–7

——, 'It Still Isn't a Science: A Reply to George Bernard Shaw', *Saturday Review of Literature*, 1949, 11–12, 61

——, 'Observations of Biological Science in Russia', *Scientific Monthly*, 16 (1923), 539–52

——, *Out of the Night: A Biologist's View of the Future* (Vanguard Press, 1935)

——, 'The Dominance of Economics over Eugenics', *Scientific Monthly*, 37 (1933), 40–7

——, 'The Measurement of Gene Mutation Rate in Drosophila, Its High Variability, and Its Dependence upon Temperature', *Genetics*, 13 (1928), 279–357

Nabhan, Gary Paul, *Where Our Food Comes from: Retracing Nikolay Vavilov's Quest to End Famine* (Island Press, 2008)

Nadin, Mihai, ed., *Anticipation: Learning from the Past: The Russian/Soviet Contributions to the Science of Anticipation* (Springer, 2015)

Nathans, Benjamin, *Beyond the Pale: The Jewish Encounter with Late Imperial Russia* (University of California Press, 2004).

Naumov, I. V., and David Norman Collins, *The History of Siberia* (Routledge, 2006)

Nazaretyan, Akop P., 'Big (Universal) History Paradigm: Versions and Approaches', *Social Evolution and History*, 4 (2005), 61–86

Nell, V., 'Luria in Uzbekistan: The Vicissitudes of Cross-Cultural Neuropsychology', *Neuropsychology Review*, 9 (1999), 45–52

Newman, Fred, and Lois Holzman, *Lev Vygotsky: Revolutionary Scientist* (Routledge, 1993)

Nikolaeva, Elena I., 'Alexander Luria: Creator in the Perspective of Time', in *Anticipation: Learning from the Past* (Hanse-Wissenschaftskolleg Institute for Advanced Study, 2014)

O'Connor, Timothy E., 'Lunacharskii's Vision of the New Soviet Citizen', *Historian*, 53 (1991), 443–54

Oleynikov, Pavel V., 'German Scientists in the Soviet Atomic Project', *Nonproliferation Review*, 7 (2000), 1–30

'On the 100th Anniversary of the Birth of Nikolai Petrovich Dubinin (1907–1998)', *Russian Journal of Genetics*, 43 (2007), 92–4

Oushakine, Serguei Alex., 'The Flexible and the Pliant: Disturbed Organisms of Soviet Modernity', *Cultural Anthropology*, 19 (2004), 392–428

Ozernyuk, N. D., 'Two Anniversaries of the Institute of Developmental Biology', *Herald – Russian Academy of Science*, 77 (2007), 634–9

Paul, Diane B., 'A War on Two Fronts: J. B. S. Haldane and the Response to Lysenkoism in Britain', *Journal of the History of Biology*, 16 (1983), 1–37

——, 'H. J. Muller, Communism, and the Cold War', *Genetics*, 119 (1988), 223–25

——, '"Our Load of Mutations" Revisited', *Journal of the History of Biology*, 20 (1987), 321–35

Paul, Diane B., and C. B. Krimbas, 'Nikolai V. Timofeeff-Ressovsky', 1992, http://edoc.mdc-berlin.de/9111

Pavlov, D. S., and V. S. Shishkin, 'Significance of Biological Stations for Progress of Science and Education', *Biology Bulletin of the Russian Academy of Sciences*, 30 (2003), 422–4

Pavlov, Ivan Petrovich, *Experimental Psychology and Other Essays* (Philosophical Library, 1957)

——, *Polnoe Sobranie Sochinenni [Complete Works]* (Akademii Nauk SSSR, 1951)

——, 'New Research on Conditioned Reflexes,' *Science*, 58, no. 1506 (1923), pp. 359–361.

——, *The Work of the Digestive Glands*, trans. William Henry Thompson (C. Griffin, 1910)

Pavlov, Ivan Petrovitch, and Gleb Vasilievich Anrep, *Conditioned Reflexes: An Investigation of the Physiological Activity of the Cerebral Cortex* (Oxford University Press: Humphrey Milford, 1928)

Pearson, Karl, *The Life, Letters and Labours of Francis Galton* (Cambridge University Press, 2011)

Perutz, M., 'Physics and the Riddle of Life', *Nature*, 326 (1987), 1–20

Pethybridge, Roger W., 'The 1917 Petrograd Soviet and the Centralist Issue', *Government and Opposition*, 5 (1970), 327–44

Petrov, R. V., 'The Man Who Dared: Recollection about N. V. Timofeeff-Ressovsky' (http://wwwinfo.jinr.ru/~drrr/Timofeeff/auto/petrov_e.html)

Pipes, Richard, *Russia under the Bolshevik Regime: 1919–1924* (Harvill, 1994)

——, *Russia Under the Old Regime* (Scribner, 1975), p. 84.

Piqueras, M., 'Meeting the Biospheres: On the Translations of Vernadsky's Work', *International Microbiology: The Official Journal of the Spanish Society for Microbiology*, 1 (1998), 165–70

Podrabinek, Alexander, *Punitive Medicine* (Karoma, 1980)

Poglazov, Boris Fedorovic, et al. (eds), *Evolutionary Biochemistry and Related Areas of Physicochemical Biology* (Bach Institute of Biochemistry and Russian Academy of Sciences, 1995)

Polikarpov, G. G., 'Sketches to the Portrait of Nikolai Timofeeff-Ressovsky' (http://wwwinfo.jinr.ru/~drrr/Timofeeff/auto/polikarpov_e.html)

Pollock, Ethan, *Stalin and the Soviet Science Wars* (Princeton University Press, 2006)

Pomper, Philip, *Trotsky's Notebooks, 1933–1935: Writings on Lenin, Dialectics, and Evolutionism* (iUniverse, 1999)

Popovsky, Mark A., 'The Last Days of Nikolai Vavilov', *New Scientist*, 80 (1978), 509–11

——, *Manipulated Science: The Crisis of Science and Scientists in the Soviet Union Today* (Doubleday, 1979)

——, *The Vavilov Affair* (Archon Books, 1984)

Pribram, Karl H., 'In Memory of Alexander Romanovitsch Luria', *Neuropsychologia*, 16 (1978), 137–9

Pringle, Peter, *The Murder of Nikolai Vavilov: The Story of Stalin's Persecution of One of the Great Scientists of the Twentieth Century* (Simon & Schuster, 2008)

Pudovkin, Vsevolod, *Mechanics of the Brain* (film; Mezhrabpom-Rus, 1926)

Putrament, A., 'The Grim Heritage of Lysenkoism: Four Personal Accounts. III. How I Became a Lysenkoist', *Quarterly Review of Biology*, 65 (1990), 435–45

Qualset, Calvin O., 'Jack R. Harlan (1917–1998): Plant Explorer, Archaeobotanist, Geneticist, and Plant Breeder', in *The Origins of Agriculture and Crop Domestication* (International Plant Genetic Resources Institute, 1997)

Radzikhovskii, L. A., and E. D. Khomskaya, 'A. R. Luria and L. S. Vygotsky: Early Years of Their Collaboration', *Russian Social Science Review*, 23 (1982), 34–52

Radzinskii, Edvard, *Alexander II: The Last Great Tsar* (Free Press, 2005)

Rapoport, Yakov L'vovic, *The Doctors' Plot* (Fourth Estate, 1991)

Rappaport, Helen, *Joseph Stalin: A Biographical Companion* (ABC-CLIO, 1999)

Bibliography

Rassweiler, Anne D., *The Generation of Power: The History of Dneprostroi* (Oxford University Press, 1988)

Ratner, Vadim A., 'Nikolay Vladimirovich Timofeeff-Ressovsky (1900–1981): Twin of the Century of Genetics', *Genetics*, 158 (2001), 933–9

Reich, Walter, 'The World of Soviet Psychiatry', *New York Times* (magazine, 30 January 1983)

Rendle, Matthew, 'Revolutionary Tribunals and the Origins of Terror in Early Soviet Russia', *Historical Research*, 84 (2011), 693–721

Reznik, Semen, *Nikolai Vavilov* (Mol. gvardiia, 1968)

Richter, B. S., 'Nature Mastered by Man: Ideology and Water in the Soviet Union', *Environment and History*, 3 (1997), 69–96

Richter, Jochen, 'Pantheon of Brains: The Moscow Brain Research Institute 1925–1936', *Journal of the History of the Neurosciences*, 16 (2007), 138–49

Riehl, Nikolaus, *Stalin's Captive: Nikolaus Riehl and the Soviet Race for the Bomb*, trans. Frederick Seitz (Chemical Heritage Foundation, 1996)

Roach, Joseph, 'The Future That Worked', *Theater*, 28 (1998), 19–26

Roberts, C. E. Bechhofer, *Through Starving Russia: Being a Record of a Journey to Moscow and the Volga Provinces, in August and September, 1921* (Methuen, 1921)

Rogers, James Allen, 'Darwinism, Scientism, and Nihilism', *Russian Review*, 19 (1960), 10–23

——, 'Russian Opposition to Darwinism in the Nineteenth Century', *Isis*, 65 (1974), 487–505

——, 'The Reception of Darwin's Origin of Species by Russian Scientists', *Isis*, 64 (1973), 484–503

Rogger, Hans, 'Amerikanizm and the Economic Development of Russia', *Comparative Studies in Society and History*, 23 (1981), 382–420

Roginsky, Arsenii B., Felix F. Perchenok, and Vadim M. Borisov, 'Community as the Source of Vernadsky's Concept of Noosphere', *Configurations*, 1 (1993), 415

Rokityanskij, Yakov G., 'N. V. Timofeeff-Ressovsky in Germany (July, 1925– September, 1945)', *Journal of Biosciences*, 30 (2005), 573–80

Roll-Hansen, Nils, 'A New Perspective on Lysenko?', *Annals of Science*, 42 (1985), 261–78

——, *The Lysenko Effect: The Politics of Science* (Humanity Books, 2005)

——, 'Wishful Science: The Persistence of T. D. Lysenko's Agrobiology in the Politics of Science', *Intelligentsia Science: The Russian Century 1860–1960* ed. Michael D. Gordin (*Osiris* 23, 2008), 166–88

Rossianov, Kirill O., 'Beyond Species: Il'ya Ivanov and His Experiments on Cross-Breeding Humans with Anthropoid Apes', *Science in Context*, 15 (2002), 277–316

——, 'Editing Nature: Joseph Stalin and the "New" Soviet Biology', *Isis*, 84 (1993), 728–45

——, 'Stalin as Lysenko's Editor: Reshaping Political Discourse in Soviet Science', *Configurations*, 1 (1993), 439–56

Rubenstein, Joshua, and V. P. Naumov, *Stalin's Secret Pogrom: The Postwar Inquisition of the Jewish Anti-Fascist Committee* (Yale University Press/United States Holocaust Memorial Museum, 2001)

'Russia: Collective Congress', *Time* (25 February 1935)

'Russian Admits Ape Experiments', *New York Times* (19 June 1926)

Safonov, V., *Land in Bloom* (Foreign Languages Publishing House, 1951)

Sakharov, Andrei D., *Memoirs* (Knopf, 1990)

Santiago-Delefosse, M. J., and J.-M. O. Delefosse, 'Spielrein, Piaget and Vygotsky: Three Positions on Child Thought and Language', *Theory and Psychology*, 12 (2002), 723–47

Schmerling, Louis, *Vladimir Nikolaevich Ipatieff: November 21, 1867–November 29, 1952* (National Academy of Sciences, 1975)

Schrödinger, Erwin, *What Is Life? The Physical Aspect of the Living Cell* (The University Press, 1945)

Schultz, R. S., 'Industrial Psychology in the Soviet Union', *Journal of Applied Psychology*, 19 (1935), 265

Scott, John, *Behind the Urals: An American Worker in Russia's City of Steel* (Indiana University Press, 1989)

Sechenov, Ivan M., *Reflexes of the Brain*, ed. Kh. S. Koshtoyants, trans. S. Belsky (Massachusetts Institute of Technology, 1970)

——, *Selected Psychological and Physiological Works* (E. J. Bonset, 1968)

Service, Robert, *Lenin: A Biography* (Harvard University Press, 2000)

——, *The Russian Revolution, 1900–1927* (Palgrave Macmillan, 2009)

Shernock, Stanley Kent, 'Continuous Violent Conflict as a System of Authority', *Sociological Inquiry*, 54 (1984), 301–29

Shoenberg, D., 'Piotr Leonidovich Kapitza. 9 July 1894–8 April 1984', *Biographical Memoirs of Fellows of the Royal Society*, 31 (1985), 327–74

Shtil'mark, F. R, and Geoffrey Uil'yamovich Harper, *History of the Russian Zapovedniks, 1895–1995* (Russian Nature Press, 2003)

Siddiqi, Asif A., 'Imagining the Cosmos: Utopians, Mystics, and the Popular Culture of Spaceflight in Revolutionary Russia', *Intelligentsia Science*, ed. Michael D. Gordin et al., 2008

——, 'The Sharashka Phenomenon' Russian History Blog (http://russianhistoryblog.org/2011/03/the-sharashka-phenomenon)

Siegelbaum, Lewis H., 'Soviet Norm Determination in Theory and Practice, 1917–1941', *Soviet Studies*, 36 (1984), 45–68

——, *Stakhanovism and the Politics of Productivity in the USSR, 1935–1941* (Cambridge University Press, 1988)

Siegel, Katherine A. S., 'Technology and Trade: Russia's Pursuit of American Investment, 1917–1929', *Diplomatic History*, 17 (1993), 375–98

Simms, James Y., 'The Crop Failure of 1891: Soil Exhaustion, Technological Backwardness, and Russia's "Agrarian Crisis"', *Slavic Review*, 41 (1982), 236–50

——, 'The Economic Impact of the Russian Famine of 1891–92', *Slavonic and East European Review*, 60 (1982), 63–74

Sirotkina, Irina E., 'Mental Hygiene for Geniuses: Psychiatry in the Early Soviet Years', *Journal of the History of the Neurosciences*, 16 (2007), 1–2

——, 'The Ubiquitous Reflex and its Critics in Post-Revolutionary Russia', *Berichte Zur Wissenschaftsgeschichte*, 32 (2009), 70–81

——, 'When Did "Scientific Psychology" Begin in Russia?', *Physis; Rivista Internazionale Di Storia Della Scienza*, 43 (2006), 1–2

Sirotkina, Irina E., and Elena V. Biryukova, 'Futurism in Physiology: Nikolai Bernstein, Anticipation, and Kinaesthetic Imagination', in Mihai Nadin (ed.) *Anticipation: Learning from the Past: The Russian/Soviet Contributions to the Science of Anticipation* (Springer, 2015)

Smil, Vaclav, *The Earth's Biosphere: Evolution, Dynamics, and Change* (MIT Press, 2002)

Smith, Roger, *Inhibition: History and Meaning in the Sciences of Mind and Brain* (University of California Press, 1992)

Sobolev, Sergei L., and Aleksei A. Liapunov, 'Kibernetika i estestvoznanie' ['Cybernetics and science'], in *Filosofskie problemy sovremennogo estestvoznaniia: Trudy Vsesoiuznogo soveshchaniia po filosofskim voprosam estestvoznaniia [Philosophical Problems of Modern Science: Proceedings of the All-Union Conference on the Philosophical Problems of the Natural Sciences]*, ed. P. Fedoseev (Izd-vo Akademii nauk SSSR, 1959), pp. 251–2.

Sochor, Zenovia A., *Revolution and Culture: The Bogdanov–Lenin Controversy* (Cornell University Press, 1988)

——, 'Soviet Taylorism Revisited', *Soviet Studies*, 33 (1981), 246–64

Solomon, Susan Gross, *Doing Medicine Together: Germany and Russia between the Wars* (University of Toronto Press, 2006)

Solzhenitsyn, Aleksandr Isaevich, *The Gulag Archipelago, 1918–1956: An Experiment in Literary Investigation, I–II, III–IV, V–VII* (Harper & Row, 1974)

'Soviet Backs Plans to Test Evolution', *New York Times*, 17 June 1926

Soyfer, Valery N., *Lysenko and the Tragedy of Soviet Science* (Rutgers University Press, 1994)

——, 'New Light on the Lysenko era', *Nature*, 339 (1989), 415–20

——, 'Radiation Accidents in the Southern Urals (1949–1967) and Human Genome Damage', *Comparative Biochemistry and Physiology Part A: Molecular and Integrative Physiology*, 133 (2002), 715–31

——, 'The Consequences of Political Dictatorship for Russian Science', *Nature Reviews Genetics*, 2 (2001), 723–9

——, 'Tragic History of the VII International Congress of Genetics', *Genetics*, 165 (2003), 1–9

——, *Vlast'i nauka [Power and Science]* (Lazur, 1993)

Spencer, L. J., 'Memorial of Alexander Evgenievich Fersman', *American Mineralogist*, 31 (1946), 173–8

Stack, Megan, 'Research Monkeys Languish in a State of Limbo', *Latimes.com*, 12 April 2008

Stalin, Joseph Vissarionovich, *Address Delivered in the Kremlin Palace to the Graduates from the Red Army Academies, May 4, 1935* (Foreign Languages Publishing House, 1955)

——, 'Dizzy with Success: Concerning Questions of the Collective-Farm Movement', *Pravda*, 1930

——, *Problems of Leninism* (Foreign Languages Publishing House, 1947)

——, *Works*, Vol. 10: *1921–1927* (Lawrence & Wishart, 1954)

——, *Works*, Vol. 11: *1928–Mar 1929* (Lawrence & Wishart, 1955)

——, *Works*, Vol. 12: *April 1929–June 1930* (Lawrence & Wishart, 1955)

——, *Works*, Vol. 13: *July 1930–January 1934* (Lawrence & Wishart, 1955)

Stanchevici, Dmitri, *Stalinist Genetics: The Constitutional Rhetoric of T. D. Lysenko* (Baywood, 2012)

Starbuck, Dane, *The Goodriches: An American Family* (Liberty Fund, 2001)

Stites, Richard, *Revolutionary Dreams: Utopian Vision and Experimental Life in the Russian Revolution* (Oxford University Press, 1991)

——, 'Trial as Theatre in the Russian Revolution', *Theatre Research International*, 23 (1998), 7

Stockdale, Melissa, 'The Russian Experience of the First World War', in Abbott Gleason (ed.), *A Companion to Russian History* (Wiley-Blackwell, 2009), 311–34

Stone, David R., 'Russian Civil War (1917–1920)', in *The Encyclopedia of War* (Blackwell, 2011)

Suny, Ronald Grigor, 'Stalin and His Stalinism: Power and Authority in the Soviet Union, 1930–1953', in David L. Hoffmann (ed.), *Stalinism* (Blackwell, 2003), pp. 9–34

Suvorov, N. F., and V. N. Andreeva, 'Problems of the Inheritance of Conditioned Reflexes in Pavlov's School', *Neuroscience and Behavioral Physiology*, 21 (1991), 8–16

Svidersky, V. L., 'Orbeli and Evolutionary Physiology', *Journal of Evolutionary Biochemistry and Physiology*, 38 (2002), 513–36

Swanson, James M., 'The Bolshevization of Scientific Societies in the Soviet Union: An Historical Analysis of the Character, Function, and Legal Position of Scientific and Scientific-Technical Societies in the USSR, 1929–1936' (Indiana University) 1968

Swianiewicz, S., 'Coercion and Economic Growth', *The Political Quarterly*, 31 (1960), 453–65

Talis, Vera L., 'New Pages in the Biography of Nikolai Alexandrovich Bernstein', *Anticipation: Learning from the Past* (Hanse-Wissenschaftskolleg Institute for Advanced Study, 2014)

Targulian, O. M., *Controversial questions on genetics and selection: Transactions of the 4th Session of the Academy, 19–27 Dec. 1936* (Moscow, Leningrad, 1937)

Thornstrom, Carl-Gustaf, and Uwe Hossfeld, 'Instant Appropriation: Heinz Brücher and the SS Botanical Collecting Commando to Russia 1943', *Plant Genetic Resources Newsletter*, 2002, 54–7

Thurston, Robert W., 'The History of Torture Shows It Does Not Work', *History News Network*, 2009

Timofeev-Ressovsky, N. V., *Vospominaniia: Istorii, rasskazannye im samim, s pis'mami, fotografiiami i dokumentami [The Stories Told by Himself with Letters, Photos and Documents]* (Soglasie, 2000)

Tirard, S. P., 'Origin of Life and Definition of Life, from Buffon to Oparin', *Origins of Life and Evolution of the Biosphere*, 40 (2010), 215–20

Todes, Daniel P., 'Darwin's Malthusian Metaphor and Russian Evolutionary Thought, 1859–1917', *Isis*, 78 (1987), 537–51

——, *Darwin without Malthus: The Struggle for Existence in Russian Evolutionary Thought* (Oxford University Press, 1989)

——, *Ivan Pavlov: A Russian Life in Science* (Oxford University Press, 2014)

——, 'Pavlov and the Bolsheviks', *History and Philosophy of the Life Sciences*, 17 (1995), 379–418

——, *Pavlov's Physiology Factory: Experiment, Interpretation, Laboratory Enterprise* (Johns Hopkins University Press, 2001)

Tromly, Benjamin, 'The Leningrad Affair and Soviet Patronage Politics, 1949–1950', *Europe–Asia Studies*, 56 (2004), 707–29

Trotsky, Leon, *My Life: An Attempt at an Autobiography* (Pathfinder, 1970)

——, *Problems of Everyday Life: And Other Writings on Culture and Science* (Pathfinder Press, 1998)

Vaingurt, Julia, 'Poetry of Labor and Labor of Poetry: The Universal Language of Alexei Gastev's Biomechanics', = *Russian Review*, 67 (2008), 209–29

Valkova, Olga, 'The Conquest of Science: Women and Science in Russia, 1860–1940', *Osiris*, 23 (2008), 136–65

Vavilov, N. I., *Five Continents*, trans. L. E. Rodin (IPGRI, 1997)

——, *The Origin, Variation, Immunity and Breeding of Cultivated Plants: Selected Writings* (Chronica Botanica, 1951)

——, 'The Problem of the Origin of the World's Agriculture in the Light of the Latest Investigations', in *Science at the Crossroads* (Frank Cass, 1931)

——, *Vavilov, Nikolai Ivanovich. The Scientific Legacy in His Letters*, Vol. 5, *1936–1937* (Nauka-M, 2002)

Vein, Alla A., 'Science and Fate: Lina Stern (1878–1968), A Neurophysiologist and Biochemist', *Journal of the History of the Neurosciences*, 17 (2008), 195–206

Vein, Alla A., and Marion L. C. Maat-Schieman, 'Famous Russian Brains: Historical Attempts to Understand Intelligence', *Brain*, 131 (2008), 583–90

Veresov, N., 'Nikolai Bernstein: The Physiology of Activeness and the Psychology of Action', *Journal of Russian and East European Psychology*, 44 (2006), 3–11

Vernadsky, Vladimir I., *150 Years of Vernadsky: The Biosphere: Volume 1*, ed. Jason A. Ross, trans. Meghan K. Rouillard (CreateSpace, 2014)

——, *Geochemistry and the Biosphere*, ed. Frank B. Salisbury, trans. A. L. Yanshin (Synergetic Press, 2007)

——, *The Biosphere*, trans. David B. Langmuir (Springer-Verlag, 1997)

Vernadsky, Vladimir I., and Meghan K. Rouillard, *150 Years of Vernadsky: The Noösphere: Volume 2*, ed. Jason A. Ross (CreateSpace, 2014)

Vöhringer, Margarete, 'Pudovkin's "Mechanics of the Brain" – Film as Physiological Experiment', *The Virtual Laboratory* (Max Planck Institute for the History of Science, 2001

Volkov, Solomon, *St Petersburg: A Cultural History* (Free Press, 1995)

Van Dongen, Jeroen, 'On Einstein's Opponents, and Other Crackpots', *Studies in History and Philosophy of Science Part B: Studies in History and Philosophy of Modern Physics*, 41 (2010), 78–80

Van Voren, Robert, 'Political Abuse of Psychiatry – an Historical Overview', *Schizophrenia Bulletin*, 36 (2010), 33–5

——, 'Psychiatry as a Tool of Repression Against Dissidents in the USSR', *Deeds and Days (Darbai Ir Dienos)*, 55 (2011), 29–42

Vorontzov, Dimitri, 'Michurin, Dir. by Alexander Dovzhenko, 1948', *Vorontzov*, 2011 (http://vorontzov.com/michurin-dir-by-alexander-dovzhenko-1948)

Vucinich, Alexander, *Darwin in Russian Thought* (University of California Press, 1988)

——, *Einstein and Soviet Ideology (Stanford Nuclear Age)* (Stanford University Press, 2000)

——, *Empire of Knowledge: The Academy of Sciences of the USSR (1917–1970)* (University of California Press, 1984)

——, 'Ivan Pavlov: Science, Philosophy and Ideology', *Texas Reports on Biology and Medicine*, 32 (1974), 107–20

——, *Science in Russian Culture, 1861–1917* (Stanford University Press, 1970)

——, *Social Thought in Tsarist Russia: The Quest for a General Science of Society, 1861–1917* (University of Chicago Press, 1976)

——, 'Soviet Marxism and the History of Science', *Russian Review*, 41 (1982), 123–43

Vygodskaya, Gita L., 'His Life', *School Psychology International*, 16 (1995), 105–16

Vygodskaya, Gita. L., and T. M. Lifanova, 'Lev Semenovich Vygotsky Part 1: Life and Works', *Journal of Russian and East European Psychology*, 37 (1999), 3–31

——, 'Lev Semenovich Vygotsky Part 2: Through the Eyes of Others', *Journal of Russian and East European Psychology*, 37 (1999), 32–90

Vygotsky, L. S., 'Letters to Students and Colleagues', *Journal of Russian and East European Psychology*, 45 (2007), 11–60

——, 'Soznanie kak problema psikhologii povedeniia' ['Consciousness as a problem of Behavioural Psychology'; 1925], *Sobranie sochinenii*, vol. 1 (Pedagogika, 1982), pp. 78–98.——, *Mind in Society: The Development of Higher Psychological Processes*, ed. Michael Cole (Harvard University Press, 1978)

——, *The Collected Works of L. S. Vygotsky. Including the Chapter on the Crisis in Psychology Volume 3*, ed. R. W. Rieber and Jeffrey L. Wollock, trans. René van der Veer (Springer, 1997)

——, *The Psychology of Art* (MIT Press, 1971)

——, *The Vygotsky Reader*, ed. René van der Veer and Jaan Valsiner (Blackwell, 1994)

Walker, Barbara, 'Kruzhok Culture: The Meaning of Patronage in the Early Soviet Literary World', *Contemporary European History*, 11 (2002), 107–23

Walker, Mark, interview with Nikolaus Riehl, 1984, American Institute of Physics (http://bit.ly/27PWtat)

——, Interview with Nikolaus Riehl, 1985, American Institute of Physics (http://bit.ly/1WbMtEo)

Walker, Shaun, 'Stalin's Space Monkeys – Science, News', *Independent*, 15 April 2008

Wazeck, Milena, *Einstein's Opponents: The Public Controversy about the Theory of Relativity in the 1920s* (Cambridge University Press, 2014)

Weiner, Douglas R., *A Little Corner of Freedom: Russian Nature Protection from Stalin to Gorbachev* (University of California Press, 1999)

——, 'Community Ecology in Stalin's Russia: "Socialist"and "Bourgeois" Science', *Isis*, 75 (1984), 684–96

——, 'Dzherzhinskii and the Gerd Case: The Politics of Intercession and the Evolution of "Iron Felix" in NEP Russia', *Kritika*, 7 (2006), 759

——, *Models of Nature: Ecology, Conservation, and Cultural Revolution in Soviet Russia* (Indiana University Press, 1988)

——, 'The Historical Origins of Soviet Environmentalism', *Environmental Review*, 6 (1982), 42–62

——, 'The Roots of "Michurinism": Transformist Biology and Acclimatization as Currents in the Russian Life Sciences', *Annals of Science*, 42 (1985), 243–60

Weismann, August, Edward Bagnall Poulton, S. Schönland and A. E. Shipley, *Essays upon Heredity and Kindred Biological Problems* (Clarendon Press, 1891)

Weiss K. M., '"There Is No Intra-Specific Struggle in Nature": Can We Inherit the Lessons of Lysenko's Time, in Our Own Time?', *Evolutionary Anthropology*, 18 (2009), 50–4

Wetter, Gustav A., *Dialectical Materialism: A Historical and Systematic Survey of Philosophy in the Soviet Union* (Praeger, 1959)

——, 'Ideology and Science in the Soviet Union – Recent Developments', *Daedalus*, 89 (1960), 581–603

White, James D., 'Alexander Bogdanov's Conception of Proletarian Culture', *Revolutionary Russia*, 26 (2013), 52–70

——, 'Bogdanov in Tula', *Studies in Soviet Thought*, 22 (1981), 33–58

Williams, Robert C., 'Collective Immortality: The Syndicalist Origins of Proletarian Culture, 1905–1910', *Slavic Review*, 39 (1980)

Windholz, George, 'Pavlov's Conceptualization of Paranoia within the Theory of Higher Nervous Activity', *History of Psychiatry*, 7 (1996), 159–66

——, 'Soviet Psychiatrists under Stalinist Duress: The Design for a "New Soviet Psychiatry" and Its Demise', *History of Psychiatry*, 10 (1999), 329–47

——, 'The 1950 Joint Scientific Session: Pavlovians as the Accusers and the Accused', *Journal of the History of the Behavioral Sciences*, 33 (1997), 61–81

Wolfe, Audra J., 'What Does It Mean to Go Public? The American Response to Lysenkoism, Reconsidered', *Historical Studies in the Natural Sciences*, 40 (2010), 48–78

Wolfe, Ross, 'The Ultra-Taylorist Soviet Utopianism of Aleksei Gastev, *The Charnel-House* (7 December 2011), http://thecharnelhouse.org

Wren, Daniel A., and Arthur G. Bedeian, 'The Taylorization of Lenin: Rhetoric or Reality?', *International Journal of Social Economics*, 31 (2004), 287–99

Wynne, Clive D. L., 'Kissing Cousins', *New York Times* (12 December 2005)

——, 'Rosalia Abreu and the Apes of Havana', *International Journal of Primatology*, 29 (2008), 289–302

Yakovlev, Y. A., *Red Villages: The 5-Year Plan in Soviet Agriculture* (M. Lawrence, 1931)

Yashin, D. I., *Experiments in the Revival of Organisms* (Soviet Film Agency, 1940).

Yasnitsky, Anton, 'Vygotsky Circle during the Decade of 1931–1941: Toward an Integrative Science of Mind, Brain, and Education' (University of Toronto, 2009)

Yasnitsky, Anton, and M. Ferrari, 'From Vygotsky to Vygotskian Psychology: Introduction to the History of the Kharkov School', *Journal of the History of the Behavioral Sciences*, 44 (2008), 119–45

Yassour, Avraham, 'Bogdanov-Malinovsky on Party and Revolution', *Studies in Soviet Thought*, 27 (1984), 225–36

——, 'The Empiriomonist Critique of Dialectical Materialism: Bogdanov, Plekhanov, Lenin', *Studies in Soviet Thought*, 26 (1983), 21–38

Young, Robert M., 'Evolutionary Biology and Ideology: Then and Now', *Science Studies*, 1 (1971), 177–206

——, 'Malthus and the Evolutionists: The Common Context of Biological and Social Theory', *Past and Present*, 1969, 109–45

Zajicek, Benjamin, 'Scientific Psychiatry in Stalin's Soviet Union: The Politics of Modern Medicine and the Struggle to Define "Pavlovian" Psychiatry, 1939–1953' (University of Chicago, 2009)

Bibliography

Zakharov, I. K., L. D. Kolosova and V. K. Shumny, 'Raisa L'vovna Berg (March 27, 1913–March 1, 2006)', *Russian Journal of Genetics*, 42 (2006), 1470–3

Zamyatin, Evgenij Ivanovic, *We*, trans. Bernard Guilbert Guerney and Michael Glenny (Penguin, 1972)

Zhinkin, L. N., and V. P. Mikhailov, 'On "The New Cell Theory": Two Soviet Authors Critically Review Recent Soviet Work on the Origin of the Cell', *Science*, 128 (1958), 182–86

Ziegler, Charles E., *Environmental Policy in the USSR* (University of Massachusetts Press, 1987)

Zimmer, Karl Günter, 'N. W. Timoféeff-Ressovsky 1900–1981', *MUT Mutation Research – Fundamental and Molecular Mechanisms of Mutagenesis*, 106 (1982), 191–93

Zimmer, Karl Günter, Max Delbrück, and N. V. Timofeev-Resovskii, *Creating a Physical Biology: The Three-Man Paper and Early Molecular Biology*, ed. Phillip R. Sloan and D. Brandon Fogel (University of Chicago Press, 2011)

Zirkle, Conway, *Death of a Science in Russia: The Fate of Genetics as Described in Pravda and Elsewhere* (University of Pennsylvania Press, 1949)

Zuberi, Matin, 'Stalin and the Bomb', *Strategic Analysis*, 23 (1999), 1133–53

Zvereva, G. V., 'Professor I. I. Ivanov – Osnovopolozhnik Iskusstvennogo Osemeneniia Zhivotnykh', *Veterinariia*, 7 (1970), 88–90

Index

Index

epigenetics, 222

eugenics, 142–8, 157–8; recast as 'human genetics', 147–8; denounced as fascist, 268–9, 274, 284

evolution, Lamarckian model of, 115–17; efforts to reconcile with genetics, 132–5; 'tested' by I. Ivanov's hybridisation experiments, 156

factories, management of, 43, 71, 73, 78–81, 167–8; peasant workers in, 80, 169; effect of Shakhty trial on, 172–3; Stakhanovites in, 237

famine, 3; of 1891–2, 16–18; of 1905–6, 22; of 1921–2, 48–9, 55–7, 167, 200; of 1932–3, 221, 225; of 1946–47, 350–1

farming *see* agriculture

Fedotov, Dmitry Dmitryevich (1908–82), 317–18

Fersman, Alexander Evgenievich (1883–1945), 25, 53, 67, 129, 156

'The Fields in Winter' (*Pravda*), 206–7

Filipchenko, Yuri Alexandrovich (1882–1930), 147–8

First Five Year Plan, 153, 160, 172, 174, 177, 202, 221, 238, 239, 424

First World War, 25–6, 37, 38, 42

'fly room' (Columbia University), 148

Flyorov, Georgy Nikolayevich (1913–90), 329, 332

Foerster, Otfrid (1873–1941), 136

Fofanova, Margarita Vasilyevna (1883–1976), 197–8, 199

food: surpluses required for capitalism, 3; city shortages of, 5, 38, 52, 58; withheld by farmers, 38, 51; forcibly requisitioned, 51, 213

forestry: V. Dokuchaev and, 387–8; T. Lysenko and, 390; mismanaged 431–2

'founding fathers' campaign, 354–5, 360, 375

Frenkel, Iakov Ilich (1894–1952), 251, 439n16

Freud, Sigmund (1856–1939), 98, 107

fruit fly, *see* Drosophila

Fedorov, Nikolai Fedorovich (1829–1903), 416–17

Fuchs, Klaus (1911–88), 330

Galton, Francis (1822–1911), 115, 142–3, 421

Ganike, Evgeny (?–?), 127

Gastev, Alexei Kapitonovich (1882–1939), 22–3, 27, 68, 70, 81, 422, 424; literary influences, 69; revolutionary activity, 69;

shop-floor experience, 69, 70, 76–7, 78; 'The Accursed Question' (1904), 70; poetry, 70, 73–5, 77, 81; Syndicalist sympathies, 71; in Paris, 69–70, 71–2; 'Express: a Siberian Fantasy' (1924), 72–3; union secretary, 73; *Poetry of the Factory Floor* (*Poézija Rabodego Udara*), 74, 165–6, 167; founds the Central Institute of Labour, 77

genes: nature of, 132, 150; dominant and recessive, 114, 135; measurement of, 262–3. *See also* mutation

genetics, 115, 119, 120, 121, 130, 131, 133–4, 195, 201, 277–8, 288; of behaviour, 126–7; of populations, 132, 134–5; of poultry, 130, 140, 178; international character of, 133, 263, 282–3, 297–8, 361; and medicine, 119–20, 151–2; denounced as fascist, 269–70, 274, 275, 278, 279; Marxists defend, 191, 192–3, 272–3; within the Academy of Sciences, 284–6; Western campaigners defend, 292–3, 342–4, 369–70; overseas contacts renewed after Second World War, 345; extinguished, 367–9; nuclear physics and, 336, 401; cybernetics and, 404–5. *See also* epigenetics, International Genetics Congress

Genetics and Plant Breeding, 1929 Congress, 202, 208–9

Germany: scientific institutions in, 24; trade with Russia, 26, 171; eugenics in, 157–8; under National Socialism, 264; invades Soviet Union, 296, 300–1, 305–7; technology seized by allied forces, 331–2, 334

Golder, Frank Alfred (1877–1929), 56, 57

Gorbunov, Nikolai Petrovich (1892–1938), secretary to V. Lenin, 46; scientific patron, 138, 157; commissar of agriculture; 173, 202, 211; and N. Vavilov, 299

Gorky, Maxim (born Alexei Maximovich Peshkov, 1868–1936), 27, 146, 246, 418, 425–6; and Lenin, 35; as patron, 62–3; founds TsEKUBU, 63; leaves for Italy, 64–5; support for N. Koltsov, 130, 178; returns to Soviet Union, 233; as propagandist, 234–5; hatred of rural life, 234–5; utopianism, 235; and Stalin, 240; isolated from leadership, 241–2; controversy surrounding death of, 242

Gorshkov, Igor Sergeyevich (1896–1965), 187–8

grain, improved varieties, 198, 205, 216, 226,

Index

pleads for L. Landau's freedom, 256; letters to Soviet leadership, 256–7; advises on uranium production, 328, 339; and liquid oxygen production, 337–8, 339; and atom bomb project, 337, 338–9 454n29

Karpechenko, Georgy Dmitrievich (1899–1942), 299

Karpinsky, Alexander Petrovich (1847–1936), 53

Keller, Boris Alexandrovich (1874–1945), 284

KEPS, 129

Kerkis, Julius, (?–?), 289–90

Khariton, Yulii Borisovich (1904–96), 319, 329, 392

Khrushchev, Nikita Sergeyevich (1894–1971), 340, 351, recollections concerning J. Stalin 387, 394; made General Secretary, 399–400; supports Lysenko 410–13; projects to transform nature, 431–2

Khvat, Alexander Grigorievich (1907–?), 298–300

Kirov, Sergei Mironovich (1886–1934), 202, 240; death of, 241

Kliueva, Nina Georgievna (1898–1971), 347–50

Kogan, Boris Borisovich (1896–1968), 158

Kol, Alexander Karpovich (?–?), 215

Kolbanovsky, Vladimir Alexandrovich (?–?), 289, 406–7

Kolchak, Alexander Vasilyevich (1874–1920), 48

Kolman, Ernst (1892–1979): *Life and Technology in the Future* (1928), 154; and B. Hessen 248–9, 250; edits *Under the Banner of Marxism*, 268, 289; and cybernetics, 405–6

Koltsov, Nikolai Konstantinovich (1872–1940), 127–8, 139, 200, 275; criticises I. Pavlov's work on the inheritance of acquired behaviours, 126–7 founds Institute of Experimental Biology, 128–9; steers development of Soviet genetics, 130, 159; arrested over Tactical Centre affair, 129–30; and eugenics, 63, 145, 146–7, 160, 175, 284–5; 'Betterment of the Human Race' (1922), 146–7, 279, 280; official hostility towards, 176, 177–8, 191, 279–80; and collectivisation, 202; defends genetics at 4th session of the Lenin Academy, 274, 278; nominated as member of the Academy of Sciences, 283–4; work investigated by Academy of Sciences, 284–6. *See also* Institute of Experimental Biology

Koltsov, Maria Polievktovna (born Sadovnikova, 1872–1940), 284, 286

Koltushi (Institute of Genetics of Higher Nervous Activity), 127, 374, 377, 415

Komarov, Vladimir Leontyevich (1869–1945), 285, 302

Kornilov, Konstantin Nikolaevich (1879–1957), 97, 98, 108

Korogodin, Vladimir Ivanovich (1929–2005), 402

Korolev, Sergei Pavlovich (1907–66), 313, 408, 425

Koshtoyants, Khachatur Sedrakovich (1900–61), 285

Kostov, Doncho Stoianov (1897–1949), 271

KR affair, 347–50, 361

Krasin, Leonid Borisovich (1870–1926), 421

Krasnoselska-Maksimova, Tatiana Abramovna (?–?), 223

Kronstadt rebellion, 52

Kropotkin, Petr Alekseevich (1842–1921), 138, 139, 144

Krupskaya, Nadezhda Konstantinovna ('Nadya') (1869–1939), 31, 65, 79, 105

Krylenko, Nikolai Vasilyevich (1885–1938), 172

Krylov, Alexei Nikolaevich (1863–1945), 25

kulaks, 220–1

Kurchatov, Igor Vasilyevich (1903–60), 328, 329, 391; constructs first Soviet nuclear reactor, 331, 337, 459n20; develops atomic weapons, 393, 408

Kuznetstroi (coal fields), 234

Kyshtym (nuclear waste dump), 433

Laboratory No. 2, 331, 334

Lamarck, Jean-Baptiste (1744–1829), 115–16

Lamarckism, 115–16, 117, 121, 125, 151; eugenic implications of, 274; experimental support for, 117, 120, 122; and Marxism, 117–18, 123–4, 125, 127, 192, 274, 420; and medicine, 119–20; objections to, 274, 444n2; Stalin's support for, 190–1, 354, 359. *See also* Kammerer, P., Studentsov, N.

Landau, Leo Davidovich (1908–68), 252, 253–5, 256

Lebedev, Sergey Alexeyevich (1902–74), 408

Lekhnovich, Vadim Stepanovich (1902–?) 297, 310

Lenin, Vladimir Ilyich (born Ulyanov, 1870–1924), 2, 17, 21, 27, 30–1, 41, 42, 44,